Aflatoxin in Maize

A Proceedings of the Workshop

El Batan, Mexico, April 7-11, 1986

Sponsored by: CIMMYT, UNDP and USAID

The International Maize and Wheat Improvement Center (CIMMYT) is an internationally funded, nonprofit scientific research and training organization. Headquartered in Mexico, CIMMYT is engaged in a worldwide research program for maize, wheat, and triticale, with emphasis on food production in developing countries. CIMMYT is one of 13 nonprofit international agricultural research and training centers supported by the Consultative Group for International Agricultural Research (CGIAR). The CGIAR is sponsored by the Food and Agriculture Organization (FAO) of the United Nations, the International Bank for Reconstruction and Development (World Bank), and the United Nations Development Programme (UNDP). The CGIAR consists of 40 donor countries, international and regional organizations, and private foundations.

CIMMYT receives support through the CGIAR from a number of sources, including the international aid agencies of Australia, Austria, Brazil, Canada, China, Denmark, Federal Republic of Germany, France, India, Ireland, Italy, Japan, Mexico, the Netherlands, Norway, the Philippines, Saudi Arabia, Spain, Switzerland, United Kingdom and the USA, and from the European Economic Commission, Ford Foundation, Inter-American Development Bank, International Bank for Reconstruction and Development, International Development Research Centre, OPEC Fund for International Development, Rockefeller Foundation, and the United Nations Development Programme. Responsibility for this publication rests solely with CIMMYT.

Correct Citation: Zuber, M.S., E.B. Lillehoj, and B.L. Renfro, eds. 1987. Aflatoxin in Maize: A proceedings of the workshop. CIMMYT, Mexico, D.F.

ISBN 968-6127-12-7

Acknowledgements

We are thankful to several organizations, groups and individuals for making significant contributions to the success of the workshop on the aflatoxin problem in maize and the publication of this proceedings. We are grateful to the participants for the preparation of papers and for their deliberations which resulted in the recommendations included in the proceedings. The US University Collaborative Maize Research Working Group is to be commended for their proposal that this workshop be held and for their continuous support since that time.

Support for travel and living costs is gratefully acknowledged from CIMMYT, the United Nations Development Programme and USAID. The three of us also value the encouragement and support given to us by our directing staffs. The Visitor Services of CIMMYT deserve special thanks for making the participants comfortable and for handling the workshop organization and activities so efficiently.

The skillfull editorial work by Elizabeth Cuéllar, as well as the editorial assistance of Kelly Cassaday, is valued highly. We also thank CIMMYT Communications staff members Patricia Martínez, Maricela A. de Ramos and Silvia Bistrain for typesetting; Miguel Mellado E., Rafael de la Colina, José Manuel Fouilloux, and Bertha Regalado for providing the design and layout of this proceedings, and Alma McNab for overseeing the translations of papers and abstracts.

M.S. Zuber
E.B. Lillehoj
B.L. Renfro
Workshop Organizers

Table of Contents

4
Country Reports

1

Preface

A program for developing collaborative maize research was initiated by US maize scientists in 1977 to bring together CIMMYT maize improvement scientists, scientists in the developing countries and those from the USA. This effort was started in response to initiatives arising from the US Foreign Assistance Act of 1961, Title 12 amendment. Many formal meetings were held among scientists from 15 US universities that had active maize research programs. A cooperative research program for maize was developed and endorsed by the Board for International Food and Agricultural Development (BIFAD) in 1981, but it was not funded. Discussions were continued among representatives of BIFAD, USAID, CIMMYT and the universities, with the result being the US Universities-CIMMYT Maize Conference, which was held at CIMMYT in August 1984.

The broad purpose of the conference was to initate dialogue among maize scientists of CIMMYT, the USA and developing countries. Specific objectives were to:

- Identify specific collaborative research, technical assistance and training activities in maize that would have high pay-off potential, and
- Provide for a mechanism which would foster and support the organization and operation of long-term maize research programs in problem areas of mutual concern.

At the conference nearly 100 maize scientists and administrators discussed issues in major subject areas relating to the CIMMYT global maize program. Proceedings of the conference were published in July 1986.

The Ad Hoc working group appointed at the conference held its first meeting in September 1986. At that meeting several specific collaborative activities were selected for further consideration. Among these was a proposal for cooperative research on aflatoxin, which was submitted by M.S. Zuber and L.L. Darrah of the University of Missouri.

Since aflatoxin contamination of maize grain is a major problem in the tropics and subtropics, often resulting in serious health problems among humans and livestock, the proposal was recommended for immediate action. Therefore, an aflatoxin workshop was planned to be held at CIMMYT in April 1986, with many scientists from CIMMYT, the USA and developing countries invited to present papers. The workshop was unique in that most of the scientists involved had extensive research experience in aflatoxin contamination. Many areas of potential collaboration on the problem of aflatoxin contamination of maize were identified, and the foundation laid for future cooperative research on a most important problem that affects the lives of many people of the world.

M.S. Zuber
E.B. Lillehoj
B.L. Renfro
Technical Editors

The CIMMYT Maize Improvement Program Today and Tomorrow

R.P. Cantrell, Director, Maize Program, CIMMYT, Mexico

In August 1984, a meeting was held at CIMMYT for maize researchers from US universities. That meeting was an important landmark for CIMMYT, pointing the way to a future of greater cooperation with US universities in the exchange of maize germplasm and scientific knowledge. This type of cooperation is already taking place through our joint efforts in the improvement of various temperate gene pools, and through cooperative work on insect pest resistance. Other such projects were proposed during the US universities meeting, and some of those are now becoming a reality; one of them is this aflatoxin workshop.

Projects like these are a vital source of new technology for the CIMMYT Maize Program. They add significantly to the body of ideas and techniques that can be drawn upon for the development of improved germplasm and other research products. By taking full advantage of technology generated in the US universities, it is hoped that products of even greater usefulness may be developed, and that in delivering those products to our clients, some of the vigor and innovation of your research will be channeled into the national maize programs of Third World countries.

The connections CIMMYT has and is forming with US universities also serve as our window to the future of agricultural research. By staying in touch with those of you who are working with biotechnology, for example, CIMMYT scientists can better anticipate the implications that this research has for their work. Although it is outside the CIMMYT mandate to engage in that research, it is well within its responsibilities to know what results are coming to light and how they can be applied to crop improvement.

The basis for cooperation between CIMMYT and universities in the USA must, of course, be a thorough and mutual understanding. It is hoped that this paper will contribute to that understanding by presenting an overview of the maize program as it is has evolved over the past few years and as it may look in the future. Several possibilities for joint research projects between CIMMYT and US universities will be mentioned, and it is to be hoped that opportunities for others will be seen.

The CIMMYT Mandate

CIMMYT is a nonprofit research and training organization, dedicated to improving the efficiency of maize, wheat and triticale production, mainly in developing countries. It is one of 13 such organizations that are supported by the Consultative Group on International Agricultural Research. This group, the CGIAR, is sponsored by the Food and Agriculture Organization, the World Bank and the United Nations Development Programme. The CGIAR consists of representatives from 45 donors, including individual countries, international and regional organizations and private foundations. CIMMYT's funding comes from the international aid agencies of a total of 20 countries and from a number of other groups involved with international agricultural development.

The objective of CIMMYT's maize program is to facilitate maize research in Third World countries. This is done by providing national maize programs with improved germplasm and other products and services, such as research procedures, training and technical information. These are developed and delivered by 15 to 20 scientists working at the research stations in Mexico and an equal number of maize researchers involved in bilateral and regional programs outside of Mexico.

The ultimate aim of CIMMYT research is an increase in the options available to farmers for raising the productivity of the resources they commit to maize production. In working toward that goal, the center cooperates closely with national programs and gives highest priority to CIMMYT-national program relationships. An active interest is also taken in the work of other groups, such as universities and private seed companies, whose research results and products can help farmers increase the efficiency of maize production.

Population improvement and variety development

One important part of the CIMMYT Maize Program that many of you will have heard about is the population improvement and variety development scheme. This system has a funnel-shaped structure very similar to that of any other large breeding program for a major crop. At the wide end of the funnel is the germplasm bank, in which over 10,000 accessions are maintained.

Drawing on these genetic resources, 50 gene pools or complexes have been built over the past 15 years or so. Each gene pool is adapted to one of several large regions (such as the lowland tropics) that are termed mega-environments. These encompass many smaller maize-growing niches that are distinct from one another in some respects, but similar in elevation, climate and other features that affect maize germplasm requirements. The gene pools, subdivided by maturity group, grain color and grain type, are improved in the backup unit by means of half-sib family selection under fairly mild selection pressure.

From the best fraction of the pools, populations have been developed. Within the advanced unit, these are placed under more intense selection pressure in a modified full-sib recurrent selection scheme that includes international testing. In the first stage of testing, selected full-sib progenies are evaluated at five or six different locations. On the basis of these results, experimental varieties are formed and are then distributed to national maize researchers for evaluation at 30 to 50 locations. The best performers in these trials, which are termed elite varieties, are then tested at 60 to 80 sites.

The varieties coming out of the breeding program are by and large intermediate rather than finished products. Generally, they are not adapted to a particular ecological niche, but rather to a mega-environment. For that reason, they must undergo a certain amount of adaptive testing before they can be released to farmers.

Improved research and techniques

In the development and delivery of intermediate research products, using the most appropriate techniques can be almost as important as having the best available germplasm. For that reason, a sizeable portion of time and resources are committed to the development of improved research techniques. These make CIMMYT's program more efficient and can also help national programs in conducting their research and in applying its results more effectively to meet their farmers' needs. Some examples of these research products are techniques for mass rearing and artificial infestation with insect pests for resistance screening, and breeding methodologies for developing, maintaining and producing improved varieties and hybrids. Another set of techniques that have proven especially useful to national programs are methods developed by CIMMYT scientists for on-farm research. With this type of research, national scientists examine farmers' circumstances, identify production problems and carry out experiments to resolve those problems under the conditions of representative farmers.

Training

At CIMMYT ideas are conveyed about on-farm research and other research techniques by several means, but in large part this is done through in-service and in-country training programs. The majority of maize program trainees participate in one of the five- to six-month in-service courses, production agronomy research, maize improvement, protein evaluation or experiment station management, which are held at CIMMYT headquarters in Mexico. Other trainees come to the center for as long as a year as visiting scientists to work on some research problem in which they and CIMMYT are mutually interested. A small number of predoctoral fellows who have finished their university course work conduct their thesis research under the supervision of CIMMYT scientists. Positions are also available for postdoctoral fellows, who become more closely involved in the research program while still working independently on problems that are of special interest to them.

Candidates for training at CIMMYT headquarters are usually identified by senior national maize researchers or administrators with whom the regional staff are working closely. Many candidates will already have taken part in training activities organized in their own countries as part of one of CIMMYT's regional maize programs. These activities include short courses on specific topics, such as on-farm research and regional workshops in which members of national programs discuss their plans and share research results.

The regional programs

The extremely varied work of each of the regional maize programs, of which there are six around the world (Figure 1), is carried out by a team of two or three regional maize specialists. Each one, while working within his own discipline in a complementary relationship with other members of the team, also maintains a broad interest in almost all aspects of maize research and production.

The principal task of the regional maize specialists is to assist national researchers in managing and improving their scarce resources and sometimes in acquiring new resources from CIMMYT or other organizations. This assistance is provided mainly through frequent and timely visits to researchers in national programs. The specialists join national program staff in their experimental plots at critical times during the growing season and accompany them on trips to farmers' fields. In addition, they are frequently present at planning meetings, where they are full participants in the discussion of research priorities.

The approach to research planning that CIMMYT is trying to promote in its regional programs is similar to that used by private seed companies. First, national researchers and administrators are encouraged to identify specific product needs. Since on-farm research is the best tool for identifying those needs, much time is spent helping national scientists make this type of work a permanent feature of their research programs. Once product needs have been identified, the regional staff work with national scientists in planning research and allocating resources needed to come up with products required by the nation's farmers.

Future Maize Program Plans

It would hardly be appropriate for CIMMYT maize researchers to preach such a doctine as this if they did not follow it to the letter themselves. To better practice what they preach, program staff are now in the process of gathering more detailed information about the maize germplasm requirements of major maize-growing environments in the Third World. For selected countries, Ghana, for example, mega-environments are being delineated and general information compiled about maize production in each of those areas. This information, most of it supplied by regional staff and their colleagues in national programs, includes items such as the amount of land in maize production and the maturity time of maize produced on that land.

In addition to recording the general features of maize production in a given mega-environment, CIMMYT is trying to learn more about the extent and severity of specific problems, such as diseases and insect pests. The ultimate goal of this exercise is to gain a clearer picture of germplasm needs throughout the developing world.

The next step will be to make certain adjustments in CIMMYT's maize research program aimed at improving its capacity to meet germplasm needs with greater accuracy. It remains to be seen what the exact details of some of those adjustments will be, since they are still under consideration. But changes can be expected or are already being made in four areas of research:

- The germplasm bank;
- The backup unit, which is concerned with the development and improvement of gene pools;
- The advanced unit, which has responsibility for population improvement; and
- The new hybrid maize program.

Other additions to the program are being planned as well, including greater emphasis on stress tolerance and some work on the efficiency of nitrogen utilization.

The germplasm bank

The germplasm bank will be discussed first, because the genetic diversity it contains is in a historical sense the foundation of the maize improvement program. This collection of maize seed, one of the oldest in the CGIAR system, was established in the 1940s with collections gathered by breeders who were then sampling the genetic variation of maize in Mexico, the Caribbean and Guatemala. The bank became more international in scope between 1967 and 1975, as collections from Brazil, Bolivia and Peru were added to it. CIMMYT took an active role during that period in planning collection expeditions with the International Board of Plant Genetic Resources in areas of the world where there were unique maize genetic resources and for which collections either had not been made or were not representative of existing variation.

During all of this time, the main function of the bank has been to supply useful materials for maize improvement work. Although it will, of course, continue to perform that function, a somewhat broader role for the germplasm bank is now envisioned.

For several reasons it is important that the CIMMYT Maize Program more fully exercise its stewardship over the valuable genetic resources contained in the bank. First is the success with which the program has met in accelerating the spread of improved varities. That success, ironically, has promoted genetic erosion of landraces and created the need for gene conservation by seed banks. A second reason is that CIMMYT possesses the only international maize bank in the

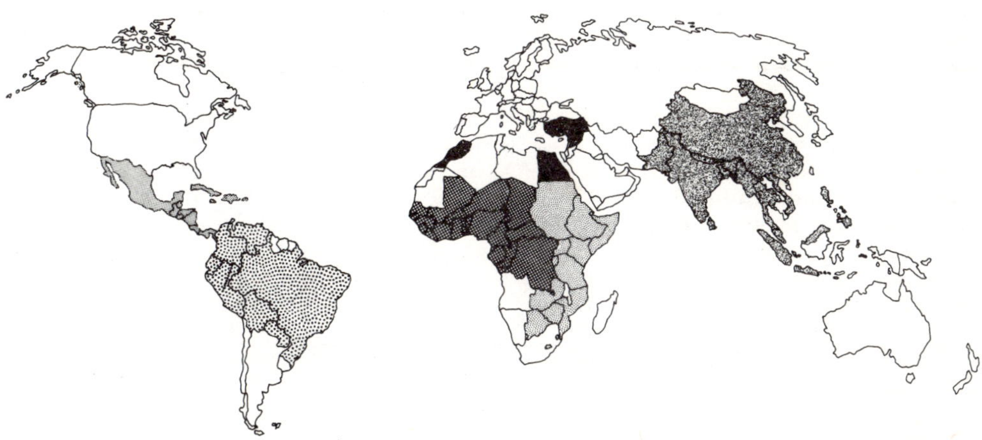

Figure 1. CIMMYT regional maize programs

Americas that has accepted entire national collections and has the capacity to regenerate and evaluate its accessions. Thirdly, CIMMYT is located at the center of origin of maize and in the area where its wild relatives are found in greatest genetic diversity.

This broader role for the maize germplasm bank has been made possible by additional funding. With that support, it was decided to make two important improvements in the bank's facilities. The first, completed in 1985, was the modification of one of the cold storage rooms so that temperature could be maintained at -15°C. This improvement is expected to at least double the lifetime of the seed, and thus increase its lifespan and reduce the frequency of regenerations, from 20 years under previous storage conditions to as much as 100 years under long-term storage. This reduction in the interval between regenerations, each of which causes some genetic drift, should keep the total change in the genetic composition of the accessions to a minimum.

The second improvement, still underway, is the development of a computerized data management system for cataloging and describing accessions. At present, no gene bank in the Americas, including CIMMYT's, can produce a passport catalog of its maize germplasm holdings. But this year, with the new system in place, it is hoped that a catalog of CIMMYT's holdings can be produced and that it will provide better access to this germplasm. That task and the storage of CIMMYT's active maize collection head the list of major goals for the germplasm bank. Another objective will be to evaluate the collections, with the aim of reducing duplication of certain accessions.

An additional goal is the support of research on the landraces of maize in the Americas. This research, particularly studies on the amount of genetic erosion that has already taken place among the landraces, would benefit greatly from the involvement of institutions possessing facilities for laboratory analysis not available at CIMMYT. Leading universities would be likely partners for cooperative reseach and work involving teosinte and *Tripsacum*. This would greatly improve the understanding, documentation and preservation of the unique genetic variation that exists in these wild relatives of maize, and it would provide universities with opportunities to pursue their growing interest in the biochemistry and subcellular structure of these materials.

Another function that could be performed is coordination of a world maize collection, somewhat like the rice network that the International Rice Research Institute has helped establish. Among the primary activities would be the addition of plant genetic resources to the list of topics covered in CIMMYT's crop improvement course and the provision of information on the materials available in the germplasm bank. The passport catalog previously mentioned would be a good first step in creating a better awareness of what is in the bank. Afterwards, a pamphlet would be made available to explain how to request materials.

Underlying the five bank functions outlined is one central aim, a more efficient management of resources so that they are of maximum utility to germplasm users in national maize programs and other research institutions. This ambition is also the driving force behind changes being considered in the backup and advanced units.

The backup unit

The backup unit, which handles introduction nurseries and is responsible for the gene pools, has traditionally had two roles, the maintenance of genetic diversity and the provision of the advanced unit with improved genotypes for further refinement. Although both functions are necessary, the pursuit of one can limit efforts to accomplish the other. For example, if excessive selection pressure is applied in the improvement of materials in the backup unit, there is a risk of reducing the diversity in that material, until finally there is little difference between the pools and their corresponding advanced populations.

One way out of this dilemma, and the course now being considered, would be to reduce the total number of pools and divide them into two groups, each having a different objective. One group, consisting of pools that include the whole range of types from a particular region in the world, would be maintained within that region in equilibrium with its environmental conditions and changing pest populations and under very low selection pressure. These pools could even be rotated, under the coordination of one of the regional programs, among various national programs having common germplasm requirements. Within such a system, both CIMMYT

The central aim of CIMMYT's maize germplasm bank is to ensure that its holdings are of maximum use to germplasm users in national maize programs and other research institutions.

and national program breeders would have ready access to valuable sources of genetic diversity.

The second group would consist of pools carrying a high concentration of genes for one or more major traits, such as resistance to certain diseases or insect pests. The information now being gathered on mega-environments would provide a basis for deciding which traits should receive the most attention. Intense selection pressure would be applied to each of these pools for the desired trait and only mild pressure for other characteristics.

These pools would serve as sources of vital characteristics needed in the materials that are improved by the advanced unit and distributed to national maize research programs through the international testing network. As with the first group, this second group of pools would also be of service to CIMMYT's regional maize staff and to clients in national programs. All of those scientists could be responsible for incorporating traits from the pools into adapted materials with high yield potential and other characteristics required of a final product.

Dividing the pools into groups and reducing their total number should offer several benefits besides those already mentioned. As it is currently organized, the backup unit works with some 30 gene pools, giving roughly the same emphasis to each one, even though materials from some pools are in much greater demand than others. By concentrating more exclusively on high-priority traits in a select group of pools, more rapid progress could be made in improving those traits and meeting the germplasm needs of developing countries with much greater precision.

The advanced unit

Many similar opportunities for improving efficiency have arisen in the advanced unit, and to take advantage of them an approach is being considering which is roughly parallel to the one described for the backup unit. As with the gene pools, a large number of populations are currently being improved in the advanced unit. From each population, experimental varieties are developed and tested internationally. All populations, like the pools, are given about the same amount of emphasis and are handled according to the same breeding methodology.

The new approach being considered is to reduce the total number of populations and divide them into two groups, infinite-life and finite-life populations. The infinite-life populations, of which there would be one for each ecology/grain type and color/maturity category, would include populations already in existance and perhaps some new ones. As their name implies, these populations would be open-ended. For the next ten years or so, or until better materials were found to replace them, they would continue to receive new germplasm, and they would undergo improvement according to the same modified full-sib recurrent selection scheme presently in use and would be tested internationally within the International Progeny Testing Trial (IPTT) system. By working with a smaller number of the infinite-life populations, more attention could be devoted to the solution of problems, such as diseases and insect pests, that are specific to particular regions of the world.

The finite-life populations would be the primary means of focusing on the above problems. The original development and improvement of the finite-life materials would be carried out either by national program scientists, regional maize staff, the backup unit or other units within the CIMMYT Maize Program. These groups would employ whatever methodology seemed most appropriate for solving the region-specific problems at hand. If the materials produced showed sufficient promise for solving the problems, they could be proposed as finite-life populations, and if accepted, enter the IPTT system. These populations would be improved within the IPTT system for no more than two cycles and then be returned to their originators. The chief advantages of this approach are that it would sharpen the focus on region-specific problems, give greater flexibility in the use of various breeding methodologies, more fully involve national scientists in the development of improved germplasm and improve the mechanism for delivering that germplasm to national programs.

The new hybrid maize program

All of the changes described here involve modifications in systems already in progress in the CIMMYT Maize Program. The last topic to be covered is an entirely new effort at headquarters to serve national researchers who are interested in the development of maize hybrids.

In the past, the Maize Program has had no systematic means of assisting those researchers. But for a number of reasons, it is considered appropriate and necessary to expand CIMMYT's capacity to support hybrid work. Foremost among those reasons is that a growing number of national programs have now become interested in and have acquired the capacity to develop maize hybrids. In cases where hybrid development is considered as potentially successful, the aim for the future will be to help national programs go about this task in an efficient and cost-effective way.

CIMMYT has now established its own hybrid program (Figure 2), which will cater to the needs of national hybrid development efforts in several ways. One of those is to compile information about inbreeding depression and

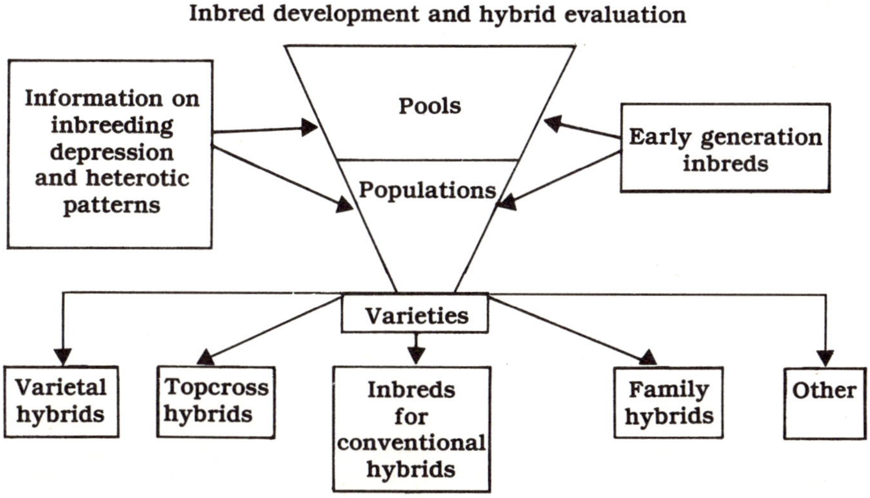

Figure 2. The production of early generation, inbred maize lines for increase by national programs for hybrid development, CIMMYT, Mexico

heterotic patterns for both the gene pools and populations. Another is to select superior materials during the improvement of pools and populations. They will be taken through a couple of generations of inbreeding and then be made available to national programs. Plans are also underway to provide national researchers with detailed information and training in various techniques of hybrid development.

For two reasons, much of CIMMYT's effort in this area will be devoted to the so-called nonconventional hybrids. First, there is not much published material on the development of family, topcross and variety hybrids, and it would be worthwhile to try to fill this information gap. Second, these types of hybrids, because they are much easier to produce than the conventional ones, should be a better option for many developing countries that have started or expect to start hybrid programs.

Conclusion

This overview of the program pretty much reflects its operations today, but is not necessarily the description CIMMYT staff members would have given ten years ago nor is it likely to be the one they give several years from now.

Important adjustments have been made in the CIMMYT Maize Program and will continue to be made as called for by changing circumstances and new opportunities for achieving greater efficiency. Thus, what is contemplated now is not a different program, but only the next step in its development, which has been going on for 20 years. Among the important developments that took place in the past were the initiation of the international testing network in 1974 and the creation of the first regional program during that same year. It is hoped that the changes now under consideration will be just as beneficial in making the CIMMYT Maize Program a more effective tool for agricultural development in the Third World.

3

The Aflatoxin-In-Maize Problem: The Historical Perspective

E.B. Lillehoj, Southern Regional Research Center, Agricultural Research Service, US Department of Agriculture, New Orleans, Louisiana, USA

Abstract

The discovery of aflatoxin in agricultural commodities introduced a new area of research—mycotoxicology. This review summarizes the historical development of aflatoxin problems in maize and the attendant responses in the agricultural research community. Originally, studies were focused on the stored commodity, since the two toxin-producing species, Aspergillus flavus *and* A. parasiticus, *were considered storage fungi. Laboratory investigations identified major factors influencing development of the toxin-producing species in storage, such as moisture, temperature, aeration and substrate. Various modifications of key components of the storage environment were developed to control the fungi. The discovery of preharvest infection of developing maize kernels by* A. flavus/ A. parasiticus *and subsequent production of aflatoxin introduced a new facet of mycotoxicology. The occurrence of storage fungi in the field required a radical reorientation of scientific thinking. The same environmental factors that influence fungi in storage appear to affect the processes of field infection/ contamination to varying degrees. However, in the developing maize ecosystems, agricultural activities dramatically influence interacting species, including host plants, attendant microbes, arthropods and other pests. In spite of the complexity of the preharvest contamination process, carefully conceived experiments have identified several important aspects of the problem. Contributing factors are assessed along with strategies for general control of aflatoxin-contaminated maize, with special emphasis on preharvest problems.*

Resumen

El descubrimiento de las aflatoxinas en los productos agrícolas introdujo una nueva área de investigación: la micotoxicología. En este trabajo se resume el desarrollo histórico de los problemas causados por las aflatoxinas y las respuestas correspondientes de la comunidad de investigación agrícola. En un principio, las observaciones se concentraron en el producto almacenado, ya que las dos especies productoras de toxinas, Aspergillus flavus *y* A. parasiticus, *se consideraban hongos de almacenamiento. Las investigaciones efectuadas en laboratorios identificaron los principales factores que afectan el desarrollo de las especies productoras de toxinas durante el almacenamiento, tales como humedad, temperatura, aereación y substrato. Se llevaron a cabo diversas modificaciones de los elementos del almacenamiento más importantes para combatir los hongos. El descubrimiento de la infección, en la fase de precosecha, con* A. flavus/A. parasiticus *de los granos de maíz en desarrollo y la subsecuente producción de aflatoxinas introdujeron una nueva faceta de la micotoxicología. La presencia de hongos de almacenamiento en el campo exigió una reorientación radical del pensamiento científico. Al parecer, los mismos factores ambientales que afectan a los hongos en condiciones de almacenamiento, afectaban en diversos grados los procesos de infección/contaminación en el campo. No obstante, en los ecosistemas del maíz en desarrollo, las actividades agrícolas influyen en forma drástica en las especies que interactúan, incluyendo plantas huésped, microbios correspondientes, artrópodos y demás plagas. A pesar de la complejidad del proceso de contaminación en la fase de precosecha, ciertos experimentos diseñados con sumo cuidado han identificado diversos aspectos importantes del problema. Los factores que contribuyen se evalúan con estrategias destinadas al control general del maíz contaminado con aflatoxinas, poniendo énfasis en los problemas relacionados con la fase de precosecha.*

Fungal contamination of agricultural crops has plagued farmers since the beginning of agriculture. The first crop plants were domesticated in relatively natural habitats, thereby maintaining competition in heterogeneous populations of biotypes. Human activities at first had miniscule effects on the environment, but as agriculture became established, new technology created the possibility of controlling the environment to a greater extent. An axiom of history is the profound change introduced into all aspects of human life by the Industrial Revolution, but nowhere has its impact been greater than in agriculture. With large-scale mechanization, plant breeding and soil fertilization practices, the age of intensive agriculture began, increasing dramatically after World War II. The most important agroecological aspect of these changes has been the homogenization of crop types and associated biota. This change resulted primarily from such factors as monocropping, pesticide use and an associated reduction in biological heterogeneity in large ecosystems.

Aflatoxin Discovery

The current epoch of fungal research in food/feed safety emerged as a result of an outbreak of disease in turkeys in England in 1960 (8). Fortunately, the English penchant for thorough and detailed explanations initiated a search for the causative agent of the disease; this effort was the beginning of a new area of agricultural research that has been labeled mycotoxicology. Initial histological examination of tissues from the diseased birds demonstrated an acute hepatic necrosis associated with bile-duct proliferation. Since the disease was not infectious, the possibility of poison was considered. Examination of bird rations showed that a common factor in disease outbreaks was the use of a Brazilian peanut meal (8). Heterogeneous distribution of toxin-tainted subsamples in larger lots indicated the possibility of microbial involvement. Subsequent tests showed the common occurrence of fungal isolates associated with the *Aspergillus flavus* group (8). Two closely related species, *A. flavus* and *A. parasiticus*, have since been identified as the toxin-producing species (17); these species are ubiquitous saprophytes that produce greenish-yellow spores and no ascospores. Although a trained mycologist can readily identify microscopic morphological differences between *A. flavus* and *A. parasiticus*, their macroscopic appearance is almost identical.

The toxic activity of fluorescent compounds in samples of contaminated peanut meals was identified in an innovative duckling bioassay system. Four closely related compounds were characterized and were generically named aflatoxins B_1, B_2, G_1 and G_2 (B = blue fluorescence; G = green fluorescence) with subscripts identifying relative chromatographic mobilities (Figure 1). Subsequent studies of aflatoxin production demonstrated the ability of *A. flavus* to produce exclusively B_1 and B_2, whereas *A. parasiticus* exhibited the capacity to produce all four toxins. Investigations of *A. flavus* populations obtained from toxin-contaminated commodities routinely provided single spore isolates that lacked the ability to produce toxin (8,9).

In contrast, *A. parasiticus* isolates with few exceptions were toxin producers. Once isolated, a significant fraction of *A. flavus* isolates lose the ability to produce aflatoxin in laboratory culture, whereas the toxin-producing trait in *A. parasiticus* is relatively stable. These observations indicate a profound difference in the genetic determinants of toxin production between the two closely related species. The mystery of this variation has been markedly increased by observations that aflatoxin-contaminated commodities generally contain predominantly B_1 and B_2, but a limited fraction of samples contain all four toxins (10,16,28,51,53). The results challenge the research-oriented intellect because they indicate distinct differences between processes that control infections and/or toxin production by the two toxin-producing species. Unravelling the mystery will provide provocative new information on the natural infection/contamination process.

Plant pathology/mycotoxicology

One aspect of the evolution of mycotoxicology has been the cooperation among mycotoxin researchers and traditional plant pathologists. Since plant pathologists generally study pathogens in the gene-for-gene context, the idea that a saprophyte could be an organism of interest was considered antithetical to generally accepted views of plant pathology. Some background of the history of this discourse is enlightening. Biologists with broad phenomenological backgrounds have attempted to explain the differences between parasites and saprophytes from various viewpoints. To avoid misconceptions, microbial activity in an ecosystem has been broadly grouped into two categories, biotrophy or parasitism (deriving nutrients from living material) and necrotrophy or saprotrophy (deriving nutrients from nonliving material) (61). Lewis (22) has expanded the description with a series of definitions that precisely describe these processes by subdividing them into obligate saprotrophy and obligate or facultative necrotrophy. Numerous mycotoxin-producing fungi can be saprophytic, but they also occur in living tissues and are defined by Lewis as facultative necrotrophs, i.e., species that are usually saprotrophic, but which can also function as parasites

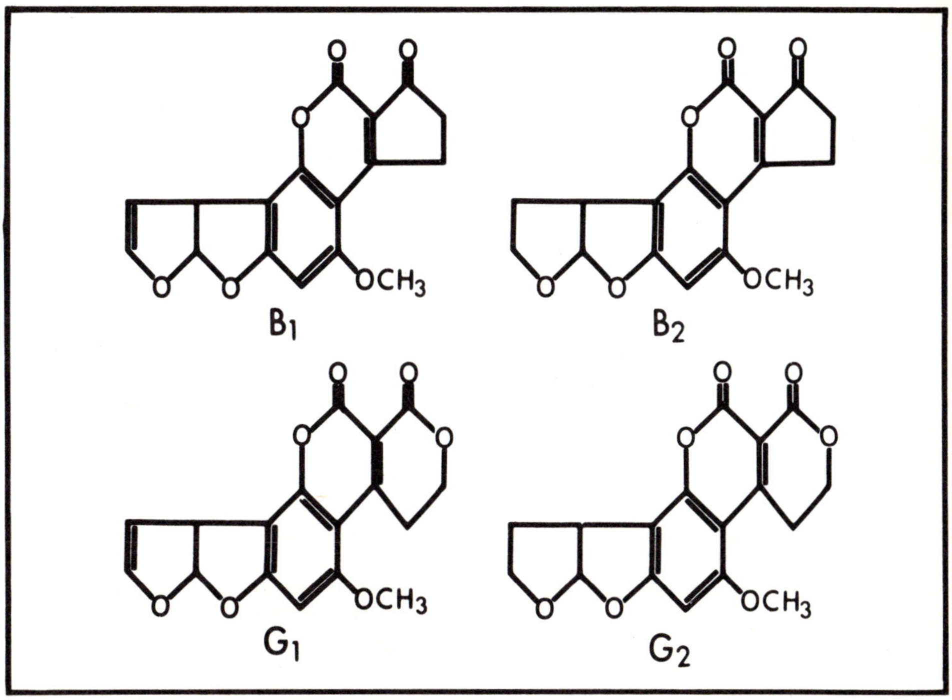

Figure 1. Structural formulae of the various aflatoxins

(8). Lewis presents a convincing argument that the established grouping of biological activity is arbitrary because the activities depend on the environment; individual responses in a species may range from extremes of parasitism to independent, saprotrophic development.

Plant pathologists have intensively considered the evolutionary origin of crop-plant pathogens. Nelson (38) has reviewed the coevolution of plant pathogens and their host plants and offers imaginative interpretations of this process. He contends that the gene-for-gene association between a microbe and a plant evolved in a stepwise progression, with gradual accumulation of genes for resistance in the plant and for virulence in the microbe. The evolutionary plateau was characterized by a balance between potential pathogens and hosts, a balance which could be upset by natural disasters that radically modified the environment. Harper (14) states, "Epidemic diseases (like epidemic outbreaks of pest populations) are not seen in natural vegetation—except after some major disturbance." Monocropping and other intensive agricultural practices can be characterized as typical of a natural catastrophe, with concomitant disequilibrium of the pertinent biota (23). The continual adaptation of crop lines to parasite challenge by stepwise incorporation of resistance to a multigene dependence represents an intriguing parallel to the evolutionary logic. A comprehensive view of contemporary agricultural practices clearly reveals the development of unique agroecosystem niches that are particularly suitable for microbes such as *A. flavus* and *A. parasiticus*.

Aflatoxin in Postharvest Maize

Storage fungi

Although the aflatoxin discoveries introduced a new era of interest in fungal toxins, the activity of fungi in reducing food/feed quality in storage has been an area of concern for a long time. The development of storage fungi in a postharvest commodity is determined by a number of factors, such as availability of inoculum, physical integrity of seed, moisture, temperature, aeration and nature of the substrate. Among the variables, moisture is clearly a dominant factor. Storage fungi, principally *Aspergillus* and *Pencillium* spp., are commonly found in maize stores at 13 to 18% moisture (6,7). The *A. glaucus* group predominates at 13 to 15% moisture, but above 15% other microbes appear, including the toxin-producing species *A. flavus, A. ochraceus* and *A. versicolor*. Lopez and Christensen (33) reported that *A. flavus* did not invade starchy grains below 17.5% moisture. In response to observations relating moisture levels to fungal growth, the US Department of Agriculture (USDA) assumed a very conservative position, describing drying techniques for control of mycotoxins in postharvest maize and recommending reduction of moisture to 13% within 24 hours after harvest (62).

Storage fungi water requirements

In defined media, an optimum a_W (available water) of 0.91 to 0.99 has been observed for growth of *A. flavus* and *A. parasiticus* (24,39,40). Although an a_W of 0.87 did not dramatically reduce fungal growth, aflatoxin production was restricted. The tests demonstrated an enhanced sensitivity of aflatoxin production to moisture levels in comparison with fungal growth processes. The studies also identified the xerotolerant rather than xerophilic properties of the toxin-producing fungi. The degree of xerotolerance in a competitive microbial environment of diminished water availability can be a key determinant in successful establishment of a specific fungus. The aflatoxin-producing species are relatively xerotolerant, but other

Aspergilli, particularly in the *A. glaucus* group, exhibit a greater degree of xerotolerance.

The moisture requirements in intact seeds clearly differ from submerged fermentations. In mature maize kernels, *A. flavus* does not routinely exhibit extensive growth below an a_W of 0.85 (33,46). However, at slightly higher levels (a_W/ 0.87), the fungus grows and produces aflatoxin (73). Although the observations indicate a precise requirement for available water, environmental factors in a storage bin would certainly influence the process. Competing microbial flora would distinctly affect the ability of a particular species to dominate. A number of microbial species have been identified as effective competitors with the aflatoxin-production strains. For example, *A. niger*, *A. oryzae* and *R. nigricans* can effectively reduce development of *A. flavus* and *A. parasiticus* (3,34,66). Microbial competition through augmentation of the indigenous microbes has been considered as a strategy for control of specific fungal species in stored commodities (D. Sauer, personal communication).

Moisture in stored maize

Although *A. flavus* appears to require at least 17.5% moisture for development on a starchy grain (33), the moisture distribution within a stored lot is critical. Moisture levels of high-moisture and low-moisture grain fractions equilibrate after blending (15). However, moisture contents of the fractions vary at equilibrium depending on whether the grain was adsorbing or desorbing moisture during equilibration (4). Moisture retention in high-moisture grain relative to drier grain has been attributed to hysteresis (15, 26). High-moisture (27 to 28%) and low-moisture (10%) maize blends that have mean moisture levels of 14% or less will support *A. flavus* development and aflatoxin production (26). Prerequisite conditions for spore germination and germ tube growth, such as relative humidity, temperature and spore densities, may determine critical aspects of initial *A. flavus* development, with a subsequent decrease in environmental restrictions by the developing mycelium. Christensen and Kaufmann (7) state, "Once the fungi have invaded grain, they will continue to grow in it at a lower moisture content than they otherwise would."

Fungi in stored maize

A detailed study was carried out on kernel deterioration in a bin of stored maize in southeastern Missouri, USA; the mean moisture level was in the safe range of 12 to 15% moisture. Microbial profiles of kernels from charred areas of apparent hot spots were compared with normal kernels from the bin (Table 1) (25). *Aspergillus flavus* was the most common *Aspergillus* species detected; incidence ranged from 10 to 80%. A similar pattern of *A. flavus* development was detected in earlier studies by Quasem and Christensen (42). *Aspergillus terreus* Thom was commonly detected in discolored kernels, but was rare in sound maize. Although previous investigations had identified the presence of *A. terreus* and *A. flavus* in deteriorating maize (12), the Missouri study elucidated the exclusive, simultaneous occurrence of the two fungi in highly discolored kernels. *Aspergillus ochraceus, Penicillium* spp., *Rhizopus* spp. and *Fusarium* spp. were widely detected in the better quality kernels, whereas *Absidia* spp., *Mucor* spp. and *Cephalosporium* spp. exhibited increased incidence in discolored kernels. A single sample was contaminated with aflatoxin; it also exhibited the highest incidence of *A. flavus* kernel infection (80%). The results demonstrated that fungi can develop in localized areas of high moisture in a stored commodity, even though the overall mean moisture is within the safe range.

Temperature effect on storage fungi

In conjunction with moisture, temperature plays an important role in the development of storage fungi. Generally, fungi grow readily between 20° and 30°C, with a restrictive range of 0° to 60°C (8). *Aspergillus flavus* exhibits optimum growth between 36° and 38°C, but ranges from 6° to 46°C (39,40,47,48). In laboratory media, maximum aflatoxin production has been observed at 25°C, with no toxin biosynthesis below 7.5°C or above 40°C (8,47). Cyclic temperature variation during development appears to increase aflatoxin yields (24,48), apparently due to the catabolic degradation of toxin by the producing organism. Since the fungus appears to simultaneously synthesize and degrade the aflatoxins, subtle environmental changes could dramatically vary the final levels of toxin. In the microenvironment of stored maize, the establishment of a fungal infection not only increases moisture levels through elaboration of respiration water but also increases temperature since the process is exothermic. Although aeration can reduce the formation of pockets of fungal activity, it also provides ideal air levels for the fungus. Developing new procedures for circulating inert atmospheres in stored commodities could accomplish the diffusion of hot spots without increasing oxygen availabilty (72). However, if moisture conditions are high in an anaerobic environment, a lactic acid fermentation routinely develops in the ensiling process. Accumulation of lactic acid in stored maize might be considered undesirable, since it changes the organoleptic properties of the commodity.

Other factors affecting storage fungi

The microbial profile of a freshly harvested crop influences subsequent competitive interactions. Damaged grain provides an opportunity for a fungus to circumvent the natural protection of the integuments and

Table 1. Incidence of microbes on samples of white maize stored in southeastern Missouri, 1974

Microbe	Incidence (%)[a]					
	1[b]	2	3	4	5	6
Aspergillus flavus Link ex Fries	27	23	80	50	30	10
Aspergillus terreus Thom	60	30	14	0	0	0
Aspergillus fumigatus Fresenius	0	3	3	0	0	0
Aspergillus niger van Tieghem	0	0	10	0	3	7
Aspergillus glaucus group	0	0	0	3	3	0
Aspergillus ochraceus Wilhelm	0	0	0	0	0	17
Penicillium spp.	0	0	3	20	3	27
Rhizopus spp.	0	0	14	37	0	0
Other *Mucorales* (*Absidia* and *Mucor* spp.)	27	37	43	17	17	0
Fusarium spp.	0	0	20	87	37	63
Cephalosporium spp.	27	10	0	0	6	0

a/ Incidence based on plating 30 surface-sterilized kernels from original samples on nutrient agar and enumerating the microbes after 4 days at 28°C. Percentages may total more than 100% since some kernels were infected by more than one microorganism. Samples 1-5 were from hot spots with varied moisture levels that ranged from 21-23% MC. Sample 6 represented the bulk commodity at an average MC of 14.5%.

b/ 1-5 discolored, 6 not discolored

Source: Lillehoj *et al.* (25)

establish infection sites in the vulnerable interior (60). Aeration can also be a particularly critical factor for storage microbes since fungi are aerobic. Reduced oxygen or increased carbon dioxide levels reduce fungal activity and toxin production (24). Development of inert atmospheres for maize storage has been considered as a practical approach (72). Breeding to develop grain with increased storage attributes has also been considered. Koehler (21) provided evidence for a relationship between genetically transmitted traits in kernel pericarp thickness and susceptibility to fungal infection; subsequent studies have identified other genetically mediated differences in stored maize kernels to invasion by storage fungi (5,36).

Aflatoxin in Preharvest Maize

Stored maize surveys/ inoculum source

In response to reports of aflatoxin occurrence in peanut meal, a number of surveys were carried out on US agricultural commodities to determine the extent of aflatoxin contamination. Examination of 1311 maize samples from the 1964 and 1965 crop years demonstrated that 2.3% contained the toxin (50). A similar examination of the 1967 crop year showed that 2.1% of the samples contained toxin (52). In a subsequent study of 293 samples of export maize collected at 10 ports, eight samples contained aflatoxin B_1 at levels ranging from 3 to 27 ppb (53). In a limited examination of 1969 and 1970 maize from Alabama, North Carolina, South Carolina, Tennessee and Virginia, 21 of 60 samples contained aflatoxin and 12 samples exceeded 20 ppb of B_1 (51). Because knowledge at the time limited *A. flavus/A. parasiticus* to storage, the results of the surveys were interpreted solely as postharvest development of the fungi. The limited information suggested that maize from the southern USA contained a higher incidence of the toxin.

The major questions that emerged from the surveys concerned the origin of the aflatoxin-producing fungal inoculum and regional variations. Conventional theory held that field fungi occurring in seed above 22% moisture included *Alternaria, Helminthosporium, Fusarium* and *Cladosporium* spp. (6,7) and indicated that relatively xerotolerant species such as *A. flavus/A. parasiticus* would not be able to compete with the less xerotolerant field fungi in developing maize. Therefore, the presence of a large inoculum of the *Aspergillus* spp. on kernels from the field appeared contradictory.

Joint USDA-industry research

In 1971, the US Food and Drug Administration (FDA) seized a product made from maize grown in southeastern Missouri (2). At the time, identification of poor storage conditions and those responsible for such conditions was required, and for this reason industrial representatives became involved in the problem. The American Corn Millers Federation aroused concern in the technical community about the hazards inherent in the aflatoxin problem. A particularly active research group at Quaker Oats, under the direction of H.W. Anderson, W.R. Wichser and E.W. Nehring, became directly involved. In response to industry concern, the Agricultural Research Service (ARS) of the USDA initiated a joint venture with Quaker Oats to acquire information on the nature of the aflatoxin contamination. Since the maize in question was a closely monitored commodity targeted exclusively for food use, industry representatives were confident that its postharvest handling was unimpeachable and chances of the toxin accumulating after removal from the field were negligible. In joint meetings among interested scientists, research approaches were developed to acquire fundamental information on the origin of the aflatoxin-producing fungi in developing maize. The Quaker Oats group concentrated on maize at locations producing white varieties, and the ARS team at the Northern Regional Research Center (NRRC), Peoria, Illinois, examined the developing crop in southeastern Missouri and southern Illinois.

Confirmation of aflatoxin contamination in preharvest maize

In 1972, initial cooperative projects were carried out to learn more about aflatoxin contamination of maize. The Quaker Oats team utilized earlier observations of the occurrence of *Aspergilli* in preharvest maize (56) to examine development of *A. flavus*. Ears were inoculated with spores of the fungus and some were sprayed with insecticides. Kernels inoculated between two weeks after flowering to maturity yielded aflatoxin with little effect from insecticide treatment (1). Visual observations of maize at various locations in the USA identified the presence of bright greenish-yellow fluorescence (BGYF) on preharvest kernels in Georgia. The fluorescence had initially been observed by Marsh and co-workers (35) in cotton fibers in association with *A. flavus* development. Investigation of the origin of BGY-fluorescing material demonstrated that it was not aflatoxin but a derivative of kojic acid, another relatively unique fungal metabolite. Bright greenish-yellow fluorescence in maize kernels was adapted as a presumptive test for presence of fungi in the *A. flavus* group (9,24). With the 1972 information as a guide, the Quaker Oats team carried out a field inoculation experiment in Georgia in 1973 that involved examination of environmental factors in the fungal invasion process. They observed a distinct association between occurrence of BGY fluorescence and kernel insect damage (1). They also noted increased aflatoxin contamination under stressed growing conditions produced by dense plant populations or reduced fertility. Individual kernels demonstrated aflatoxin levels exceeding 400 ppm.

The NRRC group developed two general approaches for examining preharvest aflatoxin occurrence in maize. The first was the establishment of field incidence of *A. flavus* and its toxic metabolites in a region covering several hundred square miles; the second was an intensive survey of harvested maize from a limited number of fields to determine the actual extent of contamination within fields. Since the FDA had seized maize grain produced in southeastern Missouri (2), the studies were carried out in the area with two specific objectives, the determination of the association of BGY fluorescence and insect damage with aflatoxin levels, and the elucidation of factors that might contribute to preharvest fungal infection and toxin formation.

Sixty ears from each of 60 fields in a four-county area of southeastern Missouri, 600 ears from two Missouri fields and 750 ears from five fields in southern Illinois were collected. Mycological studies of kernels demonstrated an average *A. flavus* incidence of about 5%, with elevated occurrence in kernels from insect-damaged ears (10). Although earlier reports had identified the presence of *A. flavus* in preharvest maize (43,44,58,59), the 5% incidence exceeded prior observations. Morphological tests identified elevated occurrence of *A. flavus* relative to *A. parasiticus* in kernels and insects (10,16,28,31). After shelling, drying and cracking, 237 samples of the 3600 ears in the general survey and 12 of 1350 ears in the intensive study exhibited BGY fluorescence (29). Aflatoxin tests showed that 120/3600 in the general survey and 6/1350 in the intensive study contained aflatoxin levels exceeding 20 ppb. Distinct regional and field-to-field differences were observed in toxin occurrence.

European corn borer [*Ostrinia nubilalis* (Hubner)] and corn earworm [*Heliothis zea* (Boddie)] caused extensive damage on test ears and the incidence of aflatoxin-positive maize was significantly higher in insect-damaged than in undamaged ears. In addition, 195 insects collected from ears yielded a 15% incidence of *A. flavus*. Evidence was provided for insect vectoring of the fungal inoculum by larval ingestion of spores and transfer to developing kernels. The observations of the Quaker Oats and USDA studies were published in 1975 (1,29), and provided convincing evidence for the preharvest aflatoxin contamination of maize kernels.

In addition to the studies of preharvest maize, a survey was conducted on the stored commodity for the presence of aflatoxin. Approximately 0.5 million bushels (12.5 million kg) of maize grain under Commodity Credit Corporation (CCC) loan were analyzed for aflatoxin. Aflatoxin was detected in 30% of 1283 truckloads of maize: 136 trucks, <10 ppb; 93 trucks, 10 to 19 ppb; 45 trucks, 20 to 29 ppb; 91 trucks, 30 to 100 ppb and 20 trucks, >100 ppb (54). Field and postharvest surveys demonstrated a broad occurrence of aflatoxin in both developing and stored maize. The incidence and levels of aflatoxin identified a serious agricultural problem.

Regional differences in preharvest aflatoxin contamination of maize

The FDA seizure of maize in South Carolina in early 1973 (65) prompted a study by ARS to acquire more information on the nature of aflatoxin contamination. Kernel samples of 4.5 kg each were collected in the field and at country elevators from an eight-county area (Table 2). Of the 184 samples taken from the field, 92 had detectable toxin levels and 62 had toxin above 20 ppb (31). Of the 113 samples taken at elevators, 60 had detectable levels of toxin and 32 had aflatoxin above 20 ppb. Mycological studies of the 152 aflatoxin-positive samples showed that 120 had one or more kernels internally infected with *A. flavus* (16,49). Of the 297 samples,

Table 2. Distribution of BGY fluorescense, aflatoxin, and mean aflatoxin B_1 levels between test areas of South Carolina maize samples, 1973

Area	Number of samples[a/]				Mean Aflatoxin B_1 (ppb)
	Total	BGY +	aflatoxin +		
1	40	38	20		22
2 and 3	3	3	3		35
4	16	10	3		3
5	32	21	15		18
6	151	103	72		56
7	14	9	8		45
8	41	32	31		78
Total	297	216	152	Overall mean	42

a/ Shelled samples obtained from picker-shellers or trucks at elevators during harvest

Source: Lillehoj *et al.* (31)

276 had one or more kernels exhibiting *A. flavus*. In 75 samples, all 50 test kernels exhibited the fungus. Of 375 insects collected from South Carolina, 274 contained *A. flavus*, with particularly high incidence levels in maize weevils. The South Carolina study provided a convincing follow-up to the original observations of field contamination by *A. flavus* and aflatoxin. The elevated levels indicated a high risk of contamination in maize grown in high-humidity, high-temperature areas of the southern part of the Corn Belt.

Controlled field-plot studies

In response to the observations of preharvest aflatoxin contamination of maize, a cooperative venture was initiated between NRRC scientists and M.S. Zuber at the University of Missouri. Dr. Zuber was the supervisor of one of the most productive maize-breeding groups in ARS and was considered to be the best choice to direct activities of cooperating field-oriented scientists. He brought to the project an extraordinary breadth of understanding of breeding in the context of production systems, and in addition provided access to key scientists both in ARS and in the state experiment stations. The first interregional, controlled investigation was designed and carried out in 1973. Normal and opaque-2 type of white and yellow hybrids were grown at Tifton, Georgia; College Station, Texas; Columbia, Missouri; and Peoria, Illinois (32).

Ears were inoculated with a hypodermic syringe through the husk into developing kernels, dispensing a measured volume of an *A. flavus* inoculum. Ears were harvested 15, 30, 45 and 70 days after treatment. The number of aflatoxin-positive ears and toxin levels increased generally from northern to southern locations. Although a hybrid difference was detected in overall assessment of aflatoxin accumulation, no variation could be attributed to normal versus opaque types (Table 3). Most of the toxin production occurred during the first 30 days after inoculation. Twelve percent of the physically damaged ears and 4% of the untreated ears contained aflatoxin. More than 90% of the damaged and control ears contaminated with toxin came from Georgia and Texas. The study provided proof that *A. flavus* infects developing maize if the fungus is introduced into developing kernels. The results also provided convincing evidence that

Table 3. Levels of aflatoxin B_1 in maize ears inoculated with *Aspergillus flavus* and grown at diverse locations, 1973

Experimental observations	Aflatoxin B_1[a] (mean ppb)
Hybrid	
I	59.8
II	15.3
LSR[b]	1.5
Endosperm type	
Normal	24.2
Opaque-2	37.8
LSR[b]	1.5
Days after inoculation	
15	11.7
30	36.8
45	45.2
70	43.2
LSR[b]	1.9
Location	
Illinois	2.4
Missouri	22.5
Texas	114.5
Georgia	133.9

a/ Aflatoxin B_1 levels are presented as geometric means of aggregate maize samples from ears inoculated with *A. flavus*.

b/ LSR is the least significant ratio (5% level) of two means

Source: Lillehoj *et al.* (32)

regional differences might be critical in determining the extent of fungal infection and toxin synthesis in the preharvest crop.

Available information: 1974

In 1974, field studies were carried out to examine the effect of maize genotypes and insecticide applications on preharvest aflatoxin contamination. The results underscored hybrid variation in susceptibility to contamination and the difficulty of completely controlling insects on developing ears (30,67,68). By the end of the 1974 crop year, a number of facts concerning the preharvest contamination process had been established:

- Yellow and white maize were equivalent in susceptibility to fungal infection;
- A positive relationship was observed between BGY-fluorescing particles and presence of aflatoxin;
- Aflatoxin contamination varied both intra- and interregionally;
- *Aspergillus flavus* predominated in aflatoxin-contaminated maize kernels and associated insects;
- Kernel damage by insects increased the potential for aflatoxin accumulation;
- Intensive insecticide application reduced but did not eliminate preharvest toxin production;
- *Aspergillus flavus* infection occurred from two weeks after flowering to physiological maturity, with maximum infection in the late-milk to early-dough stage (20 days post-flowering);
- Variation in timing of maize maturation appeared to be linked to contamination;
- Stress factors during crop development seemed to increase susceptibility; and
- Genotypic determinants, such as enhanced husk development, were linked to reduced preharvest aflatoxin contamination.

Preharvest aflatoxin contamination in midwestern maize

Observations of BGY fluorescence were made in the 1975 Iowa maize crop. Since prior information suggested that *A. flavus* infections might be restricted to southern regions of the USA, a study was carried out to identify the extent of *A. flavus* infection and aflatoxin accumulation before harvest in the Midwest. Of 214 freshly harvested samples, BGY fluorescence and aflatoxin were detected in 47% and 17% of the samples, respectively (27). Only four samples contained more than 20 ppb of alatoxin, and the highest level was 56 ppb with distinct field-to-field variation. Mycological studies identified *A. flavus*-infected kernels in approximately 60% of the samples. The freshly harvested samples provided evidence for extensive BGY fluorescence in Iowa maize, but only limited accumulation of aflatoxin. A visual examination of standing stalks was also carried out; about 400 ears were examined in each of eight fields in western Iowa. Husks were pulled back and ears examined for the presence of *A. flavus* spores prior to harvest (27). Visible *A. flavus* spores were invariably observed in insect damage tracks of the second generation European corn borer larvae (28). Mycological tests showed that *Fusarium moniliforme* and *Penicillium oxalicum* were widely distributed, with very few other fungal species detected. *Fusarium moniliforme* did not appear to be incompatible with *A. flavus*, since the two species were often found in the same insect damage tracks (28). The discovery of broad *A. flavus* infection and aflatoxin contamination of Iowa maize modified the view that the toxin was found exclusively in the south. Examination of weather conditions in the affected areas of Iowa showed extensive drought, particularly during the flowering period, with associated higher temperatures. The results demonstrated that maize in the Midwest was vulnerable to aflatoxin contamination during periods of stress.

Preharvest moisture

Information available in the mid-1970s characterized several environmental components that directly contributed to preharvest aflatoxin accumulation in maize. Moisture was a major factor among many that affected the contamination process. Although the xerotolerance of *A. flavus* provided an opportunity for competitive development at moistures between 17 and 22%, it was apparent that the fungus was infecting kernels at 50% moisture and above. To examine the moisture-related factors of preharvest toxin contamination, an interregional study was carried out in maize grown in Illinois, Missouri and Georgia (32). Ears were inoculated with *A. flavus* spores 20 days after flowering, and subsequent ear samples were collected at designated intervals. Fifteen days after inoculation (late-dough to early-dent stage) moisture levels ranged from 46 to 65%, and low levels of aflatoxin were detected in kernels from all locations. Although aflatoxin levels increased in Missouri and Georgia samples through physiological maturity (45 days after inoculation), continued aflatoxin production occurred only at the Georgia location. Toxin levels remained low in Illinois, with intermediate levels in Missouri. The results demonstrated that early aflatoxin contamination occurred in three diverse environments, but moisture levels did not appear to independently exert a controlling influence in the process.

Preharvest temperatures

In the field, moisture and temperature are obviously confounded, since elevated temperatures increase plant development rates, evaporation and water utilization. Although laboratory temperature studies have identified *A. flavus* as a mesophile, it does not exhibit any of the properties of an authentic thermophile (4,8,39,40,47). The observations of increased aflatoxin production under elevated temperatures during ear development raised perplexing questions.

To provide definitive information on temperature effects on developing maize ears, studies were conducted in controlled-environment chambers at North Carolina State University. Kernels were inoculated at three stages of maturity (early, medium and late dough) with subsequent development at 9.5, 12.5, 14.5 and 17.5 thermal units per day (57). Kernels were assayed for aflatoxin after 365 cumulative thermal units. Aflatoxin levels were significantly higher in kernels from plants grown at the highest daily temperature. Since other environmental factors were optimum, the results provided unequivocal evidence for the singular role of temperature on aflatoxin accumulation in developing kernels. Results from controlled environments cannot be extrapolated directly to field conditions, because the temperature in a crop environment is an integral part of other key factors, such as moisture, plant stress, microbial competition and weed competition. Subsequent studies of the effects of water stress on the developing crop have provided compelling evidence for an association between water deprivation and aflatoxin contamination (11,18,19).

Preharvest substrate

The nature of the substrate is obviously an important question in understanding preharvest infection of maize kernels by *A. flavus* and subsequent aflatoxin production. Numerous studies have identified the maturity of kernels as an important aspect of the infection process (1,18,19,24,75,76). Although maturity effects could be explained by insect-fungus-plant interaction, G. Payne and colleagues at North Carolina State University have demonstrated the ability of *A. flavus* to grow on degreening silks and to infect kernels directly without insect activity (20,41). These observations have made an important contribution to understanding the breadth of metabolic capabilities of *A. flavus*. However, there appear to be some relatively strict limitations on *A. flavus* substrate requirements. Broad natural occurrence of aflatoxin in US markets has only been observed in maize, cottonseed, peanuts, grain sorghum, millet, copra, tree nuts and figs (8,63,64). Absence of aflatoxin in freshly harvested soybeans presents an intellectual challenge to mycotoxicologists (55). Since soybeans are grown in the southern USA in close proximity to aflatoxin-contaminated maize, absence of inoculum cannot explain the inability of the fungus to establish a toxin-producing presence. A number of antinutritional factors in the plant, such as phytate, trypsin inhibitor and plant volatiles, have been identified as possible inhibitory factors (13,37,45,71). However, no single factor has been identified in developing soybeans that can be directly assigned the inhibitory function.

Research since 1975

As the information on aflatoxin accumulation in preharvest maize kernels evolved, it became apparent that a quick solution to the problem would not be found. Careful assessment of the situation by breeders in the cooperative research group indicated that the best long-range strategy for control involved identification of genetically mediated resistance in the host plants. The conceptual basis for appropriate approaches was first provided by M.S. Zuber and N.W. Widstrom. Early studies had identified variation among hybrids in *A. flavus* infection of kernels. The resistance was generally attributed to morphological resistance provided by increased husk cover (75,76). With the early observations as a guide, diallel studies have been carried out to identify inbred-linked differences (69,70,74). The investigations have provided clear differences among a limited number of inbred lines. However, these studies have shown that the inoculation technique is a critical factor. Introducing inoculum into the silk area without physical damage of kernels represents an entirely different mechanism of fungal access than placing the spores into developing kernels using a hypodermic needle or pinboard. The host-plant resistance involved in each of the inoculum procedures would reflect distinct differences in the morphological/biochemical process.

Information gathered by late 1974 on preharvest aflatoxin contamination has since been modified by several observations:

- Characterization of a definite association between elevated temperature during kernel development and increased aflatoxin accumulation;
- Elucidation of a mechanism for kernel infection by *A. flavus* without insect activity and toxin contamination of intact kernels;
- Identification of *A. flavus* resistance factors in inbred lines that reflect variations based on the inoculation method;
- Characterization of a direct correlation between water stress in developing maize and susceptibility to *A. flavus* infection and toxin contamination;

- Identification of drought conditions as a cause of low-level aflatoxin contamination in Midwestern maize, and widespread, high levels of toxin in the southern USA;

- Elucidation of increased aflatoxin accumulation by introduction of an *A. flavus* inoculum demonstrated by the absence of naturally occurring inoculum levels to achieve maximum preharvest levels of the toxin;

- Characterization of a limited distribution of fungal species in preharvest-contaminated maize with common occurrence of *A. flavus* and *F. moniliforme*; and

- Identification of naturally occurring factors in maize that inhibit *A. flavus* development.

Summary

Scientists gathered under the banner of mycotoxicology have shared unique experiences during the past 15 years. They have participated in the evolution of a new discipline. Creating a new area of inquiry can be controversial and the study of toxic fungal metabolites is no exception. The fundamental dilemma in mycotoxicology is its multidisciplinary nature; the scope and the diversity of professional interests make it difficult to establish a single discipline. The work inherently requires the expertise of microbiologists, plant pathologists, plant physiologists, veterinarians, entomologists, mycologists, agronomists, plant breeders, soil scientists, toxicologists, immunologists, oncologists, biochemists, chemists, public health scientists, epidemiologists, climatologists and nutritionists. In mycotoxicology the dialogue among practitioners has often resulted in recognizing common research interests and continuously learning new skills. For example, one complex question is the evolutionary logic for the acquisition of genetic information by microbes such as *A. flavus* for transformation of relatively innocuous, initial polyketides through a series of steps to the final carcinogenic, polycyclic hydrocarbon, aflatoxin. Without the interest of oncologists, pertinent structure-function questions might not have been raised among microbiologists. This example demonstrates the virtues of creative multidisciplinary exchanges. The spectrum of approaches in mycotoxicology appeals to scientists with wide-ranging views of nature and a compulsion to piece together bits of information that, to other scientists, may appear unrelated.

Reviewing studies on aflatoxin in maize provides a good example of the processes involved in mycotoxicology. At the outset, plant pathologists considered investigations of saprotrophs a somewhat pedestrian pursuit. Therefore, other scientists, particularly applied microbiologists, began to examine *A. flavus* invasion of living plant tissue and the elaboration of toxic secondary metabolites. Microbiologists challenged the established dogma of plant pathology and integrated information into new theories that included the ability of facultative necrotrophs, such as *A. flavus*, to infect developing maize kernels and produce aflatoxin before harvest. This important conceptual shift involved the recognition that the agronomic situation is dynamic, and that intensive cultural practices are radically altering the agroecosystem. The long-range importance of these observations is linked to the recognition that contamination of developing crop plant tissues by mycotoxins is not unique. The environmental factors responsible for the selection of toxin-producing fungi are in place and will continue to function, probably more aggressively. This premise is based on the continued narrowing of genetic diversity as greater amounts of land are committed to agriculture. The ever-increasing difficulty of producing safe food/feed commodities emphasizes the need to understand the basic aspects of the contamination process; understanding the process of natural contamination represents a great challenge to mycotoxicology.

A clear idea of the processes involved in preharvest aflatoxin contamination of maize kernels requires a detailed knowledge of the microbial activities involved in genetic regulation of secondary metabolite synthesis in the substrate of maturing seed. The survival advantage that aflatoxin production provides *A. flavus* is not clear. Without a definitive biological explanation of this phenomenon, the problem will remain relatively intractable. The cycle of *A. flavus*, from saprophytic development on decaying plant parts to facultative necrotrophy in living tissues, must be explained, with particular emphasis on build-up of inoculum potential.

Information on the competitive microflora in monocropping of maize or dicropping of maize/soybeans should be identified and procedures developed to augment natural competitors of *A. flavus*. The activity most likely to offer a rapid solution to *A. flavus* infection of developing maize kernels is the identification of genetic resistance in host plants and the incorporation of the character(s) into commercial cultivars. Problems in breeding relate to the opportunistic nature of *A. flavus*, which is not highly adapted for development on maize (compared with traditional maize pathogens). Therefore, the resistance will require innovative introduction of combinations of characters to achieve desired objectives. Breeders should tailor research strategies to the development of hybrids that have desirable traits in specific ecosystems. For example, hybrids with the specific ability to support nontoxin-producing microbes that would effectively compete with *A. flavus* should be developed. The incorporation of certain predator-control antinutritionals into hybrids should also be considered, with the parallel development of postharvest technologies to reduce the negative aspects of the substances in foods/feeds. Heat inactivation of the antinutritionals in soybeans, a process that provides a food-quality meal, could serve as a model. The future problem areas in mycotoxicology do not represent opportunities for the timid. However, the profession is not noted for timidity, and the scientists in the field are a particularly creative group who will provide exciting and provocative information in a new era of ecological understanding of agriculture.

References

1. Anderson, H.W., E.W. Nehring and W.R. Wichser. 1975. Aflatoxin contamination of corn in the field. Journal of Agricultural and Food Chemistry 23:775-782.

2. Anonymous. 1971. FDA recalls corn meal, bread mix allegedly tainted by toxin. Southwestern Miller 50:26.

3. Ashworth, L.J., H.W. Schroeder and B.C. Langley. 1965. Aflatoxins: Environmental factors governing occurrence in Spanish peanuts. Science 148:1228-1229.

4. Brockington, S.F., H.C. Dorin and H.K. Howerton. 1949. Hygroscopic equilibria of whole kernel corn. Cereal Chemistry 16:166-173.

5. Cantone, F.A., J. Tuite, L.F. Bauman and R. Stroshine. 1983. Genotypic differences in reaction of stored corn kernels to attack by selected *Aspergillus* and *Penicillium* spp. Phytopathology 73:1250-1255.

6. Christensen, C.M. 1957. Deterioration of stored grains by fungi. Botanical Review 23:108-134.

7. Christensen, C.M., and H.H. Kaufmann. 1969. Grain storage: The role of fungi in quality loss. University of Minnesota Press, Minneapolis, Minnesota, USA.

8. Detroy, R.W., E.B. Lillehoj and A. Ciegler. 1971. Aflatoxin and related compounds. *In* Microbial Toxins, vol. 6, A. Ciegler, S. Kadis and S.J. Ajl, eds. Academic Press, New York, N.Y., USA. Pp. 3-178.

9. Fennell, D.I., R.J. Bothast, E.B. Lillehoj and R.E. Peterson. 1973. Bright greenish-yellow fluorescence and associated fungi in white corn naturally contaminated with aflatoxin. Cereal Chemistry 50:404-414.

10. Fennel, D.I., E.B. Lillehoj and W.F. Kwolek. 1975. *Aspergillus flavus* and other fungi associated with insect-damaged field corn. Cereal Chemistry 52:314-321.

11. Fortnum, B.A., and A. Manwiller. 1985. Effects of irrigation and kernel injury on aflatoxin production in selected maize hybrids. Plant Disease 69:262-265.

12. Goodman, J.J., and C.M. Christensen. 1952. Grain storage studies. XI. Lipolytic activity of fungi isolates from stored corn. Cereal Chemistry 29:299-308.

13. Halim, A.H., C.E. Wassom, H.L. Mitchell and L.K. Edmunds. 1973. Suppression of fungal growth by isolated trypsin inhibitors of corn grain. Journal of Agricultural and Food Chemistry 21:1118-1119.

14. Harper, J.L. 1977. Population Biology of Plants. Academic Press, New York, N.Y., USA.

15. Hart, J.R. 1967. A method for detecting mixtures of artificially dried corn with high moisture corn. Cereal Chemistry 44:601-606.

16. Hesseltine, C.W., O.L. Shotwell, W.F. Kwolek, E.B. Lillehoj, W.K. Jackson and R.J. Bothast. 1976. Aflatoxin occurrence in 1973 corn at harvest. II. Mycological studies. Mycologia 68:341-353.

17. Hesseltine, C.W., O.L. Shotwell, M. Smith, J.J. Ellis, E. Vandegraft and G. Shannon. 1970. Production of various aflatoxins by strains of the *Aspergillus flavus* series. *In* Proceedings of the First US-Japanese Conference on Toxic Microorganisms, M. Herzbery, ed. US Department of the Interior, Washington, D.C., USA. Pp. 202-210.

18. Jones, R.K. 1981. Effect of nitrogen fertilizer, planting date, and harvest date on aflatoxin production in corn inoculated with *Aspergillus flavus*. Plant Disease 65:741-744.

19. Jones, R.K., H.E. Duncan and P.B. Hamilton. 1981. Planting date, harvest date, and irrigation effects on infection and aflatoxin production by *Aspergillus flavus* in field corn. Phytopathology 71:810-816.

20. Jones, R.K., H.E. Duncan, G.A. Payne and K.J. Leonard. 1981. Factors influencing infection by *Aspergillus flavus* in silk-inoculated corn. Plant Disease 64:859-863.

21. Koehler, B. 1938. Fungus growth in shelled corn as affected by moisture. Journal of Agricultural Research 56:291-307.

22. Lewis, D.H. 1973. Concepts in fungal nutrition and the origin of biotrophy. Biological Review of the Cambridge Philosophical Society 48:261-278.

23. Lillehoj, E.B. 1982. Evolutionary basis and ecological role of toxic microbial secondary metabolites. Journal of Theoretical Biology 97:325-332.

24. Lillehoj, E.B. 1983. Effect of environmental and cultural factors on aflatoxin contamination of developing corn kernels. *In* Aflatoxin and *Aspergillus flavus* in Corn, U.L. Diener, R.L. Asquith and J.W. Dickens, eds. Southern Cooperative Series Bulletin 279. Alabama Agricultural Experiment Station, Auburn, Alabama, USA. Pp. 27-34.

25. Lillehoj, E.B., D.I. Fennell and S. Hara. 1975. Fungi and aflatoxion in a bin of stored white maize. Journal of Stored Products Research 11:47-51.

26. Lillehoj, E.B., D.I. Fennell and C.W. Hesseltine. 1976. *Aspergillus flavus* infection and aflatoxin production in mixtures of high-moisture and dry maize. Journal of Stored Products Research 12:11-18.

27. Lillehoj, E.B., D.I. Fennel and W.F. Kwolek. 1976. Aflatoxin and *Aspergillus flavus* occurrence in 1975 corn at harvest from a limited region of Iowa. Cereal Chemistry 54:366-372.

28. Lillehoj, E.B., D.I. Fennell and W.F. Kwolek. 1976. *Aspergillus flavus* and aflatoxin in Iowa corn before harvest. Science 193:495-496.

29. Lillehoj, E.B., W.F. Kwolek, D.I. Fennell and M.S. Milburn. 1975. Aflatoxin incidence and association with bright greenish-yellow fluorescence and insect damage in a limited survey of freshly harvested high-moisture corn. Cereal Chemistry 52:403-412.

30. Lillehoj, E.B., W.F. Kwolek, A. Manwiller, J.A. DuRant, J.C. LaPrade, E.S. Horner, J. Reid and M.S. Zuber. 1976. Aflatoxin production in several corn hybrids grown in South Carolina and Florida. Crop Science 16:483-485.

31. Lillehoj, E.B., W.F. Kwolek, G.M. Shannon, O.L. Shotwell and C.W. Hesseltine. 1975. Aflatoxin occurrence in 1973 field corn. I. A limited survey in the southeastern U.S. Cereal Chemistry 52:603-611.

32. Lillehoj, E.B., W.F. Kwolek, E.E. Vandegraft, M.S. Zuber, O.H. Calvert, N.W. Widstrom, M.C. Futrell and A.J. Bockholt. 1975. Aflatoxin production in *Aspergillus flavus* inoculated ears grown at diverse locations. Crop Science 15:267-270.

33. Lopez, L.C., and C.M. Christensen. 1967. Effect of moisture content and temperature on invasion of stored corn by *Aspergillus flavus*. Phytopathology 57:588-590.

34. Maing, I.Y., J.C. Ayres and P.E. Koehler. 1973. Persistence of aflatoxin during fermentation of soy sauce. Applied Microbiology 25:1015-1017.

35. Marsh, P.B., M.E. Simpson, R.J. Ferretti, G.V. Merola, J. Donoso, G.O. Craig, M.W. Trucksess and P.S. Work. 1969. Mechanism of formation of a fluorescence in cotton fiber associated with aflatoxin in the seeds at harvest. Journal of Agricultural and Food Chemistry 17:462-466.

36. Moreno-Martinez, E., and C.M. Christensen. 1971. Differences among lines and varieties of maize in susceptibility to damage by storage fungi. Phytopathology 61:1498-1500.

37. Nagarajan, V., and R.V. Bhat. 1972. Factor responsible for varietal differences in aflatoxin production in maize. Journal of Agricultural and Food Chemistry 20:911-914.

38. Nelson, R.R. 1979. Some thoughts on the coevolution of plant pathogenic fungi and their hosts. *In* Host-Parasite Interfaces, B.B. Nickol, ed. Academic Press, New York, N.Y., USA. Pp. 17-25.

39. Northolt, M.D., and L.B. Bullerman. 1982. Prevention of mold growth and toxin production through control of environmental conditions. Journal of Food Protection 45:519-526.

40. Northolt, M.D., C.A.H. Verhulsdonk, P.S.S. Soentoro and W.E. Paulsch. 1976. Effect of water activity and temperature on aflatoxin production by *Aspergillus parasiticus*. Journal of Milk and Food Technology 39:170-174.

41. Payne, G.A. 1983. Epidemiology of aflatoxin formation by *A. flavus*. *In* Aflatoxin and *Aspergillus flavus* in Corn, U.L. Diener, R.L. Asquith and J.W. Dickens, eds. Southern Cooperative Series Bulletin 279. Alabama Agricultural Experiment Station, Auburn, Alabama, USA. Pp. 16-20.

42. Quasem, S.A., and C.M. Christensen. 1958. Influence of moisture content, temperature and time on the deterioration of stored corn by fungi. Phytopathology 48:544-549.

43. Rambo, G.W., J. Tuite and R.W. Caldwell. 1974. *Aspergillus flavus* and aflatoxin in preharvest corn from Indiana in 1971 and 1972. Cereal Chemistry 51:848-853.

44. Rambo, G.W., J. Tuite and P. Crane. 1974. Preharvest inoculation and infection of dent corn ears with *Aspergillus flavus* and *Aspergillus parasiticus*. Phytopathology 64:797-800.

45. Reed, C., and D. Penner. 1978. Peptidases and trypsin inhibitor in developing endosperm of opaque-2 and normal corn. Agronomy Journal 70:377-380.

46. Sauer, D.B., and R. Burroughs. 1980. Fungal growth, aflatoxin production, and moisture equilibration in mixtures of wet and dry corn. Phytopathology 70:515-521.

47. Schindler, A.F., J.G. Palmer and W.V. Eisenberg. 1967. Aflatoxin production by *Aspergillus flavus* as related to various temperatures. Applied Microbiology 15:1006-1009.

48. Schroeder, H.W., and H. Hein, Jr. 1968. Effect of diurnal temperature cycles on the production of aflatoxin. Applied Microbiology 16:988-990.

49. Shotwell, O.L., M.L. Goulden, E.B. Lillehoj, W.F. Kwolek and C.W. Hesseltine. 1976. Aflatoxin occurrence in 1973 corn at harvest. III. Aflatoxin distribution in contaminated, insect-damaged corn. Cereal Chemistry 54:620-626.

50. Shotwell, O.L., C.W. Hesseltine, H.R. Burmeister, W.F. Kwolek, G.M. Shannon and H.H. Hall. 1969. Survey of cereal grains and soybeans for the presence of aflatoxin. II. Corn and soybeans. Cereal Chemistry 46:454-463.

51. Shotwell, O.L., C.W. Hesseltine and M.L. Goulden. 1973. Incidence of aflatoxin in southern corn, 1969-1970. Cereal Science Today 18:192-196.

52. Shotwell, O.L., C.W. Hesseltine, M.L. Goulden and E.E. Vandegraft. 1970. Survey of corn for aflatoxin, zearalenone, and ochratoxin. Cereal Chemistry 47:700-707.

53. Shotwell, O.L., C.W. Hesseltine, E.E. Vandegraft and M.L. Goulden. 1971. Survey of corn from different regions for aflatoxin, ochratoxin, and zearalenone. Cereal Science Today 16:266-270.

54. Shotwell, O.L., W.F. Kwolek, M.L. Goulden, L.K. Jackson and C.W. Hesseltine. 1975. Aflatoxin occurrence in white corn under loan, 1971. I. Incidence and level. Cereal Chemistry 52:373-380.

55. Shotwell, O.L., E.E. Vandegraft, and C.W. Hesseltine. 1978. Aflatoxin formation on sixteen soybean varieties. Journal of the Association of Official Analytical Chemists 61:574-577.

56. Taubenhaus, J.J. 1920. A study of black and yellow molds of ear corn. Texas Agricultural Experiment Station Bulletin 270:3-38.

57. Thompson, D.L., E.B. Lillehoj, K.J. Leonard, W.F. Kwolek and M.S. Zuber. 1980. Aflatoxin concentration in corn as influenced by kernel development stage and post-inoculation temperature in controlled environments. Crop Science 20:609-612.

58. Tuite, J. 1961. Fungi isolated from unstored corn seed in Indiana, 1956-1958. Plant Disease Reporter 45:212-215.

59. Tuite, J., and R.W. Caldwell. 1971. Infection of corn seed with *Helminthosporium maydis* and other fungi in 1970. Plant Disease Reporter 55:387-389.

60. Thompson, R.A., and G.H. Foster. 1963. Stress cracks in artificially dried corn. US Department of Agriculture Marketing Research Report 631:1-24.

61. Thrower, L.B. 1966. Terminology for plant parasites. Phytopathologische Zeitschrift 56:258-261.

62. US Department of Agriculture. 1968. Preventing mycotoxins in farm commodities. US Department of Agriculture Research Service Report 20-16:1-7.

63. US Food and Drug Administration. 1968. Notices of judgment. FDA, Washington, D.C., USA. P. 37.

64. US Food and Drug Administration. 1969. Seizures and postal service cases. FDA, Washington, D.C., USA. P. 31.

65. US Food and Drug Administration. 1973. Seizures and postal service cases. FDA Consumer 7:32.

66. Weckbach, L.W., and E.H. Marth. 1977. Aflatoxin production in a competitive environment. Mycopathologia 62:39-45.

67 Widstrom, N.W., A.N. Sparks, E.B. Lillehoj and W.F. Kwolek. 1975. Aflatoxin production and lepidopteran insect injury on corn in Georgia. Journal of Economic Entomology 68:855-856.

68. Widstrom, N.W., A.N. Sparks, E.B. Lillehoj and W.F. Kwolek. 1976. Corn earworm damage and aflatoxin B_1 on corn ears protected with insecticide. Journal of Economic Entomology 69:677-679.

69. Widstrom, N.W., B.R. Wiseman, W.W. McMillian, W.F. Kwolek, E.B. Lillehoj, M.D. Jellum and J.H. Massey. 1978. Evaluation of commercial and experimental three-way corn hybrids for aflatoxin B_1 production potential. Agronomy Journal 70:986-988.

70. Widstrom, N.W., D.M. Wilson and W.W. McMillian. 1984. Ear resistance of maize inbreds to field aflatoxin contamination. Crop Science 24:1155-1157.

71. Wilson, D.M., R.C. Gueldner and R.A. Hill. 1983. Effects of plant metabolites on the *Aspergillus flavus* group and aflatoxin production. *In* Aflatoxin and *Aspergillus flavus* in Corn, U.L. Diener, R.L. Asquith and J.W. Dickens, eds. Southern Cooperative Series Bulletin 279. Alabama Agricultural Experiment Station, Auburn, Alabama, USA. Pp. 13-16.

72. Wilson, D.M., and E. Jay. 1975. Influence of modified atmosphere storage on aflatoxin production in high-moisture corn. Applied Microbiology 29:224-228.

73. Winn, R.T., and G.T. Lane. 1978. Aflatoxin production on high moisture corn and sorghum with a limited incubation. Journal of Dairy Science 61:762-764.

74. Zuber, M.S., O.H. Calvert, W.F. Kwolek, E.B. Lillehoj and M.S. Kang. 1978. Aflatoxin B_1 in an eight-line diallel of *Zea mays* infected with *Aspergillus flavus*. Phytopathology 68:1346-1349.

75. Zuber, M.S., O.H. Calvert, E.B. Lillehoj and W.F. Kwolek. 1976. Preharvest development of aflatoxin B_1 in corn in the United States. Phytopathology 66:1120-1121.

76. Zuber, M.S., and E.B. Lillehoj. 1979. Status of the aflatoxin problem in corn. Journal of Environmental Quality 8:1-5.

Biology of *Aspergillus flavus* and *A. parasiticus*

U.L. Diener and N.D. Davis, Department of Botany, Plant Pathology and Microbiology, Alabama Agricultural Experiment Station, Auburn University, Auburn, Alabama, USA

Abstract

Aspergillus flavus *and* A. parasiticus *are the only fungi that produce aflatoxin. Several media have been developed for identifying toxin formation without chemical analysis. Natural substrates are superior to artificial media for quantitative production of the aflatoxins. Pure cultures of fungi grown on yeast extract (2%) with various amounts of sucrose produce high yields of secondary metabolites. Aflatoxin formation is optimal at relatively high temperatures of 28° to 35°C. The amount of aflatoxin formed is also related to the fungal strain.* Aspergillus flavus *and* A. parasiticus *are capable of seed invasion under certain conditions, are pathogenic to insects, and grow as saprophytes on crop debris in the field and soil. Another source of primary inoculum may be the sclerotia of* A. flavus, *which form in damaged and intact maize kernels, and may overwinter in kernels dispersed onto the soil by combines.*

Resumen

Aspergillus flavus *y* A. parasiticus *son los únicos hongos que producen aflatoxinas. Se han creado diversos medios para identificar la formación de toxinas sin necesidad de recurrir al análisis químico. Los substratos naturales son mejores que los medios artificiales para la producción cuantitativa de aflatoxinas. Los cultivos puros de hongos cultivados en extracto de levadura (2%) con diversas cantidades de sucrosa producen altos rendimientos de metabolitos secundarios. La formación de aflatoxinas es óptima a temperaturas relativamente altas, 28° a 35°C. La cantidad de aflatoxina que se forma también se relaciona con la cepa del hongo.* Aspergillus flavus *y* A. parasiticus *pueden invadir las semillas en determinadas condiciones, son patógenos para los insectos y crecen en forma de saprofitos en los desechos de las cosechas que se encuentran en el campo y en el suelo. Otra fuente de inóculo primario suelen ser los esclerocios de* A. flavus, *que se forman en los granos dañados y sanos de maíz y que pueden sobrevivir al invierno en granos que han sido dispersados sobre el suelo por las máquinas segadoras.*

Aspergillus flavus Link ex Fries and *A. parasiticus* Speare, two closely related fungi, occur worldwide in the air and soil, and cause preharvest aflatoxin contamination in maize, peanuts, cottonseed and tree nuts. These seed-inhabiting fungi also contaminate a wide variety of crops after harvest, during handling, in storage and during processing. *Aspergillus flavus* and *A. parasiticus* are the only two fungal species that produce aflatoxin. However, not all of their isolates or strains produce aflatoxin in the field or in laboratory culture.

The two aflatoxin-producing species can be readily distinguished morphologically and by chemical analysis of metabolites (Table 1). Aflatoxin in maize is almost exclusively produced by *A. flavus*, which exhibits biseriate conidiophores and compact ivy-green colonies on culture media. Chemical analysis of *A. flavus*-contaminated maize typically reveals the presence of only aflatoxin B_1 and aflatoxin B_2. *Aspergillus flavus* appears to be the dominant fungal species in maize, cottonseed, tree nuts and, according to some authorities, peanuts. *Aspergillus parasiticus* has been isolated more frequently from peanuts than from any other crop. Some strains of *A. flavus* produce cyclopiazonic acid (CPA) in addition to aflatoxin, whereas no strain of *A. parasiticus* has been reported to produce CPA (10). Other research (11) has characterized *A. flavus* as proteolytic and *A. parasiticus* as lipolytic.

Media for Identification of Aflatoxin-Producing Fungi

An *Aspergillus* differential medium (ADM) has been developed for rapid detection of aflatoxigenic species; the medium contains ferric citrate, which induces orange-yellow (cadmium yellow) reverse coloration in agar plate colonies of *A. flavus* and *A. parasiticus* (1). Although toxigenic species of *Fusarium* and *Penicillium* fail to produce the pigment, other species of the *A. flavus* group and several *A. ochraceus* group species also produce the pigment. Recognizing the limited usefulness of ADM, streptomycin sulfate and Botran (2,6-dichloro-4-nitro-aniline) were incorporated into ADM (BSAD) to suppress the growth of bacteria and fungi (12); the BSAD medium was used for the detection of *A. flavus* in cottonseed following incubation for five days at 28°C. After 30 isolates of 11 *Aspergillus* species and 24 other fungal genera were screened, it was reported that only *A. flavus*, *A. parasiticus* and *A. oryzae* formed the orange-yellow pigmentation. Modifying ADM with yeast extract (2%), peptone (1%), ferric ammonium citrate (0.05%), and agar (1.5%) resulted in large colony growth and characteristic orange-yellow reverse pigmentation, enabling ready recognition of *A. parasiticus* and *A. flavus* after 42 hours of incubation at 30°C (16). After microbial inhibitors were added, this medium was named *Aspergillus flavus* and *parasiticus* agar (AFPA). Ferric ions and organic nitrogen were essential for color production. This medium has been used extensively with peanuts and soils, and results may be reproduced and compared with other enumeration media incubated for longer periods.

Another method for detecting aflatoxin-producing fungi utilizes the ultraviolet (UV) fluorescence of aflatoxin produced in a modified Czapek agar containing

Table 1. Key characteristics of the aflatoxin-producing fungi *Aspergillus flavus* and *A. parasiticus*

Characteristic	*A. flavus*	*A. parasiticus*
Conidiophore arrangement	Consistently biseriate	Mostly uniseriate, sometimes mixed
Conidia	Almost smooth to slightly roughened	Distinctly verruculose
Colony color	Yellow green	Ivy green
Colony surface	Irregular, some aerial hyphae	Compact, velvety
Chemical analysis	AFB_1[a/] + AFB_2 and cyclopiazonic acid (proteolytic)	AFB_1 + AFB_2 + AFG_1 + AFG_2 + AFM_1 (lipolytic)

[a/] AF = aflatoxin

maize steep liquor, $HgCl_2$, and $(NH_4)_2H_2PO_4$ (instead of $NaNO_3$); cultures are incubated at 28°C for seven to ten days (13). The presence of aflatoxin was confirmed by thin-layer chromatography (TLC) of chloroform extracts of the fluorescing agar. In Brazil, an agar medium containing commercial coconut extract (CAM) was developed to rapidly detect aflatoxin formation of *Aspergillus* spp. (15). On this medium, aflatoxin-positive isolates showed a characteristic blue or blue-green fluorescence in agar surrounding the colonies under UV light. Fluorescence appeared in 32 hours at 24° to 28°C with high aflatoxin-producing isolates, increasing in intensity with prolonged incubation. Incubation for three days was sufficient for detection of low aflatoxin production.

Media for *A. Flavus* Growth and Aflatoxin Production

Mycotoxin production by fungi in liquid media is usually substantially lower than on natural substrates. When a large quantity of a mycotoxin is needed for long-term feeding studies with experimental or domesticated animals, it is usually produced on natural substrates. The use of semisynthetic media permits the investigator to have precise control over several variables in the production of secondary metabolites by a specific fungus. With chemically-defined media, the nutrient requirements for maximum production of a mycotoxin, as well as specific aspects of the biochemical pathways of metabolite synthesis of the fungus, may be determined. Isolation of metabolites produced in the filtrate or in the mycelium is less complicated in liquid media, since there are fewer interfering substances than with most natural substrates.

Aflatoxin accumulation in liquid media is seldom comparable to that obtained in natural substrates (18,19). However, the combination of a high-yielding, aflatoxin-producing strain of *A. flavus* or *A. parasiticus* and media amended with maize steep liquor or yeast extract increased yields tenfold as compared to those first reported (9). Initially, aflatoxins were produced in Czapek medium with added zinc sulfate or yeast extract in Raulin-Thom medium (9). The first liquid medium developed by Auburn University investigators (8) for stationary culture was named SMKY and contained technical grade sucrose (200 g), magnesium sulfate (0.5 g), potassium nitrate (3.0 g) and Difco yeast extract (7.0 g). An Alabama isolate of *A. flavus* produced 1.0 to 1.5 ppm of aflatoxin B_1 at 25°C in ten days on SMKY medium; an isolate of *A. parasiticus* formed 1.3 ppm of aflatoxin B_1 and 2.5 ppm of aflatoxin G_1 at 30°C in 15 days (unpublished data).

A 2% yeast extract and 15 to 20% sucrose medium (YES) was developed that greatly increased the yields of aflatoxin B_1 to 150 to 340 ppm with strains of the *A. flavus* group in stationary liquid cultures (5). Technical grade sucrose was superior to chemically pure sucrose, probably because it contained additional iron and other nutrients needed for maximum aflatoxin formation. Difco yeast extract gave higher yields of aflatoxin than other brands tested. The YES medium apparently provided all of the nutrients required for aflatoxin production by several isolates of the *A. flavus* group (Table 2). The medium is widely used by mycotoxicologists for large-scale production of fungal toxins (2,6,7).

Research with a chemically-defined medium (4) revealed that glucose and sucrose were excellent carbon sources for aflatoxin production. Ribose, xylose and glycerol were also good carbon sources (3). The growth of *A. flavus* is supported by numerous organic and inorganic nitrogen sources, but aflatoxin synthesis reached comparatively high levels only with the addition of yeast extract or peptone. Inorganic nitrogen, such as KNO_3, was ineffective unless an amino acid was included in the medium. Aspartate, glycine, glutamine and glutamate were good sources of organic nitrogen for aflatoxin production. Magnesium, zinc, iron and possibly molybdenum were determined to be essential for aflatoxin formation, whereas manganese appeared to reduce aflatoxin yields. The optimum concentration for magnesium sulfate was 0.5 g/liter, the optimum level of zinc sulfate was 5 ppm, and that for ferrous sulfate was 1 ppm (4).

Optimum Temperature and Time for Aflatoxin Production

The optimum temperature for growth and aflatoxin B_1 production by *A. flavus* on both SKMY medium and sterilized peanuts was 25°C, whereas the optimum for *A. parasiticus* was 30° to 35°C (8). Optimum temperature for aflatoxin G_1 formation was 25°C on both SMKY medium and peanuts, the levels being five times greater on peanuts. Maximum aflatoxin levels were noted after nine days for *A. flavus* and after seven days for *A. parasiticus*. Rabie (17), using a Czapek medium plus 1% yeast extract, observed maximum *A. flavus* growth after 12 days at 18°C, whereas maximum aflatoxin B_1 was produced at 24°C and maximum aflatoxin G_1 at 30°C. Maximum production of aflatoxin B_1 in sterilized maize kernels occurred at 30°C for 21 days at 99 + 1% relative humidity (Diener and Davis, unpublished data). The results relate closely to the optimum temperatures for silk infection by *A. flavus* in the phytotron studies by Jones (14).

Variation in Isolates of the *A. flavus* Group

Only two species of the *A. flavus* group are widely recognized as aflatoxin producers, *A. flavus* and *A. parasiticus*, but Murakami has identified *A. toxicarius* as a toxin-producing species that other taxonomists consider to be *A. parasiticus* (20). Isolates of the two aflatoxin-producing species vary widely in growth rate, sclerotial production, enzyme production and production of secondary metabolites. There are nontoxigenic isolates of both fungi, high and low toxin producers and others that produce cyclopiazonic acid, aflatrem, kojic acid and aspergillic acid, as well as a variety of aflatoxins. Aflatoxin production by an *A. flavus* isolate can be lost or greatly reduced with successive agar slant transfers. Unless cultures are lyophilized and/or maintained in soil, under oil or on natural substrates, the toxin-producing ability may be lost or becomes erratic. In general, *A. parasiticus* isolates have

Table 2. Production of aflatoxins by selected isolates of *Aspergillus flavus* grown in YES medium[a/]

Isolate	Mycelial dry weight (g/100 ml)	Aflatoxin production (mg/100 ml)		
		B_1	G_1	Total (B+G)
Ala-2	2.6	3.8	3.2	7.0
Ala-6	4.6	17.1	14.4	31.5
Ala-8	3.7	15.2	1.4	16.6
NRRL 2999	4.3	24.7	20.8	45.5
ATCC 15517	5.3	28.5	24.0	52.5
ATCC 15548	6.5	34.2	28.8	63.0
ATCC 15547	2.1	0.1	0.1	0.2

a/ 15-20% sucrose medium

been more stable in culture and do not lose aflatoxin-producing ability as rapidly as *A. flavus* isolates.

Sources of Primary Inoculum in the Field

Since *A. flavus* lacks a sexual stage and moistened conidia lose viability in 21 to 60 days, the principal source of primary inoculum has been thought to be conidia produced from mycelium of the fungus that overwintered in plant debris and litter in the soil (14). In addition, it has been suggested that airborne spores can be dispersed by high altitude wind currents or rise from ventilating ducts on farm buildings and grain storage bins.

Sclerotia are one of the principal sources of primary inoculum of several important plant pathogenic fungi and may be significant in the life cycle (Figure 1) of *A. flavus* in its infection of

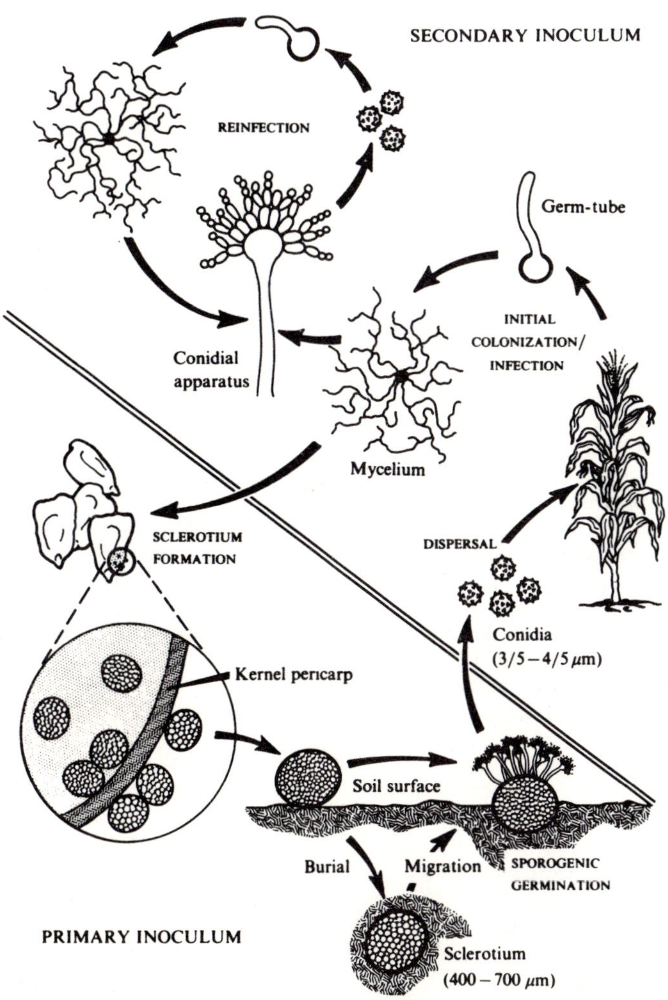

Figure 1. Schematic diagram (not to scale) showing the relationship between primary and secondary inoculum in the life cycle of *A. flavus*.

maize (22). Sclerotia of about 100 *A. flavus* isolates from both warm and cool latitudes formed readily in culture on potato-dextrose-yeast extract agar and were found to contain aflatoxins and three major indole metabolites: cyclopiazonic acid, aflatrem and dihydroxyaflavinine (21). Cyclopiazonic acid was detected primarily in sclerotia of isolates from the warmer latitudes, whereas aflatrem and dihydroxyaflavinine were detected in 85% of all strains examined. Sclerotia of *A. flavus* were produced during moist chamber incubation of maize kernels that were either naturally infected with the fungus or artificially inoculated (25). Sclerotia also formed on sterilized inoculated kernels incubated in the dark for 21 days at 25°C on sterile sand, sterilized soil and nonsterile garden soil. Exposure to light and incubation on nonsterile soil resulted in fewer sclerotia.

The first reported natural occurrence of sclerotia of *A. flavus* and immature scleroid cleistothecia of *Eupenicillium ochrosalmoneum* was from samples of insect-damaged and moldy maize ears collected from October to December of 1981 from standing, unharvested maize (24). In 1982, the moldiest maize contained the most sclerotia, both before harvest and in soil debris. Propagule density of *A. flavus* in the soil increased following harvest, indicating dispersal during combine harvesting. Sclerotium formation in preharvest maize appeared to be associated with kernel damage by biological agents (insects) or by mechanical means (23).

Sclerotial germination in *A. flavus* and *A. parasiticus* is sporogenic, with conidiophores and conidia produced directly from exposed sclerotium surfaces (22). Five of seven *A. flavus* strains and two of three *A. parasiticus* strains formed conidial heads on surface-sterilized sclerotia after 48 to 72 hours of incubation on moist sand. Sclerotia incubated on nonsterile field soil also germinated sporogenically. Sclerotia buried for ten days in nonsterile soil produced conidia within 72 hours after being washed to the soil surface. Most strains of *A. parasiticus*, which do not form sclerotia readily in culture, appear to be infrequent contaminants of maize in the field. It has not yet been established whether this fungus forms sclerotia on peanut pods and kernels and plant residues.

Aspergillus flavus sclerotia form naturally in standing maize and can be dispersed into field soils during combine harvesting. Sclerotia probably represent an important source of primary inoculum of *A. flavus* in maize, particularly in fields where maize is frequently grown. In addition to the primary inoculum sources previously mentioned, *A. flavus* and *A. parasiticus* are pathogenic to insects and sporulate on their dead bodies (22). If sclerotia are important as primary inoculum, then farm machinery that picks and shells maize could dislodge sclerotia from kernels and disperse them onto the soil. Also, with minimum tillage where the soil is not cultivated, sclerotia or fungus-infested kernels would remain on or near the soil surface, where they could produce conidia. Management practices designed to eliminate the importance of sclerotia as sources of infective inoculum may be essential for the control of alfatoxin and *A. flavus* in maize.

References

1. Bothast, R.J., and D.I. Fennell. 1974. A medium for rapid identification and enumeration of *Aspergillus flavus* and related organisms. Mycologia 66:365-369.

2. Davis, N.D., D.K. Dalby, U.L. Diener and G.A. Sansing. 1975. Medium-scale production of citrinin by *Penicillium citrinum* in a semisynthetic medium. Applied Microbiology 29:118-120.

3. Davis, N.D., and U.L. Diener. 1968. Growth and aflatoxin production by *Aspergillus parasiticus* from various carbon sources. Applied Microbiology 16:158-159.

4. Davis, N.D., U.L. Diener and V.P. Agnihotri. 1967. Production of aflatoxins B_1 and G_1 in chemically defined medium. Mycopathologia et Mycologia Applicata 31:251-256.

5. Davis, N.D., U.L. Diener and D.W. Eldridge. 1966. Production of aflatoxins B_1 and G_1 by *Aspergillus flavus* in a semisynthetic medium. Applied Microbiology 14:378-380.

6. Davis, N.D., G.A. Sansing, T.V. Ellenburg and U.L. Diener. 1972. Medium-scale production and purification of ochratoxin A, a metabolite of *Aspergillus ochraceus*. Applied Microbiology 23:433-435.

7. Davis, N.D., J.W. Searcy and U.L. Diener. 1969. Production of ochratoxin A by *Aspergillus ochraceus* in a semisynthetic medium. Applied Microbiology 17:742-744.

8. Diener, U.L., and N.D. Davis. 1966. Aflatoxin production by isolates of *Aspergillus flavus*. Phytopathology 56:1390-1393.

9. Diener, U.L., and N.D. Davis. 1969. Aflatoxin formation by *Aspergillus flavus*. *In* Aflatoxins, L.A. Goldblatt, ed. Academic Press, New York, N.Y., USA. Pp. 13-54.

10. Dorner, J.W., R.J. Cole and U.L. Diener. 1984. The relationship of *Aspergillus flavus* and *Aspergillus parasiticus* with reference to production of aflatoxins and cyclopiazonic acid. Mycopathologia 87:13-15.

11. El-Gendy, S.M., and E.H. Marth. 1980. Protoeolytic and lipolytic activities of some toxigenic and nontoxigenic Aspergillis and Penicillia. Journal of Food Protection 43:354-355.

12. Hamsa, T.A.P., and J.C. Ayres. 1977. A differential medium for the isolation of *Aspergillus flavus* from cottonseed. Journal of Food Science 42:449-453.

13. Hara, S., D.I. Fennell and C.W. Hesseltine. 1974. Aflatoxin-producing strains of *Aspergillus flavus* detected by fluorescence of agar medium under ultraviolet light. Applied Microbiology 27:1118-1123.

14. Jones, R.K. 1979. The epidemiology and management of aflatoxins and other mycotoxins. *In* Plant Disease, vol. IV, J.G. Horsfall and E.B. Cowling, eds. Academic Press, New York, N.Y., USA. Pp. 381-392.

15. Lin, M.T., and J.C. Dianese. 1976. A coconut-agar medium for rapid detection of aflatoxin production by *Aspergillus* spp. Phytopathology 66:1466-1469.

16. Pitt, J.I., A.D. Hocking and D.R. Glenn. 1983. An improved medium for the detection of *Aspergillus flavus* and *A. parasiticus*. Journal of Applied Bacteriology 54:109-114.

17. Rabie, C.J. 1965. Influence of temperature on the production of aflatoxin by *Aspergillus flavus*. *In* Symposium on Mycotoxins in Foodstuffs: Agricultural Aspects, Pretoria, Republic of South Africa. Pp. 18-29.

18. Shotwell, O.L., C.W. Hesseltine, R.D. Stubblefield and W.G. Sorenson. 1966. Production of aflatoxin on rice. Applied Microbiology 14:425-428.

19. Stubblefield, R.D., O.L. Shotwell, C.W. Hesseltine, M.L. Smith and H.H. Hall. 1967. Production of aflatoxin in wheat and oats: Measurement with a recording densitometer. Applied Microbiology 15:186-190.

20. Wicklow, D.T. 1983. Taxonomic features and ecological significance of sclerotia. *In* Aflatoxin and *Aspergillus flavus* in Corn, U.L. Diener, R.L. Asquith and J.W. Dickens, eds. Southern Cooperative Series Bulletin 279. Alabama Agricultural Experiment Station, Auburn, Alabama, USA. Pp. 6-12.

21. Wicklow, D.T., and R.J. Cole. 1982. Tremorgenic indole metabolites and aflatoxins in sclerotia of *Aspergillus flavus*: An evolutionary perspective. Canadian Journal of Botany. 60:525-528.

22. Wicklow, D.T., and J.E. Donahue. 1984. Sporogenic germination of sclerotia in *Aspergillus flavus* and *A. parasiticus*. Transactions of the British Mycological Society 82:621-624.

23. Wicklow, D.T., and B.W. Horn. 1984. *Aspergillus flavus* sclerotia form in wound-inoculated perharvest corn. Mycologia 76:503-505.

24. Wicklow, D.T., B.W. Horn, W.R. Burg and R.J. Cole. 1984. Sclerotia dispersal of *Aspergillus flavus* and *Eupenicillium ochrosalmoneum* from maize during harvest. Transactions of the British Mycological Society 83:299-303.

25. Wicklow, D.T., B.W. Horn and R.J. Cole. 1982. Sclerotium production by *Aspergillus flavus* on corn kernels. Mycologia 74:398-403.

Conditions that Affect Growth of *Aspergillus flavus* and Production of Aflatoxin in Stored Maize

D.B. Sauer, US Grain Marketing Research Laboratory, Agricultural Research Service, US Department of Agriculture, Kansas State University, Manhattan, Kansas, and J. Tuite, Department of Plant Pathology, Purdue University, West Lafayette, Indiana, USA

Abstract

The aflatoxin-producing fungus, Aspergillus flavus, *grows in maize in the field and also in stored grain after harvest. The most important factors affecting fungal growth and toxin production are moisture content and temperature; drying and cooling of the grain are the principal means of control. Other factors that affect* A. flavus *are oxygen and carbon dioxide concentration, physical damage to the grain, initial levels of mold contamination, insect activity and genetic differences in the maize.* Aspergillus flavus *does not grow at relative humidities below 85% or moisture contents below 16%. As moisture increases even slightly above these levels, aflatoxin risk increases greatly. If moisture is adequate, aflatoxin can be produced at temperatures ranging from 11° to 40°C, although 25° to 35°C is the optimal range. High-temperature drying increases susceptibility to fungal growth by increasing breakage susceptibility and lowering equilibrium moisture content. Aflatoxin risk increases with increasing levels of initial infection or spore contamination by* A. flavus *when the maize is put into storage. In spite of what is known about fungal growth and toxin production, there is still no reliable means of predicting aflatoxin risk under various storage conditions.*

Resumen

Aspergillus flavus, *hongo productor de aflatoxinas, se desarrolla en el maíz en los campos y también en el grano almacenado después de la cosecha. Los principales factores que afectan el crecimiento de los hongos y la producción de toxinas son el contenido de humedad y la temperatura; el principal medio de control es el secado y enfriado de los granos. Otros factores que afectan a* A. flavus *son las concentraciones de óxigeno y bioxido de carbono, el daño físico que presenten los granos, los niveles iniciales de contaminación por moho, la actividad de los insectos y las diferencias genéticas del maíz.* Aspergillus flavus *no crece en una humedad relativa de menos de 85% ni en contenidos de humedad de menos de 16%. A medida que la humedad sobrepasa, por más ligeramente que sea, estos niveles, aumenta en grado sumo el riesgo de que se produzcan aflatoxinas. Si la humedad es adecuada, las aflatoxinas pueden producirse a temperaturas que varían de 11° a 40°C, aunque el rango óptimo de temperatura es de 25° a 35°C. El secado a altas temperaturas aumenta la sensibilidad al crecimiento fungoso, ya que incrementa la tendencia a la fractura y reduce el contenido equilibrado de humedad. El riesgo de las alfatoxinas incrementa al aumentar los niveles iniciales de infección o la contaminación por esporas de* A. flavus *cuando se almacena el maíz. A pesar de todo lo que se sabe acerca del crecimiento fungoso y de la producción de toxinas, todavía no se cuenta con un medio fiable de predecir el riesgo de la infección por aflatoxinas bajo diferentes condiciones de almacenamiento.*

Aspergillus flavus has long been known as a storage mold affecting various seeds and grains. When aflatoxin was discovered in the early 1960s, first in peanuts and later in other commodities, it was generally assumed to be a storage problem. A decade passed before aflatoxin in maize was recognized as a field or preharvest problem as well. Now the preharvest aspects receive the most attention. Nevertheless, aflatoxin can be produced in stored maize, and attempts to manage the aflatoxin problem must consider both field and storage aspects and their interactions.

The environment in which *A. flavus* and maize interact is critical in determining whether aflatoxin contamination will occur. Controlling the environmental factors in storage is more feasible than controlling them in the field; manipulating the storage environment is the primary method of controlling postharvest fungi. In this paper, the environmental factors of temperature, moisture and interseed gas composition as they relate to growth and toxin production by *A. flavus* will be discussed. Grain handling, drying and aeration as they affect the potential for aflatoxin production will be reviewed, in addition to such biological factors as inoculum levels, competition among fungal species, insect activity and genetic differences among maize genotypes. Chemical control and prospects and limitations for predicting whether aflatoxin will be produced in specific situations will also be discussed.

Environmental Factors

Temperature and moisture are the most important considerations in determining whether aflatoxin will develop in maize after harvest; these two factors will be considered separately in this paper although they interact strongly. Below a certain moisture content (MC), *A. flavus* cannot grow and cannot produce toxin. Moisture will be considered in terms of relative humidity (RH) or water activity. For practical purposes, 85% RH is the lower limit for significant activity by *A. flavus*. Spore germination has been observed at 81 to 82% RH, and sporulation has been demonstrated at 83% RH (14), but many studies have shown 85% as the point at which the fungus begins to be a problem (8,10,14,25). There have been reports of growth and toxin production in rice and maize with moisture levels in equilibrium with relative humidities considerably lower (4,16). This can be explained as a problem with inadequate surface sterilization of the inoculated kernels. Grains with heavy inoculum or surface contamination can give erroneous indications of internal infection when surface sterilized and plated (26).

As RH increases above 85%, the growth of *A. flavus* increases dramatically. In this range even a small increase in moisture or RH can be very significant in terms of increasing the risk of aflatoxin contamination (Table 1) (8,10,14,25). There does not seem to be an upper limit of moisture that restricts growth or aflatoxin production in pure cultures with adequate aeration, but under normal conditions, very high moisture levels may be unsuitable for *A. flavus* because of competition from other microorganisms (7). At high MCs, other fungi, yeasts and bacteria may be better adapted than *A. flavus*, or the rapid rate of microbiological activity may produce anaerobic conditions unsuitable for *A. flavus*.

In practical situations, the MC of the grain rather than its water activity or RH is measured. There are many problems, however, in relating MC to the growth of *A. flavus* and production of aflatoxin. Different lots of maize may have different MCs when in equilibrium with the same RH.

Cultivar differences can affect equilibrium moisture content (EMC) as much as one percentage point. Maize that is absorbing moisture will equilibrate to a lower EMC than the same grain which is desorbing or losing moisture. Grain dried at a high temperature will have a lower EMC than if it were air dried at a lower temperature (25,33).

There are other factors that make it difficult to predict that a given lot of maize at, for example, 16% MC, is going to be free from *A. flavus* growth. A farmer or elevator operator may have a large quantity of 16%-MC maize which he believes is safe from *A. flavus* and aflatoxin, although perhaps not from other storage molds. Such an assumption may be invalid for several reasons. Because of sampling or measurement errors, all or part of the grain may actually have an MC higher than 16%. *Aspergillus flavus* and other fungi respond to the actual MC or interseed RH to which they are exposed, rather than to the reading given by a moisture meter. Temperature differences within a grain bulk will cause moisture to migrate from warmer to cooler areas, thus raising the moisture content in portions of the bulk. Moisture may increase as a result of rain leaking or blowing into a bin, or it may increase as a result of the growth and respiration of fungi or insects.

In spite of the difficulty in setting firm MC limits for the growth of *A. flavus*, some general guidelines are useful. At moisture contents of 16% or less, growth is not likely, but 17% may be high enough for slow growth and aflatoxin production (14,25,31). At 18 to 19% MC, *A. flavus* may grow quite rapidly in maize at favorable temperatures. Several studies have shown that considerably higher moisture contents may be optimal for growth and toxin production. Aflatoxin production in barley was greater at 28% MC than at higher or lower MCs (7). Aflatoxin was produced in rice at 24 to 26% MC, but not at 20% (5). Consistently, *A. flavus* growth in freshly harvested maize ranges from 20 to 28% MC, particularly when temperatures are in the 20° to 30°C range.

Table 1. Average moisture content, interseed relative humidity and aflatoxin B_1 content of freshly harvested, *A. flavus*-inoculated blends of wet and dry maize stored at 28°C

Average MC[a/] (%)	Interseed RH[b/] (%)	Aflatoxin B_1 content (ng/g) 14 days Hybrid A[c/]	14 days Hybrid B	28 days Hybrid A	28 days Hybrid B
16.5	84.2	ND[d/]	ND	ND	ND
16.7	85.1	ND	ND	ND	25
17.3	86.0	10	30	45	1100
17.6	86.9	25	750	150	1675
18.0	87.7	140	500	600	3375

a/ MC = moisture content
b/ RH = relative humidity
c/ Hybrids A and B = the two hybrids used to make the blends
d/ ND = not detected

Source: Sauer and Burroughs (25)

Temperatures as low as 8°C may allow the slow growth of *A. flavus*, and aflatoxin has been reported at temperatures of 11°C in sterilized rice and 14°C in nonsterile peanuts (10). Fungal growth or toxin production at minimum temperatures probably occurs only when other factors are nearly optimum. In other words, if both temperature and MC are suboptimal, the fungus is not likely to grow. Temperatures of 20° to 35°C are suitable for aflatoxin production, with 37° to 43°C probably being the upper limit (10,14,22,30). Fungal growth, however, can continue up to 44° to 46°C (30).

If both temperature and moisture are favorable for *A. flavus*, aflatoxin can be produced within 48 hours (23). When 21%-MC maize was put into bins at 30°C and cooled with refrigerated air, no aflatoxin and little *A. flavus* were detected if cooling was started immediately (Table 2). But if cooling was delayed 20 to 40 hours, *A. flavus* grew and aflatoxin was produced (9).

Atmospheric gas composition can limit fungal growth or aflatoxin production, but the principle has not been used in control measures, except for ensiling high-moisture maize. Several reports indicate that high carbon dioxide and/or low oxygen concentrations inhibit aflatoxin synthesis (11,13,35). The effect seems to be greater on aflatoxin production than on fungal growth and is more pronounced when temperature and MC are suboptimal (13).

Handling and Drying Effects

Mechanical damage to maize kernels makes them much more vulnerable to invasion by storage molds, including *A. flavus* (27,32). Under any given environmental conditions fungal growth may be several times faster in damaged compared to intact kernels. Cracks and breaks in maize are caused mainly by harvesting and handling equipment, although insect feeding may also be responsible for breaks in the pericarp. Damage to the attachment area (tip cap removal) or to

Table 2. Percentage of kernels invaded by *A. flavus* and *A. niger* after one week of storage, with and without cooling delays

Storage conditions	Kernels invaded after one week (%)			
	1970		1971	
	A. flavus	*A. niger*	*A. flavus*	*A. niger*
Cooled immediately	1	1	1	1
Cooling delayed 20 hours	50	28	6	6
Cooling delayed 40 hours	87	53	25	23

Note: Maize was field-shelled at 21-22% moisture and 30°C, then cooled to 3°C

Initial condition of grain:
1970: *A. flavus* 13/800 seeds, *A. niger* 3/800 seeds, mechanical damage 19.4%
1971: *A. flavus* 0/800 seeds, *A. niger* 2/800 seeds, mechanical damage 13.7%

Source: Converse *et al.* (9)

the germ causes much higher rates of respiration and fungal growth than does damage to other parts of the kernel (27,32). The amount of damage done to kernels by mechanical harvesting has been shown to vary with different kinds of machines and their adjustment, and also to be a function of grain MC and maize genotype (20,21). Careful attention to such variables could greatly reduce the susceptibility of maize to postharvest mold growth and aflatoxin contamination.

High-temperature drying may affect the growth of *A. flavus* in maize in several ways. High-temperature drying of maize results in a lower EMC (33). This means that at any given MC the water activity or RH is higher for artificially dried than for naturally dried maize. High drying temperatures may kill the seed, thus increasing susceptibility to fungal invasion when environmental conditions are suboptimal. High-temperature drying also makes the kernels brittle and more susceptible to breakage during handling.

Some of the fungi on and in the grain before drying may be killed by the dryer heat, but this effect is difficult to evaluate. For example, if other fungi are killed more readily than *A. flavus*, which appears to be reasonably thermotolerant (15), then *A. flavus* may grow faster because of a lack of competition or antagonism from other species. If populations of all fungi, including *A. flavus*, were greatly reduced by such a partial sterilization, then it is possible that a longer lag time would be necessary before a significant fungal population could build up. In either case the grain moisture would have to be high enough after drying to permit fungal growth, or the partial sterilization effect would not make any difference. Rewetting the grain could lead to problems here.

Low-temperature or ambient air drying is an alternative to high-temperature drying; it saves fuel and avoids the increased susceptibility to damage associated with high-temperature drying. Because low-temperature drying requires days or weeks to complete, high-moisture maize is exposed to the possibility of fungal attack and to contamination by mycotoxins before it is dried to a safe moisture level. When ambient air temperatures are below 15° to 20°C there is little risk of aflatoxin production during drying with sufficient, uniform airflow, because *A. flavus* will not compete well with other fungi at those temperatures. At higher ambient or initial grain temperatures, the risk of aflatoxin contamination increases. Maize with an initial MC of 25% is more vulnerable to fungal invasion than maize harvested at lower MCs. Some attempts to assess the aflatoxin risk potential in low-temperature drying have been unnecessarily pessimistic, because they assumed that the minimum RH, temperature or time reported for aflatoxin production would prevail even if the other environmental conditions were suboptimal or limiting (23).

The factors affecting fungal growth during low-temperature drying are the same as for storage in general. Risk increases with increasing MC and grain temperature at harvest. Damaged grain and fine material increase risk because of greater susceptibility to mold growth; they also impede airflow, making drying slower or uneven. Removing fines before filling the drying bin will reduce mold growth and aflatoxin potential.

To avoid having maize with very high MC in the upper part of the bin until the drying front has moved completely through the grain, some systems employ mechanical stirring devices to mix the wet and dry grain within the bin. With such systems, kernel-to-kernel equilibration reduces the MC of the wettest grain and thereby reduces its extreme vulnerability to fungal invasion. Another way to achieve the same goal is to mix wet and dry grain together as the drying bin is being filled. Disadvantages to the latter approach are the need to have a supply of dry grain, the possibility of uneven mixing, the extra handling, and the possibility of mold spores from the dry grain inoculating freshly harvested high-moisture grain.

Chemical Control

Organic acids, primarily propionic acid, have been successful in controlling storage molds (24); sulfur dioxide and ammonia have also been used with low-temperature drying systems. Their use requires considerable management expertise, because localized moisture increase and non-uniform treatment can result in enhanced growth of *A. flavus* and aflatoxin production. Elimination of competing organisms allows *A. flavus* and other important storage molds to proliferate dramatically. Chemicals such as thiabendazole and benomyl may have potential as preservatives (19); they are not corrosive and would not affect the marketability of the grain as the acids do.

Biological Factors

Perhaps the most significant biological factor in determining whether or not *A. flavus* will grow in stored maize is the microflora initially found in and on the grain. If there is little or no *A. flavus* on the grain at first, the chances of aflatoxin development are reduced (9,14). Competing microflora also play a role, depending on which species are present, their abundance and their ability to grow and compete under the prevailing environmental conditions. Early studies with rice showed that *A. glaucus* and *A. candidus* can reduce aflatoxin production, depending on initial inoculum levels and the storage environment (2,3). *Aspergillus niger* is commonly present on maize at harvest and may compete with *A. flavus* (34). Depending on the environmental conditions, aflatoxin may or may not be produced in the presence of other fungi (1). Maize damaged by southern corn leaf blight, and containing many other fungi, was found to be more susceptible to aflatoxin contamination (12); sorghum heavily invaded by field fungi was not predisposed to storage molds (29).

Insects that feed on maize ears in the field are known to predispose kernels to *A. flavus* infection through the physical damage caused by their feeding. Likewise, insect feeding in stored maize will open the kernels to fungal invasion. Insects also may act as vectors by carrying fungal spores on their bodies and by contaminating grain as they move about. A significant insect infestation may also raise the temperature and MC of the grain, making conditions more suitable for *A. flavus* growth.

There is evidence that maize cultivars differ in their susceptibility to the growth of storage fungi, including *A. flavus* (6,18). The mechanism of resistance may involve any of several physical or chemical factors that directly affect the ability of fungi to penetrate and invade kernels. Another approach to selecting maize strains with improved storability might be to select for resistance to mechanical damage. Genotypic differences in damage susceptibility have been demonstrated (20).

Prediction of Aflatoxin Risk

There is a significant quantity of information about the basic factors that affect growth of *A. flavus* and aflatoxin production. Temperature, moisture, oxygen, carbon dioxide, pH, mechanical damage, initial inoculum and competition from other fungi affect the process. However, on a practical basis, accurate predictions of whether or not a specific bin of maize will develop aflatoxin are not possible.

The application of recommendations for storage moisture and temperature is frequently handicapped in commercial practice because the storer does not have accurate knowledge of the amount of preharvest invasion or inoculum level of *A. flavus*, amount of kernel damage, history of the grain, range of moisture and temperature throughout the grain mass or the EMC of the grain. Any prediction system, regardless of its precision, is compromised without such information.

Agricultural engineers have worked for years with computer models to predict the rate of deterioration or the safe storage period for maize, based on respiration rates at different temperatures and MCs. Such models are generally useful, but they do not account for all variables, and they cannot predict aflatoxin development. For example, aflatoxin can be produced within the period normally considered safe for storage, which is defined as the time interval when less than 0.5% dry matter is lost (27).

Certainly storage conditions exist which prohibit aflatoxin production. For example, maize can be kept safely at MCs below 13% and a uniform temperature throughout the bulk. For economic and other practical reasons, such recommendations may justifiably be ignored. In the more marginal environments where problems may occur, there is a need for more accurate predictions of risk of aflatoxin formation and general mold spoilage.

Some of the literature on aflatoxin may not be useful or realistic in helping to determine the likelihood of aflatoxin production because of the experimental conditions used. Sterile or rewetted grain may have been used, resulting in abnormally high rates of growth or toxin production. High inoculum levels of *A. flavus* or the absence of other fungi normally found in maize may have similar effects on results. The strategies for these kinds of studies may have been to simulate worst-case scenarios or to simplify the experiment. At the interface of limiting conditions for *A. flavus*, various other fungi will be much more competitive and will usually exclude *A. flavus*. More research is needed in this area to determine whether a large amount of inoculum or extensive preinvasion by a particular fungus, including *A. flavus*, will permit its development near the limits of its environmental capabilities.

References

1. Bean, G.A., and J.S. MacFall. 1982. Microbial interactions as they affect aflatoxin production. Developments in Industrial Microbiology 23:221-228.

2. Boller, R.A., and H.W. Schroeder. 1973. Influence of *Aspergillus chevalieri* on production of aflatoxin in rice by *Aspergillus parasiticus*. Phytopathology 63:1507-1510.

3. Boller, R.A., and H.W. Shroeder. 1974. Influence of *Aspergillus candidus* on production of aflatoxin in rice by *Aspergillus parasiticus*. Phytopathology 64:121-123.

4. Boller, R.A., and H.W. Schroeder. 1974. Influence of relative humidity on production of aflatoxin in rice by *Aspergillus parasiticus*. Phytopathology 64:17-21.

5. Calderwood, D.L., and H.W. Schroeder. 1968. Aflatoxin development and grade of rough rice following prolonged storage in aerated bins. US Department of Agriculture, ARS Report 52-26.

6. Cantone, F.A., J. Tuite, L.F. Bauman and R. Stroshine. 1983. Genotypic differences in reaction of stored corn to attack by selected *Aspergillus* and *Penicillium* species. Phytopathology 73:1250-1255.

7. Chang, H.G., and P. Markakis. 1981. Effect of moisture content on aflatoxin production in barley. Cereal Chemistry 58:89-92.

8. Christensen, C.M., and C.J. Mirocha. 1976. Relation of relative humidity to the invasion of rough rice by *Aspergillus parasiticus*. Phytopathology 66:204-205.

9. Converse, H.H., D.B. Sauer and T.O. Hodges. 1973. Aeration of high moisture corn. Transactions of the American Society of Agricultural Engineers 16:696-699.

10. Diener, U.L., and N.D. Davis. 1970. Limiting temperature and relative humidity for aflatoxin production by *Aspergillus flavus* in stored peanuts. Journal of the American Oil Chemists Society 47:347-351.

11. Diener, U.L., and N.D. Davis. 1977. Aflatoxin formation in peanuts by *Aspergillus flavus*. Alabama Agricultural Experiment Station Bulletin 493. Auburn, Alabama, USA.

12. Doupnik, B., Jr. 1972. Maize seed predisposed to fungal invasion and aflatoxin contamination by *Helminthosporium maydis* ear rot. Phytopathology 62:1367-1368.

13. Epstein, D., M.P. Steinberg, A.I. Nelson and L.S. Wei. 1970. Aflatoxin production as affected by environmental conditions. Journal of Food Science 35:389-391.

14. Hunter, J.H. 1969. Growth and aflatoxin production in shelled corn by the *Aspergillus flavus* group as related to relative humidity and temperature. PhD Thesis, Purdue University, West Lafayette, Indiana, USA.

15. Kingsland, G.C. 1986. Relationship between temperature and survival of *Aspergillus flavus* Link ex Fries on naturally contaminated maize grain. Journal of Stored Products Research 22:29-32.

16. Lillehoj, E.B., D.I. Fennell and C.W. Hesseltine. 1976. *Aspergillus flavus* infection and aflatoxin production in mixtures of high-moisture and dry maize. Journal of Stored Products Research 12:11-18.

17. Lopez, L.C., and C.M. Christensen. 1967. Effect of moisture content and temperature on invasion of stored corn by *Aspergillus flavus*. Phytopathology 57:588-590.

18. Moreno-Martinez, E., and C.M. Christensen. 1971. Differences among lines and varieties of maize in susceptibility to damage by storage fungi. Phytopathology 61:1498-1500.

19. Moreno-Martinez, E., and J. Ramirez. 1985. Protective effect of fungicides on corn seed stored with low and high moisture contents. Seed Science and Technology 13:285-290.

20. Paulsen, M.R., L.D. Hill, D.G. White and G.F. Sprague. 1983. Breakage susceptibility of corn-belt genotypes. Transactions of the American Society of Agricultural Engineers 26:1830-1836, 1841.

21. Racop, E., R. Stroshine, R. Lien and W. Notz. 1984. A comparison of three combines with respect to harvesting losses and grain damage. American Society of Agricultural Engineers Paper 84-3016. St. Joseph, Missouri, USA.

22. Rambo, G., J. Tuite and G.L. Zachariah. 1975. Fluorescence associated with corn infected with *Aspergillus flavus* and *A. parasiticus* in storage. Cereal Chemistry 52:757-764.

23. Ross, I.J., J.O. Loewer and G.M. White. 1979. Potential for aflatoxin development in low temperature drying systems. American Society of Agricultural Engineers Paper 78-2-216. St. Joseph, Missouri, USA.

24. Sauer, D.B. 1973. Grain preservatives for high moisture feed grain. US Grain Marketing Research Laboratory Report, Manhattan, Kansas, USA.

25. Sauer, D.B., and R. Burroughs. 1980. Fungal growth, aflatoxin production, and moisture equilibration in mixtures of wet and dry corn. Phytopathology 70:516-521.

26. Sauer, D.B., and R. Burroughs. 1986. Disinfection of seed surfaces with sodium hypochlorite. Phytopathology. (In press.)

27. Seitz, L.M., D.B. Sauer and H.E. Mohr. 1982. Storage of high-moisture corn: Fungal growth and dry matter loss. Cereal Chemistry 59:100-105.

28. Seitz, L.M., D.B. Sauer, H.E. Mohr and D.F. Aldis. 1982. Fungal growth and dry matter loss during bin storage of high-moisture corn. Cereal Chemistry 59:9-14.

29. Seitz, L.M., D.B. Sauer, H.E. Mohr and R. Burroughs. 1975. Weathered grain sorghum: Natural occurrence of alternariols and stability of the grain. Phytopathology 65:1259-1263.

30. Sorenson, W.G., C.W. Hesseltine and O.L. Shotwell. 1967. Effect of temperature on production of aflatoxin on rice by *Aspergillus flavus*. Mycopathologia et Mycologia Applicata 33:49-55.

31. Trenk, H.L., and P.A. Hartman. 1970. Effects of moisture content and temperature on aflatoxin production in corn. Applied Microbiology 19:781-784.

32. Tuite, J., C. Koh-Knox, R. Stroshine, F.A. Cantone and L.F. Bauman. 1985. Effect of physical damage to corn kernels on the development of *Penicillium* species and *Aspergillus glaucus* in storage. Phytopathology 75:1137-1140.

33. Tuite, J., and G.H. Foster. 1963. Effect of artificial drying on the hygroscopic properties of corn. Cereal Chemistry 40:630-637.

34. Wicklow, D.T., C.W. Hesseltine, O.L. Shotwell and G.L. Adams. 1980. Interference competition and aflatoxin levels in corn. Phytopathology 70:761-764.

35. Wilson, D.M., and E. Jay. 1975. Influence of modified atmosphere storage on aflatoxin production in high-moisture corn. Applied Microbiology 29:224-228.

36. Wilson, D.M., E. Jay and R.A. Hill. 1985. Microflora changes in peanuts (groundnuts) stored under modified atmospheres. Journal of Stored Products Research 21:47-52.

Aflatoxicosis in Farm Animals

P.B. Hamilton, Department of Poultry Science, North Carolina State University, Raleigh, North Carolina, USA

Abstract
Aflatoxin has many effects on farm animals, including malabsorption of various nutrients, coagulopathy, decreased tissue integrity, poor growth, poor efficiency of feed conversion, enhanced susceptibility to infection, vaccine failures, drug failures, reproductive problems in males and females and increased sensitivity to temperature extremes. Toxic residues of aflatoxin in animal products present a hazard to public health. To date, regulatory guidelines have been based on legal considerations rather than experimental data. The available estimates are mainly from experiments designed to demonstrate an effect rather than establish a safe level. Consequently, an extremely important aspect of aflatoxin, i.e., its ability to interact deleteriously with other factors, has been virtually ignored. Appropriately designed experiments and statistical approaches have revealed that it is not feasible, either practically or theoretically, to determine a true minimum effective dose (MED) for aflatoxin in the laboratory. Epidemiological studies, coupled with laboratory experiments and mathematical corrections, have yielded an estimated MED based on the safety criterion of economic loss in broiler chickens (< 10 ppb). Additional experiments to establish a theoretical basis for the interactions of aflatoxin are needed before a rational MED can be established. It may be assumed that no level of aflatoxin is free of risk.

Resumen
Las aflatoxinas producen diversos efectos en los animales domésticos, incluyendo la malabsorción de diferentes nutrientes, coagulopatía, disminución de la integridad de los tejidos, crecimiento deficiente, deficiencia en la conversión de alimentos, mayor susceptibilidad a la infección, inutilidad de las vacunas, problemas reproductivos en machos y hembras, y mayor sensibilidad a temperaturas extremosas. Los residuos tóxicos de las aflatoxinas en los productos animales constituyen un peligro para la salud pública. Hasta la fecha, las normas reguladoras se han basado más en consideraciones legales que en información experimental. Las cifras existentes son fundamentalmente el resultado de experimentos diseñados para demostrar un efecto y no para establecer un nivel de seguridad. Por consiguiente, prácticamente se ha pasado por alto uno de los aspectos más importantes de las aflatoxinas, es decir, su capacidad de interactuar perniciosamente con otros factores. Los experimentos con un diseño apropiado y los métodos estadísticos han revelado que no es posible, ni práctica ni teóricamente, determinar en el laboratorio una verdadera dosis eficaz mínima (DEM) para las aflatoxinas. Estudios epidemiológicos, combinados con experimentos de laboratorio y correcciones matemáticas, han dado por resultado una DEM estimada, que se basa en el criterio de seguridad referente a la pérdida económica con pollos de engorda (<10 ppb). Se necesitan nuevos experimentos para establecer la base teórica de las interacciones de las aflatoxinas antes de que sea posible determinar una DEM racional. Puede suponerse que ningún nivel de aflatoxina está libre de peligro.

There are three main reasons for an international meeting on aflatoxin 25 years after its discovery:

- Aflatoxin causes myriad effects and economic losses in farm animals;
- Aflatoxin is extremely carcinogenic in animals and systems commonly used to assess the risk of carcinogenicity in humans; and
- Aflatoxin occurs in feedstuffs and foodstuffs around the world, reducing the value of many agricultural products.

Although human health concerns continue to be a motivating factor for aflatoxin in research, agricultural losses have provided most of the justification for research conducted on this toxin.

The most obvious consequence of aflatoxicosis in animals is mortality, which does not require highly trained personnel to diagnose and quantify. Turkey "X" disease in England, which led to the discovery of aflatoxin, acquired its notoriety from the death of over 100,000 turkeys, as well as chickens, ducks, pheasants and swine (10). In subsequent studies, an LD_{49} of aflatoxin was observed in a flock of laying hens 48 hours after they were fed a new supply of maize (5), and an entire flock of 15,000 turkeys perished after consuming aflatoxin-contaminated maize (8).

Mortality is merely one extreme in the broad range of effects produced by aflatoxin, some of which may be barely detectable. A typical field case of aflatoxicosis is marked not by mortality but by a decline in productivity with no visible disease symptoms. For example, Mertens (13) summarized nine independent studies showing eight signs of aflatoxicosis associated with reduced milk production in dairy cows. Most of the clinical signs are not specific for aflatoxicoses; they depend on length of time and dosage, and combinations of symptoms can vary from case to case. Similar conclusions were reached earlier in a survey of 94 confirmed cases of aflatoxicosis in swine, cattle and poultry (19). A consequence of the variation in aflatoxin symptoms is that accountants and bankers, rather than farmers, scientists and veterinarians, sometimes decide whether aflatoxicosis exists.

One extremely important aspect of aflatoxicosis is the possibility that toxic residues may be found in animal products consumed by humans. This possibility is especially critical in lactating animals, which excrete aflatoxin and its metabolites in milk, a major component of the diet of babies and children. Because the young of a species generally are more susceptible than adults, and because aflatoxin is highly carcinogenic, this problem evokes highly emotional responses, captures the attention of the news media, and provokes legal action by consumer groups and regulatory agencies. The available evidence implies a potentially severe problem. Armbrecht *et al.* (1) gave sows 100 ppb aflatoxin in their feed and found 1.5 ppb in their milk and 8.0 ppb in the livers of their suckling pigs. Nine independent experiments showed about 1% of the consumed aflatoxin occurring as a toxic metabolite in milk. It was calculated that 62 to 111 ppb dietary aflatoxin would result in 0.5 ppb in milk, depending on feed consumption and milk production. The action guideline for aflatoxin in milk in the USA is 0.5 ppb.

Dose-response relationships have been established for aflatoxin in animal products, and the information from controlled experiments has shown feed:tissue ratios (20) of 14,000:1 for beef liver, 2200:1 for eggs, 1200:1 for chicken liver, 800:1 for pork liver and 300:1 for cow's milk. There are several complicating factors in such assessments. Individual animals vary markedly in their ability to concentrate and excrete aflatoxin. Aflatoxin does not occur uniformly in feed. Nevertheless, extensive surveys have demonstrated a measurable incidence of toxic residues in marketplace samples of milk from the USA (20). The foregoing information seems to support the conclusion that it is safer for humans to consume animal products than suspect plant products.

Of all studies of aflatoxicosis in animals, the disease is best documented in chickens. Most of the principles governing aflatoxicoses in farm animals have become understood through studies of chickens; for example, the interrelationship of the toxic effects of aflatoxin was found in the bruising syndrome of broiler chickens. Apparently healthy birds exhibited bruises and hemorrhaging at slaughter, which resulted in downgraded and condemned meat and in consumers becoming dissatisfied with internal bruises that were not visible to inspectors. Experiments revealed that aflatoxin increases capillary fragility and reduces the ability of supporting tissues to cushion the blood vessels against blows (22). In addition, coagulopathy, resulting from a decrease in all components of the extrinsic and common pathways of blood clotting (3), causes a ruptured blood vessel to bleed longer.

Another economically important effect of aflatoxin in animal production is impairment of mechanisms of resistance to infectious agents. Aflatoxin impairs resistance to coccidiosis, a protozoan disease, *Salmonella* infections, infectious bursal disease caused by a virus, infections with the liver fluke *Fasciola hepatica* and to the mycosis caused by *Candida albicans* (8). Aflatoxin enhances animals' sensitivity to infectious agents by impairing practically all aspects of the immune system, including cellular, humoral and passive immunity (8). Though not always immediately apparent, drug failures (21) and vaccine failures (15) can also have damaging economic consequences. It should be noted that the immunosuppression caused by aflatoxin has no economic effect unless the compromised animal is challenged by disease.

Aflatoxin toxicity may depend on the nutritional status of the animal. The ability of aflatoxin to inhibit growth in chickens is dependent on the protein level of the diet (18), and interactions between aflatoxin and vitamin nutrition are well documented (8). Dietary fat interacts in a complex fashion with aflatoxin. In chickens, inefficient utilization of dietary lipids was indicated by as much as a tenfold increase in fecal lipids during aflatoxicosis (14); this steatorrhea was associated with a decrease in pancreatic lipase, bile salt concentration and decreased conversion efficiency of feed into animal products.

An unusual consequence of malabsorption that is of economic importance in some countries is the pale bird syndrome, which can be defined as failure of chickens to realize the color potential of their diet. The yellow color of chicken skins and egg yolks is attributable to carotenoids which must be absorbed from the diet. Many consumers will pay a premium price for poultry products high in color, which is achieved at considerable expense. Aflatoxin causes pale bird syndrome (21) by interfering with the absorption, transport, deposition and metabolism of carotenoids.

The effect of aflatoxin on animal reproduction has not received the attention it merits. An infertility syndrome of swine, apparently associated with aflatoxin, is characterized by abortion, mortality and poor growth of suckling pigs and by repeat breeding (R. Behlow, personal communication). In laying hens, aflatoxin produces a fatty liver syndrome and decreased egg production (4). The eggs produced during aflatoxicosis are also smaller than normal. Aflatoxin fed to fertile hens for only 48 hours decreased hatchability without changing egg production (11). Aflatoxin also impairs semen quality in male White Leghorn chickens (16).

The presence of aflatoxin in feed can manifest itself during extremes of environmental temperatures (9). Affected birds are more sensitive to both heat and cold. A related effect is increased sensitivity of chickens to sodium chloride (9).

A central problem in the animal production industry remains the establishment of safe and unsafe levels of aflatoxin. Several factors—legalities, analytical sensitivity, presumed safety to humans and animals, and economically affordable levels of aflatoxin—have all played a role in setting safe aflatoxin levels. It is noteworthy (perhaps scandalous) that safety has not been defined, and that the guidelines that have been established do not rest on experimental data. A lack of information has forced regulatory agencies to make subjective judgments about the value of the available data. A rational scientific study of safe levels of aflatoxin is essential for establishing appropriate criteria. For example, mortality might be useful, but is 2% acceptable, or is 8% acceptable? Also, mortality is rare compared to growth inhibition. Obviously, there are many ways to consider criteria for characterization of safe levels.

In a widely used experimental model, the growth of young broiler chickens was inhibited significantly ($P < .05$) by 2500 ppb aflatoxin, but not by 1250 ppb. These data can be interpreted as showing that aflatoxin below 2500 ppb is safe (6). However, these experiments were designed to show an effect, rather than to establish a safe level. They actually show that the safe level (minimum effective dose, MED) is between 1250 and 2500 ppb. Furthermore, data in the same publications show an MED of < 625 ppb (the lowest level tested) on susceptibility to bruising, immunosuppression, serum lipids, serum iron, plasma prothrombin, bile salts and pancreatic trypsin. The question on the proper criterion of safety becomes critical. Are sensitive physiological factors more valid than growth rate, which is economically more important? Any deviation from normal will be a disadvantage under some conditions. Stated in physiological terms, stressors below minimally effective doses combine in countercurrent stress to cause an effect (17). In more common terms, apparently ineffective factors interact synergestically.

Fundamental physiological principles play a unifying role in determining and understanding safe levels of aflatoxin. Off-range data from experiments designed to show an effect can be used to estimate an MED more precise than <625 ppb (6). If a straight line relation between response and dose is assumed, the slope of the assumed line determines the change in toxin concentration required to provide a change in response equal to the least significant difference determined by analysis of variance. Applying this simple approach to six growth-inhibitory variables with MEDs <625 ppb yielded calculated MEDs ranging from about 200 to 350 ppb. If the same ratio between growth inhibitory dose and physiologically effective dose (about 10:1) applies to single interactions, then a MED of about 20 to 35 ppb can be calculated.

In experiments designed to define a safe level of aflatoxin, the number of replicates per treatment, dosage increments and method of statistical analysis influenced the MED of aflatoxin on four dependent response variables (2). Overall, the statistical approach decreased the apparent MED on four variables by a factor of four, which lowers the estimated MED from 200 to 350 ppb to 50 to 88 ppb. It should be noted that this is a mathematical correction separate from the biology of the bird. How far can this approach be taken? It was calculated that about 400 replicates per treatment were necessary to detect true differences of 1% in body weight, which in the US broiler chicken industry represents about US one hundred million dollars annually.

Can the exhorbitant costs of determining a true MED be avoided? If a continuous relationship between aflatoxin and body weight with no threshold dose is assumed, and if the number of birds is increased to 10,000 per unit (approximate numbers at risk in poultry units), the MED would be lowered by a factor of 100 because the standard error of a treatment varies inversely with the root mean of number of observations upon which the treatment mean is based. Applying the factor to broiler chickens yielded a MED of <1 ppb. Unfortunately, such numbers of birds are essentially populations, not samples on which statistics are based. Consequently, other experiments using prohibitive numbers must be run or a practical MED must be based on field observations.

The available field data on a practical MED confirms the preceding considerations. A broiler chicken operation with two mills on the same bird management regime, using the same diet formula and receiving their feed ingredients from the same sources, had a 1% difference in feed conversion and a 2% difference in growth rate depending on the mill (7). The highest level of aflatoxin in problem feed was 30 ppb and frequency of contamination was 30%, compared to 6 ppb and 2%, respectively, in good feed. Reducing the level and frequency of aflatoxin to those of the good feed solved the problem.

A survey of five independent broiler chicken operations yielded similar results (12). Growers were classified as good, mediocre or poor, based on an objective index of productivity the prior year. Good growers produced more live birds with better body weight and feed conversion and fewer condemnations at slaughter. Good growers received about 5% more in payments than mediocre ones and 10% more than poor growers. The cost of production was about 10.5¢ per chicken for each class of grower. Good growers had a contamination frequency of 18% and a mean aflatoxin level of 6.1 ppb. Corresponding values for poor growers were 31% and 14.0 ppb.

The epidemiological studies suggest that the MED for aflatoxin in broiler chickens is below 10 ppb under field conditions where aflatoxin can interact with many other factors. The value agrees reasonably with values calculated from laboratory data and theoretical corrections. At present, there do not appear to be truly safe levels of aflatoxin. Interacting factors mean that a given amount of aflatoxin will not ensure a given response under field conditions. The higher the concentration and frequency of exposure, the higher the risk. Prudence dictates that any level probably carries a risk. Delineation and control of this risk should flow from the experimental and economic arena and not from the political arena, which has been dominant.

References

1. Armbrecht, B.H., W.T. Shalkop, H.G. Wiseman, J.E. Jackson and H.E. Zimmerman. 1969. The effects produced in brood sows by feeding aflotoxin. Toxicology and Applied Pharmacology 14:649-655.

2. Dixon, R.C., L.A. Nelson and P.B. Hamilton. 1981. Dose-response relationships during aflatoxicosis in young chickens. Toxicology and Applied Pharmacology 35:437-446.

3. Doerr, J.A., R.D. Wyatt and P.B. Hamilton. 1976. Impairment of coagulation function during aflatoxicosis in young chickens. Toxicology and Applied Pharmacology 35:437-446.

4. Garlich, J.D., H.T. Tung and P.B. Hamilton. 1974. The effects of short term feeding of aflatoxin on egg production and some plasma constituents of the laying hen. Poultry Science 53:2206-2211.

5. Hamilton, P.B. 1971. A natural and extremely severe occurrence of aflatoxicosis in laying hens. Poultry Science 50:1880-1882.

6. Hamilton, P.B. 1975. Determining safe levels of mycotoxins. Journal of Food Protection 47:570-575.

7. Hamilton, P.B. 1978. Fallacies in our understanding of mycotoxins. Journal of Food Protection 41:404-408.

8. Hamilton, P.B. 1982. Mycotoxins and farm animals. Refuah Veterinarith 39:17-45.

9. Hamilton, P.B., and J.R. Harris. 1971. Interaction of aflatoxicosis with *Candida albicans* infections and other stresses in the chicken. Poultry Science 50:906-912.

10. Hesseltine, C.W. 1979. Introduction, definition, and history of mycotoxins of importance to animal production. *In* Interactions of Mycotoxins in Animal Production. National Academy of Science (USA), Washington, D.C., USA. Pp. 3-18.

11. Howarth, B., and R.D. Wyatt. 1976. Effect of dietary aflatoxin on fertility, hatchability, and progeny performance of broiler breeder hens. Applied and Environmental Microbiology 31:680-684.

12. Jones, F.T., W.M. Hagler and P.B. Hamilton. 1982. Association of low levels of aflatoxin in feed with productivity losses in commercial broiler operations. Poultry Science 61:861-868.

13. Mertens, D.R. 1979. Biological effects of mycotoxins upon rumen function and lactating dairy cows. *In* Interactions of Mycotoxins in Animal Production. National Academy of Science (USA), Washington, D.C., USA. Pp. 118-136.

14. Osborne, D.J., W.E. Huff and P.B. Hamilton. 1976. Comparative effects of aflatoxin, ochratoxin and T-2 toxin on digestion in broiler chickens. Poultry Science 55:2075.

15 Pier, A.C., and K.L. Heddleston. 1970. The effect of aflatoxin on immunity in turkeys. I. Impairment of actively acquired resistance to bacterial challenge. Avian Diseases 14:797-809.

16. Sharlin, J.S., B. Howarth and R.D. Wyatt. 1980. Effect of dietary aflatoxin on reproductive performance of mature White Leghorn males. Poultry Science 59:1311-1315.

17. Selye, H. 1973. The evolution of the stress concept. American Scientist 61:692-706.

18. Smith, J.W., C.H. Hill and P.B. Hamilton. 1971. The effect of dietary modification on aflatoxicosis in the broiler chicken. Poultry Science 50:768-774.

19. Smith, R.B., J.M. Griffin and P.B. Hamilton. 1976. Survey of aflatoxicosis in farm animals. Applied Environmental Microbiology 31:385-388.

20. Stoloff, L. 1979. Mycotoxin residues in edible animal tissues. *In* Interactions of Mycotoxins in Animal Production. National Academy of Science (USA), Washington, D.C., USA. Pp. 157-166.

21. Tung, H.T., and P.B. Hamilton. 1973. Decreased serum carotenoids during aflatoxicosis. Poultry Science 52:80-83.

22. Tung, H.T., J.W. Smith and P.B. Hamilton. 1971. Aflatoxicosis and bruising in the chicken. Poultry Science 50:795-800.

23. Wyatt, R.D., M.D. Ruff and R.K. Page. 1975. Interaction of aflatoxin with *Eimeria tenella* infection and monensin in young broiler chickens. Avian Diseases 19:730-740.

Aflatoxicosis and Immunosuppression in Mammalian Animals

A.C. Pier, DVM, Department of Veterinary Science, University of Wyoming, Laramie, Wyoming, USA

Abstract

Aflatoxicosis in mammalian animals causes a variety of disease manifestations, which are related to the ability of aflatoxin to impair protein synthesis and react with macromolecules and cellular organells and to interfere with the normal production of cellular regulators. Acute aflatoxin poisoning causes hepatic necrosis, derangement of hepatic function, coagulopathy and extensive hemorrhagic lesions, frequently resulting in the death of the animal. Subacute or chronic aflatoxicosis causes fatty changes in the liver, enlargement of the gall bladder, periportal fibrosis with proliferative changes in bile duct epithelium, icterus and impaired growth rate or production. In addition to the liver, the thymus is also a primary target organ of aflatoxin. Consumption of aflatoxin causes thymic aplasia and marked suppression of cell-mediated immune responses, as well as nonspecific factors of the native defense mechanisms, including phagocytosis by macrophages and diminished production of complement (C4). T-cell populations of the peripheral blood and antibody titers are usually normal. Immunosuppressive effects are thought to arise from effects on antigen presentation and lymphokine production. Acute aflatoxicosis has a pathognomonic symptomatology and is relatively easy to diagnose. The signs of subacute or chronic aflatoxin poisoning are less definitive and so may go unrecognized because an animal may show only a reduced growth rate or increased susceptibility to infectious diseases to which the mycotoxicosis has predisposed it. Major economic losses from aflatoxicosis in animals are associated with the subacute or chronic forms, which are the most difficult to recognize. Carcinogenesis is not an important aspect of aflatoxicosis in food-producing animals.

Resumen

La aflatoxicosis en los mamíferos produce una amplia variedad de manifestaciones patológicas, relacionadas con la capacidad que tiene la aflatoxina de obstaculizar la síntesis de proteínas y reaccionar con las macromoléculas y orgánulos celulares y de interferir en la producción normal de reguladores celulares. El envenenamiento agudo con aflatoxina causa necrosis hepática, trastornos de la función hepática, coagulopatía y extensas lesiones hemorrágicas, que a menudo dan por resultado la muerte del animal. La aflatoxicosis sabaguda o crónica ocasiona alteraciones en la grasa del hígado, agrandamiento de la vesícula biliar, fibrosis periportal con cambios proliferativos en el epitelio de los conductos biliares, ictericia y velocidad de crecimiento o producción deficientes. Además del hígado, el timo es uno de los principales órganos atacados por la aflatoxina. El consumo de aflatoxina produce aplasia del timo y una marcada supresión de las respuestas inmunológicas en las que intervienen las células, así como factores no especificados de los mecanismos de defensa naturales, incluyendo fagocitosis por parte de los macrófagos y disminución de la producción de complemento (C4). Suelen ser normales las poblaciones de células T de la sangre periférica y las cantidades de anticuerpos. Se piensa que los efectos inmunosupresores se derivan de los efectos ejercidos sobre la presentación del antígeno y la producción de linfocitos. La aflatoxicosis aguda presenta una sintomatología patognomónica y su diagnóstico es relativamente sencillo. Los signos del envenenamiento subagudo o crónico con aflatoxina son menos definidos y, por ende, pueden pasar desapercibidos, ya que un animal puede mostrar únicamente un crecimiento reducido o una mayor sensibilidad a las enfermedades infecciosas a las que lo ha predispuesto la micotoxicosis. Las pérdidas económicas más importantes causadas por aflatoxicosis en animales se asocian con las formas subaguda o crónica, que son las más difíciles de diagnosticar. La carcinogénesis no constituye un aspecto importante de la aflatoxicosis en los animales que producen alimentos.

A quarter of a century ago, following a series of dramatic field outbreaks of feed-associated illness and mortality, a "new" animal disease was recognized. The agents proved to be a family of related mycotoxins, the aflatoxins, and thus aflatoxicosis was officially named and recorded in 1961 (5). This was not a new disease; similar clinical episodes had been recorded from animals foraging moldy maize in 1953, but the causative agent was not found (16). Undoubtedly, the aflatoxins have affected both human and animal health for many years, but they have attracted notice only in certain years and in certain geographic regions, following conditions fostering their production (12). It is largely through the reactions of animals to assorted doses and exposures of aflatoxin that the present concept of the biological effects of aflatoxicosis has emerged. This paper reviews the major biological effects of aflatoxin consumption in mammalian animals (laboratory animals, companion animals and livestock), as well as aflatoxic immunosuppression.

Aflatoxicosis in Mammalian Animals

The aflatoxins are important because they cause undesirable biological effects in animals and humans who consume them. Of the 18 or more related toxins of the aflatoxin family, aflatoxin B_1, the primary subject of this paper, is the most biologically active. The biological effects of aflatoxin are related to several discrete actions at the cellular level. Aflatoxin is a notable depressant of protein formation, in that it binds to DNA and suppresses RNA synthesis. Probably all levels of protein synthesis (i.e., transcription, translation and elongation) are involved, but the major effect of massive, prolonged exposure is transcriptional suppression.

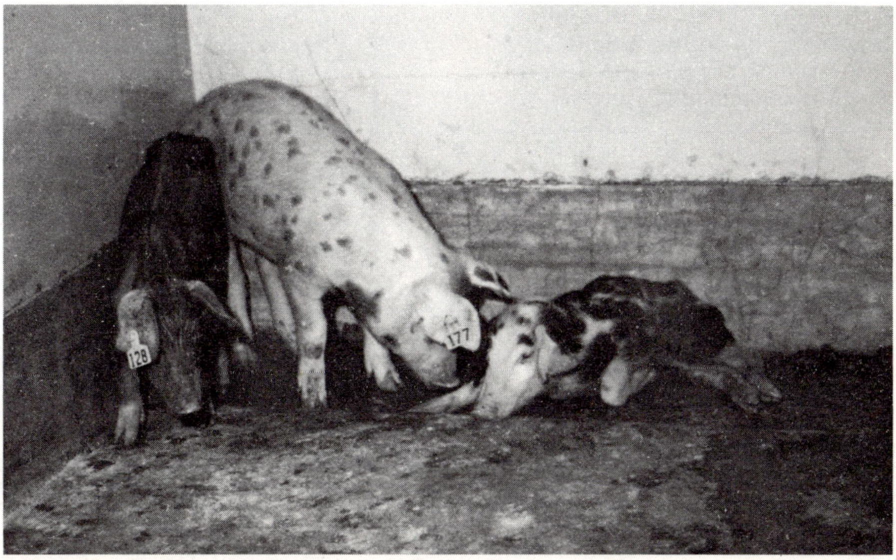

Range of symptoms in three pigs fed with the same aflatoxin-contaminated feed.

Alterations of carbohydrate and lipid metabolism and mitochondrial respiration are also known to result from aflatoxin exposure, and are apparently due to the interaction of aflatoxin with macromolecules (e.g., nucleic acids and proteins) and subcellular organelles (e.g., mitochondria and ribosomes), and to interference with the production of enzymes and other cellular regulators (2). Aflatoxin-related changes affect cellular function, cellular integrity, and ultimately animal health and productivity. In animals, clinical manifestations of aflatoxin consumption involve four general areas: acute, chronic liver damage, reduced rates of growth or productivity, impairment of immunologic responsiveness and carcinogenicity.

In aflatoxicosis, the primary target organ is the liver. Hepatic effects are changes in hepatic cellular function, cellular necrosis and proliferative repair activities. Single large doses of aflatoxin cause acute hepatitis, with swelling of the hepatic lobule and severe impairment of liver function (dry clearance) within three to six hours. Nuclear changes occur in hepatocyle within six hours and cellular necrosis within nine to 12 hours; there are elevations in serum levels of hepatocellular enzymes within 12 to 24 hours after toxin exposure (4). The acutely affected liver suffers centrilobular necrosis, congestion and hemorrhaging. Among the sequellae of altered hepatic function are icterus and impairment of clotting mechanisms. These, coupled with increased capillary permeability, result in widespread petechial hemorrhaging over mucosal and serosal surfaces and in free blood in the intestinal canal.

Following prolonged intake of lesser doses of aflatoxin, chronic hepatic involvement occurs, with a definitive pattern of fatty changes in the hepatocytes, centralobular necrosis, alteration of lobular architecture, periportal fibrosis and bile duct proliferation. The liver becomes yellow with a pebbly surface; the gall bladder is greatly enlarged and is filled with dilute bile. The affected animal is frequently icteric and has reduced hepatic function, prolonged prothrombin clotting time and elevated serum enzymes indicative of hepatic cell necrosis. Subacute or chronically affected animals have reduced growth rates (Figure 1) and milk production; generally, the animals perform poorly because of lowered feed conversion (3,8,10,13).

The susceptibility of animals to aflatoxin varies greatly. Different species of animals have different degrees of susceptibility, presumably

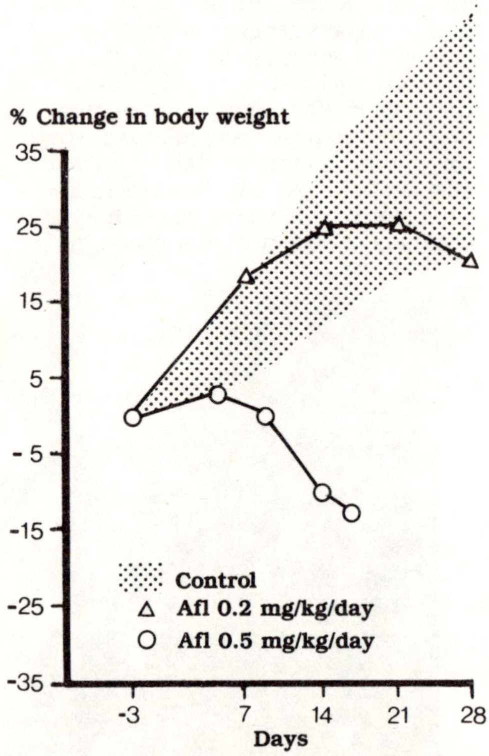

Figure 1. Effect of varying doses of aflatoxin on growth rate of calves

Source: Pier *et al.* (13)

depending on the metabolic pathways by which they are able to dispose of the toxin (6). The relative susceptibility of a number of laboratory animals, companion animals, livestock species and primates is presented in Table 1. The LD_{50} doses for cattle and horses are estimates based on limited experimental response data reported; because of the large quantities of aflatoxin involved, actual LD_{50} determinations for these species have not been made. The responses of different livestock species to continued experimental aflatoxin consumption are shown in Table 2.

Susceptibility within species is also variable. Young animals are invariably more susceptible than adults. Some data suggest differences in susceptibility between males and females of the species, but this has not been proved. Perhaps the least understood within-species variation is that found between individuals of closely related laboratory animal strains or littermates in livestock and poultry. Lethality and other biological responses (e.g., weight gain, immunological responsiveness and histopathological changes) are often highly variable among animals given the same doses of aflatoxin. In quail, within-species differences in lethality have been found to be heritable (9).

The effects of aflatoxin are also modified by interaction with other mycotoxins. Feed mixtures may include mycotoxins of different origins, and some of these are known to have

Table 1. Comparative susceptibility of animals to consumed aflatoxin

Rank	Approximate age or weight	Approximate single oral LD_{50} (mg/kg)
Highly susceptible		
Rabbits	3 months	0.30
Ducks	1 day	0.36
Cats	Adult	0.55
Swine	Weanling	0.62
Moderately susceptible		
Dogs	Adult	1.0
Horses	Adult	(0.6-1.0)[a]
Calves	1 month	(1.0-1.5)[a]
Turkeys	3 weeks	1.36
Guinea Pigs	250 g	1.4
Sheep	Adult	2.0
Baboon	5 kg	2.02
Monkeys (cynomologus)	2 years	2.2
Relatively resistant		
Chickens	1 week	6.5
Rats	Adult	5.5-17.9
Monkeys (macaque)	3+ years	7.8
Hamsters	1 month	10.2

[a] Estimated from response to experimental doses; actual LD_{50} data is not available for these species

Source: Pier (12) and Cheeke and Shull (2)

additive or synergistic effects with the aflatoxin. Rubratoxin has been shown to enhance the effects of aflatoxin in calves (13). It has been shown that the interaction of T-2 toxin with aflatoxin has an effect on lethality (7) and other biological functions (14) in laboratory animals.

In addition to causing acute, chronic disease in the liver, aflatoxin has a second target organ, the thymus. Subacute doses of aflatoxin B_1 and M_1 cause thymic aplasia in a variety of animal species (11). There is a critical depletion of cortical thymocytes, although circulating T-cell populations appear to be normal. The results of aflatoxin are immunosuppressive, with the primary effect occurring in the cell-mediated-immune (CMI) system. Among the CMI functions suppressed by aflatoxin are delayed cutaneous hypersensitivity, lymphoblastogenesis, leukocyte migration inhibition and graft-versus-host activity. Other defense mechanisms are also suppressed, including phagocytic activity of the macrophages and complement (C_4) formation. These two nonspecific factors strongly affect specific immunologic effectiveness through antigen opsonization and antigen presentation.

The final mechanism of aflatoxic immunosuppression is not clear; however, it appears to be primarily an effect on CMI rather than humoral (i.e., antibody) mechanisms. Very high

Table 2. Response of livestock to aflatoxin intake

Animal	Dose	Effect
Cattle		
Calf	0.08 mg/kg/day	Reduced weight gain
Calf	0.2 mg/kg/day	Reduced weight gain, coagulopathy
Calf	0.5 mg/kg/day	Hepatic necrosis, coagulopathy, death after 14 days
Steer	0.7 ppm ad lib.	Reduced weight gain
Steer	0.8 ppm ad lib.	Reduced delayed cutaneous hypersensitivity
Steer	1.0 ppm ad lib.	Death after 59 days
Cow	2.0 ppm ad lib.	Reduced milk yield
Cow	0.3 mg/kg/day	Inappetance, reduced milk yield, increased mastitis inflammation
Swine		
Pig	0.26 ppm ad lib.	Reduced weight gain
Pig	0.065 mg/kg/day	Immunosuppression, reduced weight gain
Pig	2-4 ppm ad lib.	Acute hepatitis, death
Horses		
Adult pony	0.075 mg/kg/day	Decreased liver function, coagulopathy, icterus, death after 37 days
Adult pony	0.15 mg/kg/day	Hepatitis, coagulopathy, death after 26 days

Source: Adapted from Brown *et al.* (1), Cysewski *et al.* (3, 4), Pier (2), Pier *et al.* (13) and Richard *et al.* (15)

doses of aflatoxin (2 to 40 ppm in feed) are known to cause diminution of immune globulins (IgA and IgG, but not IgM) and may reduce antibody titers. However, doses capable of causing reduced acquired immunity through vaccination (0.25 to 0.5 ppm) do not affect most antibody responses. Since the T-cell population in peripheral blood is unaffected, the fault is presumably a change in T-cell function, possibly associated with lymphokine production or effectiveness (11, 14). The spectrum of aflatoxin entities that affect lymphoblastogenic activity is very interesting. Aflatoxins B_2, G_1, G_2 and P_1 are relatively inactive against this function. Of particular interest is the fact that aflatoxin Q_1, which is relatively inactive toxicologically and mutagenically, is highly suppressive of lymphoblastogenesis; aflatoxicol, which is highly carcinogenic and mutagenic, is inactive in lymphoblastogenic suppression.

The clinical picture presented by animals affected by aflatoxin, therefore, is highly variable. With the sudden intake of large quantities of aflatoxin (2 to 5 ppm), which might occur with a batch of new feed or stored feed with pockets where high levels of aflatoxin have formed, an acute clinical disease typified by acute hepatic necrosis, coagulopathy and death with extensive hemorrhaging can be expected. This manifestation of the disease is easy to diagnose because of its pathognomonic symptomatology, the nature of the disease process and the dramatic nature of the effects on the primary target organ, the liver. However, this most severe form of aflatoxicosis is less frequently encountered than others.

The ingestion of smaller doses of aflatoxin (0.2 to 0.5 ppm) is more common, but results in a more protean disease whose clinical symptomatology is less definitive. Affected animals may show reduced rates of growth or production; some will be icteric and have definitive hepatic lesions, including centrilobular necrosis, periportal fibrosis with bile duct hyperplasia and enlarged gall bladders when examined at necropsy. The third clinical effect seen in animals has been called secondary mycotoxic disease, because the animal catches an infectious disease as a result of the ingestion of aflatoxin, which has suppressed its immunological responsiveness and natural defense mechanisms.

When reduced growth rate and an increased susceptibility to disease are the only clinical signs of aflatoxin, proper diagnosis is dependent on keen observation and good production records. Unfortunately, since this is the form in which aflatoxicosis in animals is most often encountered, proper diagnosis is often not made. The latter two manifestations of aflatoxicosis constitute the greatest economic loss in animals. In food-producing animals, with the exception of trout, aflatoxin-induced carcinogenicity is no problem, and in companion animals aflatoxin-induced carcinogenicity is a rarity; in laboratory animals, only rats and primates have shown a predeliction to carcinoma from aflatoxin consumption. Thus the effect of aflatoxin that constitutes the greatest public health anxiety is of little consequence in most lower animals.

The consequences of aflatoxin consumption in animals are many and varied. Unfortunately, the most frequently occurring forms of aflatoxicosis are both the most economically important and the most difficult to diagnose. More attention must be given to good clinical diagnostic technique (e.g., serum enzymes) or to chemical demonstration of aflatoxin residues in urine or other animal parts. With greater emphasis and improved technology in these areas, the true incidence of aflatoxicosis in livestock can be determined more easily.

References

1. Brown, R.D., A.C. Pier, J.L. Richard and R.E. Krogstad. 1981. Effects of dietary aflatoxin on existing bacterial intrammary infections of dairy cows. American Journal of Veterinary Research 42:927-933.

2. Cheeke, P.R. and L.R. Shull. 1985. Natural Toxicants in Feeds and Poisonous Plants. AVI Publishing Company, Westport, Connecticut, USA.

3. Cysewski, S.J., A.C. Pier, A.L. Baetz and N.F. Cheville. 1982. Experimental equine aflatoxicosis. Toxicology and Applied Pharmacology 65:354-365.

4. Cysewski, S.J., A.C. Pier, G.W. Engstrom, J.L. Richard, R.W. Dougherty and J.R. Tuhurston. 1968. Clinical pathological features of acute aflatoxicosis of swine. American Journal of Veterinary Research 29:1577-1590.

5. Goldblatt, L.A. 1969. Aflatoxin. Academic Press, New York, N.Y., USA.

6. Hsieh, D.P., Z.A. Wong, J.J. Wong, C. Michas and B. Ruebner. 1977. Comparative metabolism of aflatoxin. *In* Mycotoxins in Human and Animal Health, J.V. Rodricks, C.W. Hesseltine and M.A. Mehlman, eds. Pathotox, Park Forest South, Illinois, USA. Pp. 37-50.

7. Lindenfelser, L.A., E.B. Lillehoj and H.R. Burmeister. 1974. Aflatoxin and trichothecene toxins: Skin tumor induction and synergistic acute toxicity in white mice. Journal of the National Cancer Institute 52: 113-116.

8. Lynch, G.P., W.T. Shalkop, N.M. Jacoby, D.F. Smith and R.W. Miller. 1971. Responses of dairy calves to oral doses of aflatoxin. Journal of Dairy Science 54:1688-1698.

9. Marks, H.L., and R.D. Wyatt. 1979. Genetic resistance to aflatoxin in Japanese quail. Science 206:1329-1330.

10. National Academy of Sciences. 1979. Interactions of mycotoxins in animal production. *In* Proceedings of a Symposium. Washington D.C., USA.

11. Pier, A.C. 1986. Immunomodulation in Aflatoxicosis in Diagnosis of Mycotoxicoses, J.L. Richard and J.R. Thurston, eds. Martinus Nijhoff, Boston, Massachussetts, USA. Pp. 143-148.

12. Pier, A.C. 1981. Mycotoxins and animal health. Advances in Veterinary Science and Comparative Medicine 25:185-243.

13. Pier, A.C., S.J. Cysewski, J.L. Richard, A.L. Baetz and L. Mitchell. 1976. Experimental mycotoxicosis in calves with aflatoxin, ochratoxin, rubratoxin and T-2 toxin. *In* Proceedings of the United States Animal Health Association. Pp. 130-148.

14. Pier, A.C., M.J. Varman, R.R. Dahlgren, E.L. Belden and L.R. Maki. 1986. Aflatoxic suppression of cell mediated immune response and interaction with T-2 toxin. *In* Proceedings of the International Union of Pure and Applied Chemistry of Mycotoxins and Phytotoxins, Pretoria, South Africa. (In press.)

15. Richard J.L., A.C. Pier, R.D. Stubblefield, R.L. Lyon and R.C. Cutlip. 1983. Effect of feeding corn naturally contaminated with aflatoxin on feed efficiency, physiologic, immunologic and pathologic changes and on tissue residue in steers. American Journal of Veterinary Research 49:1294-1299.

16. Sipple, W.L., J.E. Burnside and M.B. Attwood. 1953. A disease of swine and cattle caused by eating moldy corn. *In* Proceedings of the American Veterinary Medicine Association. Pp. 174-181.

Mycotoxicoses in Farm Animals

R. Rosiles M., Pathology Department, Department of Veterinary Medicine and Animal Science, National Autonomous University of Mexico, Mexico, D.F., Mexico

Abstract
*Livestock and feed samples were analyzed for the presence of aflatoxin B_1, zearalenone, ochratoxin and scirpenols. Four mold isolates (*Penicillium, Fusarium, *and two* Aspergillus *spp.) were cultured on rice and used to inoculate feed for broiler chickens. Five groups of 20 chickens each were used; four groups were fed inoculated feed mixtures, and the fifth group was the control. Chickens showed feed-conversion deficiencies when aflatoxin B_1 levels in their feed reached 0.8 ppm. Chickens within a group showed macroscopic changes of differing intensity (e.g., changes in livers varied greatly). In 1984 and 1985 the number of aflatoxin-contaminated samples was fewer than in previous years, and the greatest demand for analysis was in chicken feed, followed by pig feed.*

Resumen
*Se examinaron ganado y muestras de alimento animal para determinar la presencia de aflatoxinas, zeralenona, ocratoxina y scirpenoles. Se cultivaron cuatro aislamientos de moho (*Penicillium, Fusarium *y dos especies de* Aspergillus*) en arroz y se utilizaron para inocular el alimento suministrado a pollos de engorda. Se emplearon cinco grupos de 20 pollos cada uno; a cuatro de ellos se les administró el alimento inoculado y el quinto grupo se usó como testigo. Los pollos presentaron deficiencias de conversión de alimentos cuando los niveles de aflatoxina en el alimento llegaron a 0.8 ppm. Los pollos de uno de los grupos presentaron cambios macroscópicos de diferente intensidad (por ejemplo, los cambios producidos en el hígado variaron considerablemente). En 1984 y 1985 el número de muestras contaminadas con aflatoxinas fue menor que en años anteriores; la mayor cantidad de solicitudes de análisis fue para el alimento para pollos, seguido por el alimento para cerdos.*

The mycotoxins produced by the fungus of the genus *Aspergillus* include a series of methoxy-benzene-bifuran derivatives, such as B_1, whose properties vary due to the addition or removal of radicals such as alcohol, phenol and hydroxyl, as well as the presence or absence of double bonds. An example is aflatoxin M_1, which is found in milk and formed by the addition of a hydroxyl radical to aflatoxin B_1. Aflatoxin P_1 is a phenolic metabolite found in the urine of animals that have consumed B_1.

The biological effects of aflatoxins on animals vary, ranging from tumors in rainbow trout to mutations and terotogenesis in rats and chicken embryos (1,3,13). The principal effect on livestock, however, is liver damage. The Peking duck is especially sensitive to the effects of aflatoxins, and for that reason is often used in aflatoxin-related biological experiments.

The trichothecenas, which are produced mainly by *Fusarium* spp., comprise another important mycotoxin group. These include the scirpenols such as the T-2 toxin and zearalenone (1,7,9). Their effect on animals varies according to the compound and the length of time and degree of exposure to it. Intraperitoneal injections of trichothecene extracts in laboratory animals cause fibrous peritonitis (1,2). The biological effect of toxin T-2 can be verified experimentally on rabbit skin tissue, which dies on contact with the scirpenol. The effects of zearalenone have been studied on virgin sows, in which it produces hyperestrogenism with degeneration and proliferation of the vaginal mucous membrane (7,12). A vegetable oil suspension of zearalenone fed to rats produces results similar to those shown by the sows (7,12). The scirpenols can be seen by using acidification combined with burning (7,10,13). Zearalenone can be identified through thin-layer chromatography by its typical bright greenish-yellow fluorescence (BGYF) (4,11,12).

Mycotoxicosis in animals can be suspected as a cause of clinical problems when there is evidence of fungal activity and growth in the feed, when the disorder is not contagious and when the clinical symptoms do not respond to antibiotics. It may also be suspected when an illness appears seasonally.

Research Materials and Methods

This UNAM study of the biological effects of aflatoxicosis in chickens is the only one that has been conducted in Mexico on any kind of farm animal. There have been other aflatoxin studies in the country, but these have focused on the chemical analysis of aflatoxins or on aflatoxin contamination of human foods.

The materials used in the study were livestock and feed samples sent to the Veterinary Medicine and Animal Science laboratory for mycotoxin analysis. A clinical report accompanied each sample, indicating that the animals consuming the feed were not producing adequately or that fungus had been found in the feed.

The samples were tested for aflatoxin B_1, zearalenone, ochratoxin and scirpenols. To test for the presence of aflatoxin B_1, it was necessary to change from long to short wave when the aflatoxins did not fluoresce, and to atomize with 20% sulfuric acid to observe the color change, or with trifluoroacetic acid to induce change in R_F (11).

Four mold isolates, *Penicillium*, *Fusarium* and two *Aspergillus* spp., were cultured on rice and used to inoculate feed for broiler chickens. One kilogram of inoculum was used for each 30 kg of feed. Five groups of 20 chickens each were used. Four groups were fed one of the inoculated feed mixtures, and the fifth group was used as the control. The chickens were fed for four weeks, with amount of feed consumed, weight gain and clinical signs recorded. The animals were then dissected and the relative weights of the heart, liver, spleen, kidney and bursa of Fabricius were recorded, along with any macro- and microscopic changes. The analyses of the samples, as well as feed consumption, weight gain and the condition of the internal organs of the chickens are shown in Tables 1 and 2.

Results

In 1984, 290 samples were received for testing by the laboratory. Forty-six (18.9%) showed aflatoxin B_1 contamination of 25 to 100 ppb. Of this number, 41 were concentrated livestock feed samples, four were sorghum and one was maize silage. In 1985, aflatoxins were found in 15 (5.4%) of the 276 samples received, 10 in concentrated feed, four in sorghum and one in wheat. The highest demand for analysis was in chicken feed, followed by pig feed.

The chickens in the experiment showed food-conversion deficiencies when aflatoxin B_1 levels in their feed reached 0.8 ppm. The relative weights of the livers and the kidneys were negatively affected. A microscopic examination showed a metamorphosis of the fat in the liver with necrosis and proliferation of the bile ducts. Degenerative changes occurred in the kidneys. There was less significant change in the other groups than in the two that were fed aflatoxin-contaminated feed.

Chickens within a group showed macroscopic changes of differing intensity; for example, changes in the livers varied greatly. Macro- and micro-level changes in the kidneys followed the same pattern. The bursa of Fabricius sometimes increased in weight and sometimes decreased within the same group. Some bone marrow changes were related to changes in fatty tissue, while others corresponded to changes in the hematopoietic (reddish) tissue, again among chickens in the same inoculation groups.

Discussion

The data for 1984 and 1985 show a drop in the number of aflatoxin-contaminated samples; in previous years, contaminated samples averaged around 30% of those sent in for analysis. The number of contaminated samples dropped to 18% of the total number received in 1984, and to 5% of the total received in 1985. This may have been due to the weather; mycotoxin contamination shows a relationship to climatic changes. The

Table 1. Results of analysis for aflatoxin B_1 in samples of livestock feed, Mexico City, 1984

Sample	Species					
	Poultry	Cattle	Swine	Trout	Dogs	Total
Sorghum	4	—	—	—	—	4
Concentrated feed	16	9	11	4	1	41
Silage	—	1	—	—	—	1
Zea concentrate	—	—	1	—	—	1
Total samples analyzed 18.0% positive						290

growing use of manufactured milled-grain feeds may lead to increased development of mycotoxicosis, since animals are unable to be selective in their choice of food (6,9,10).

References

1. Asquith, R.L. 1983. Biological effects of aflatoxins in horses. *In* Aflatoxin and *Aspergillus flavus* in Corn, U.L. Diener, R.L. Asquith and J.W. Dickens, eds. Southern Cooperative Series Bulletin 279. Alabama Agricultural Experiment Station, Auburn, Alabama, USA. Pp. 62-66.

2. Bamburg, J.R. 1972. The biological activities and detection of naturally ocurring 12, 13 epoxy-trichothecens. Clinical Toxicology 5:495-515.

3. Bodine, A.B., and D.R. Mertens. 1983. Toxicology, metabolism and physiological effects of aflatoxin in the bovine. *In* Aflatoxin and *Aspergillus flavus* in Corn, U.L. Diener, R.L. Asquith and J.W. Dickens, eds. Southern Cooperative Series Bulletin 279. Alabama Agricultural Experiment Station, Auburn, Alabama, USA. Pp. 46-50.

4. Cysewski, S.J., A.C. Pier, G.W. Engstran, J. L. Richard, R.W. Dougherty and J.R. Thurston. 1968. Clinical pathologic features of acute aflatoxicosis in swine. American Journal of Veterinary Research 29:1577-1590.

5. Edds, G.T., and R.A. Bortell. 1983. Biological effects of aflatoxin in poultry. *In* Aflatoxin and *Aspergillus flavus* in Corn, U.L. Diener, R.L. Asquith and J.W. Dickens, eds. Southern Cooperative Series Bulletin 279. Alabama Agricultural Experiment Station, Auburn, Alabama, USA. Pp. 56-61.

6. Goldblaft, C.A. 1972. Implications of mycotoxins. Clinical Toxicology 5:453-464.

Table 2. Results of analysis for aflatoxin B_1 in samples of livestock feed, Mexico City, 1985

Sample	Species			
	Poultry	Cattle	Swine	Total
Sorghum	3	—	1	4
Concentrated feed	5	4	1	10
Wheat	1	—	—	1
Total samples analyzed 5.4% positive				276

7. Harold, J.K., E.N. Malcon, G.H. Nelson, C.M. Christensen and C.J. Mirocha. 1969. Histological changes in the genital tracts of swine fed strogenic mycotoxin. American Journal of Veterinary Research 30:551-556.

8. Hoerr, F.J., and H.G. d'Andrea. 1983. Biological effects of aflatoxin in swine. *In* Aflatoxin and *Aspergillus flavus* in Corn, U.L. Diener, R.L. Asquith and J.W. Dickens, eds. Southern Cooperative Series Bulletin 279. Alabama Agricultural Experiment Station, Auburn, Alabama, USA. Pp. 51-55.

9. Rosiles, M.R., and R. Lopez. 1977. Síndrome estrogénico de origen alimenticio de cerdos. Veterinaria Mexicana 8:123-126.

10. Shreeve, B.J., D.S.P. Patterson and B.A. Roberts. 1975. Investigation of suspected cases of mycotoxicosis in farm animals in Britain. Veterinary Research 11:275-278.

11. Stoloff, L. 1972. Analytical methods for mycotoxins. Clinical Toxicology 5:465-494.

12. Thaxton, J.P., H.T. Ting and P.B. Hamilton. 1974. Immunosuppression in chickens by aflatoxin. Poultry Science 53:721-725.

13. Wong, J.J., and D.P.H. Hsieh. 1976. Mutagenicity of aflatoxin related to their metabolism and carcinogenic potential. Proceedings of the National Academy of Science 73:2241-2244.

Premature Germination of Maize and the Related Mycotoxin

M. Carvajal, Institute of Biology, National Autonomous University of Mexico, Mexico, D.F., Mexico

Abstract
Premature germination of maize (PGM) is produced by Fusarium moniliforme, *but also present are other* Fusarium *spp., such as* F. nivale, *which produces deoxynivalenol (DON). The mycotoxin associated with PGM appears to be deoxynivalenol.*

Resumen
La germinación prematura del grano de maíz es provocada por Fusarium moniliforme, *aunque también están presentes otras especies de* Fusarium, *como* F. nivale *que produce deoxinivalenol, la micotoxina que al parecer se asocia con la germinación prematura.*

Premature germination of maize (PGM) is a common disease that has been observed in the USA, France, Venezuela and Mexico (3,4,7,12). The disease was first reported in 1924 (3) and has subsequently spread significantly, particularly in central Mexico, where thousands of hectares have been affected. In Mexico, affected maize, instead of being burned or destroyed, is mixed with healthy maize and used for human consumption in *tortillas*.

Premature germination of maize consists of germination of kernels while the ears are still on the plants, witches' broom of the ears and sometimes a diminished number of kernels (Figures 1, 2 and 3). Livestock reject PGM-affected grain, and the quality of the crop is greatly reduced.

A number of causes have been suggested for PGM, such as genetic factors (2,8,10), manganese deficiency in the soil (11), *Diplodia zeae* (1) and *Fusarium moniliforme* (5,6).

Animal-Response Study

A study was conducted by the Department of Plant Pathology of the Institute of Biology of the National Autonomous University of Mexico to determine the acceptibility of PGM grain on livestock (cattle, swine, sheep, goats, chickens, ducks and geese). Two groups of grain, one sound and the other with PGM, were provided *ad libitum*. Animal responses were registered on film.

Materials and Methods

The pregerminated maize kernels were cultivated in the laboratory on potato dextrose agar (PDA), and 100 ears of five *criollo* varieties, 2 white, 2 purple and one of mixed purple and white kernels, were inoculated with a solution of spores of *F. moniliforme* (1×10^9 in 100 ml distilled water). Fifty control ears were soaked in distilled water only. The ears were wrapped in paper and each group placed in separate plastic bags and incubated at 25°C for two weeks.

Mycotoxin analysis of the pregerminated maize grain was carried out, using the method of Stoloff *et al.* (13).

Results and Discussion

All test animals initially consumed both sound and PGM grain, but after only 3 to 15 seconds or immediately after consumption of sound grain, the animals refused the PGM grain. Goats spent the most time eating PGM grain (1.5 minutes), but as soon as they ate sound grain they also rejected the PGM. From the behavior of the animals it appears that vomitoxin is responsible for the feed rejection.

Laboratory tests showed the presence in the PGM grain of the fungi *F. moniliforme, F. nivale, F. tritinctum, F. oxysporum* and occasionally *F. poae. Fusarium roseum* was not detected (Figure 4).

When ears were inoculated with *F. moniliforme*, 90% of the white *criollo* varieties, 20% of the purple and 50% of the mixed purple and white varieties exhibited PGM. Of the controls, 10% had PGM and 90% were sound. The results suggest that the white corn varieties were susceptible to the disease, the purple ones moderately resistant and those of mixed color slightly resistant. Premature germination of maize in the control cobs is explained by natural occurrence of *F. moniliforme*, as the ears were not disinfected internally.

In the Stoloff mycotoxin test (13), deoxynivalenol (DON) was obtained by thin-layer chromatography; DON is a trichothecene. The symptoms that it produces are decreased weight gain, abortion, feed refusal, vomiting, emesis in swine, digestive disorders, bloody diarrhea and hemorrhagic lesions in the stomach, heart, intestine, lungs, bladder and kidneys (9). Deoxynivalenol has been found in maize used for human consumption in the Transkei, where the incidence of esophageal cancer is high (9).

Acknowledgements

The authors wish to thank the National Commission of Science and Technology (CONACYT) for funding this study and Dr. J. Galindo for filming animal responses to feed containing PGM.

References

1. Calvert, O.H., M.S. Zuber and M.G. Neuffer. 1969. Vivipary in *Zea mays* induced by *Diplodia maydis*. Phytopathology 59:239-240.

2. Eister, W.H. 1981. Vivipary in maize. Genetics 16:574-590.

3. Eister, W.H. 1924. A primitive sporophyte in maize. American Journal of Botany 11:7-14.

4. Felix, G.R., and S. Romero. 1981. Etiología de la germinación prematura del maíz en Huamantla, Tlaxcala. Agrociencia 43:81-87.

5. Galindo, A.J. 1983. Evidencia adicional para confirmar a *Fusarium moniliforme* como causa de la germinación prematura del maíz. Collaborative agreement report, Postgraduate College of Agriculture, Chapingo, (Montecillos) and the National Autonomous University of Mexico, Mexico, D.F.

6. Galindo, A.J., and S. Romero C. 1982. *Fusarium moniliforme* as the cause of preharvest sprouting of maize in Mexico. Paper presented at the Third International Symposium on Preharvest Sprouting Damage in Cereals, Winnipeg, Canada.

7. González, N.C. 1976. Germinación de granos inmaduros en mazorcas de maíz. Agronomía Tropical 26(A):359-362.

8. Mangelsdorf, P.C. 1930. The inheritance of dormancy and premature germination in maize. Genetics 15:462-494.

9. Pathre, S.V., and C.J. Mirocha. 1979. Trichothecenes: Natural ocurrence and potential hazard. Journal of the American Oil Chemists Society 56:820-823.

10. Robertson, D.S. 1955. The genetics of vivipary in maize. Genetics 40:745-760.

11. Routchenko, W., and J.P. Soyer. 1972. Causes de la germination sur plante de grains inmatures de máiz. Données complementaires. Annales Agronomiques 23(4):445-459.

12. Routchenko, W., and J.P. Soyer. 1971. Sur les causes de la germination sur plante des grains inmatures de máiz. Résultats preliminaires. Annales Agronomiques 22(2):241-255.

13. Stoloff, L., S. Nesheim, L. Yin, J.V. Rodricks. M. Stack and A.D. Campbell. 1971. A multimycotoxin detection method for aflatoxins, ochratoxin, zearalenone, sterigmatocystin, and patulin. Journal of the Association of Official Analytical Chemists 54:91-97.

Aflatoxin B_1 in Allergenic Mold Extract Used for Immunotherapy Desensitization

M.S. Legator, V.M. Sadagopa Ramanujam, G. Kline and J.B. Ward, Jr., Division of Environmental Toxicology, Department of Preventive Medicine and Community Health, University of Texas Medical Branch, Galveston, Texas, USA

Abstract
To investigate the possibility that carcinogenic aflatoxins may be present in Aspergillus *mold extracts used for immunotherapy of patients with allergies, 12 samples from various clinics and manufacturers were obtained. Chemical analysis by thin-layer and high-pressure liquid chromatography showed the presence of various levels (0.054 to 1.08 ppm) of aflatoxin B_1 in four samples. The sample containing 1.08 ppm of aflatoxin B_1 was also found to be highly mutagenic (>1000 net revertants/100 μl) in the Ames* Salmonella typhimurium *test with strains TA100 and TA98. The findings suggest that careful screening of commercially available allergenic mold extracts is warranted. There may be a risk of iatrogenic, aflatoxin-induced illness such as Reye's syndrome from hyposensitization immunotherapy.*

Resumen
A fin de investigar la posibilidad de que ciertas afaltoxinas carcinogénicas estuvieran presentes en los extractos de moho de Aspergillus *empleados en la inmunoterapia de pacientes con alergias, se obtuvieron 12 muestras de diversas clínicas y fabricantes. El análisis químico mediante cromatografía de capa fina y cromatografía líquida de alta presión señaló la presencia de varios niveles (0.054 a 1.08 ppm) de aflatoxina B_1 en cuatro muestras. También se encontró que la muestra que contenía 1.08 ppm de aflatoxina B_1 era en extremo mutagénica (>1000 revertientes netos 100 μl) en la prueba de* Salmonella typhymurium *de Ames con las cepas TA100 y TA98. Los resultados indican que se justifica una selección cuidadosa de los extractos alergénicos de mohos que se encuentran en el mercado. Puede haber un riesgo de enfermedades iatrogénicas producidas por las alfatoxinas, tales como el síndrome de Reye, causada por la inmunoterapia de hiposensibilización.*

In injection immunotherapy (hyposensitization), mold antigens are commonly used to treat patients with rhinitis, asthma or other conditions (3,7). Injection immunotherapy is frequently employed when a patient's history and skin test suggest that such allergens as molds play an etiological role in the illness. Extracts are frequently prepared from dried ground powders, after growing the organisms on the surface of a suitable broth medium, or from the medium alone. Sometimes, the mat and medium are homogenized. The preparations are frequently standardized on the basis of protein nitrogen unit (PNU).

To investigate the possibility that carcinogenic mycotoxins, specifically aflatoxins, may be present in *Aspergillus* mold extracts used for immunotherapy of patients with allergies, 12 samples from various sources were acquired and attempts made to isolate and identify aflatoxins in the extracts (8).

Materials and Methods

Chemical analysis was conducted by thin-layer chromatography and high-pressure liquid chromatography. Mutagenicity testing was performed directly on the commercial mold extracts using the Ames bacterial bioassay with the *Salmonella typhimurium* strains TA100 and TA98.

Mold extracts

Aspergillus mold extracts (allergenic samples) were obtained from different manufacturers either directly or through allergy clinics in Galveston and Houston, Texas, USA.

Chemicals

Aflatoxins (B_1, B_2, G_1 and G_2) used as standards for thin-layer chromatography (TLC) and high-pressure liquid chromatography (HPLC) were purchased from the Sigma Chemical Company, St. Louis, Missouri. Such solvents as chloroform, methanol and water were of HPLC grade (Fisher Scientific Company, Houston, Texas). The positive control for mutagenicity assays was 2-aminoanthracene (2-AAn) (Aldrich Chemical Company, Milwaukee, Wisconsin); it was recrystallized with 50% aqueous ethanol before use.

Methods for TLC and HPLC

Thin-layer chromatography was done using the procedure of Beljaars *et al.* (4), with silica gel G glass plates (Analtech Inc., Newark, Delaware) in chloroform:acetone 90:10 (v/v). Five TLC plates were spotted with each allergenic sample (100 μl/plate). After development of TLC plates, the region very close to the origin $R_F = 0.01$ to $R_F = 0.55$, which fluoresced under ultraviolet (UV) light (aflatoxins B_1, B_2, G_1 and G_2 fluoresce under UV light) was scraped and the combined materials extracted with chloroform (2 x 100 ml). The chloroform extract was filtered, evaporated to dryness under vacuum and the residue dissolved in 500 μl of methanol for HPLC analysis. A total of 12 allergenic samples were analyzed.

High-pressure liquid chromatography was done using a Beckman Model 332 liquid chromatograph (Fullerton, California) equipped with Hitachi model 100-10 UV-visible variable wave length detector and Gilson Spectra/Glo filter fluorescence detector connected in series. An Altex Ultrasphere 5 ODS reversed phase stainless steel column was used, and the TLC extracts were eluted with methanol:water 65:35 (v/v) solvent using a flow rate of 1 ml/minute. The peak areas were calculated with an Altex C-R1A integrator.

Quantitative estimates of the allergenic sample TLC extracts separated by HPLC were made using both UV (365 and 254 nm) and fluorescence (excitation 360 nm, emission 455 nm) detectors (5,6). With UV-365 nm, the lowest detectable amount for aflatoxin B_1 is 5 nanograms; with fluorescence, amounts as low as 0.01 nanograms of B_1 can be identified. After examining HPLC profiles from each separate TLC extract and deciding whether aflatoxins were present, further HPLC quantitation was done by reporting the TLC with the allergenic samples containing aflatoxin B_1. The spot corresponding to the identified toxin was scraped and extracted, and the final amount of aflatoxin B_1 determined from its HPLC standard curve.

Ames *Salmonella typhimurium* Microsomal Assays

Strains TA100 and TA98 used in the assays came from B.N. Ames, Berkeley, California. The Ames assays of *Aspergillus* extracts were done using standard procedures (1,2,9). Four concentrations of each extract were usually assayed, sometimes with the addition of rat liver S-9 fraction. Negative controls were established by testing for spontaneous revertants for each strain with and without S-9 activation. In each experiment, positive mutagenesis controls, using 5 μg/plate of 2-acetylaminofluorene, were routinely included to confirm the reversion properties of each strain. A chemical was considered to have a positive response if the number of induced revertants was equal to or greater than twice the number of spontaneous revertants.

Results

Figure 1 shows the HPLC profile for the *Aspergillus* mix/TLC extract containing 1.08 ppm of aflatoxin B_1 with fluorescence detection. The peak, a retention time of 7.02 minutes, was identified as aflatoxin B_1 by running an aflatoxin B_1 standard under identical conditions. Further confirmation was obtained by spiking a known quantity of aflatoxin B_1 in the TLC extract corresponding to the region with R_F of aflatoxin B_1 (R_F = 0.53 in chloroform:acetone 90:10) and comparing HPLC profiles and areas of peaks. No new peak was seen, and the area of the peak identified as aflatoxin B_1 increased proportionally.

Ames test results with *Salmonella typhimurium* strains TA100 and TA98 on the *Aspergillus*-mix sample

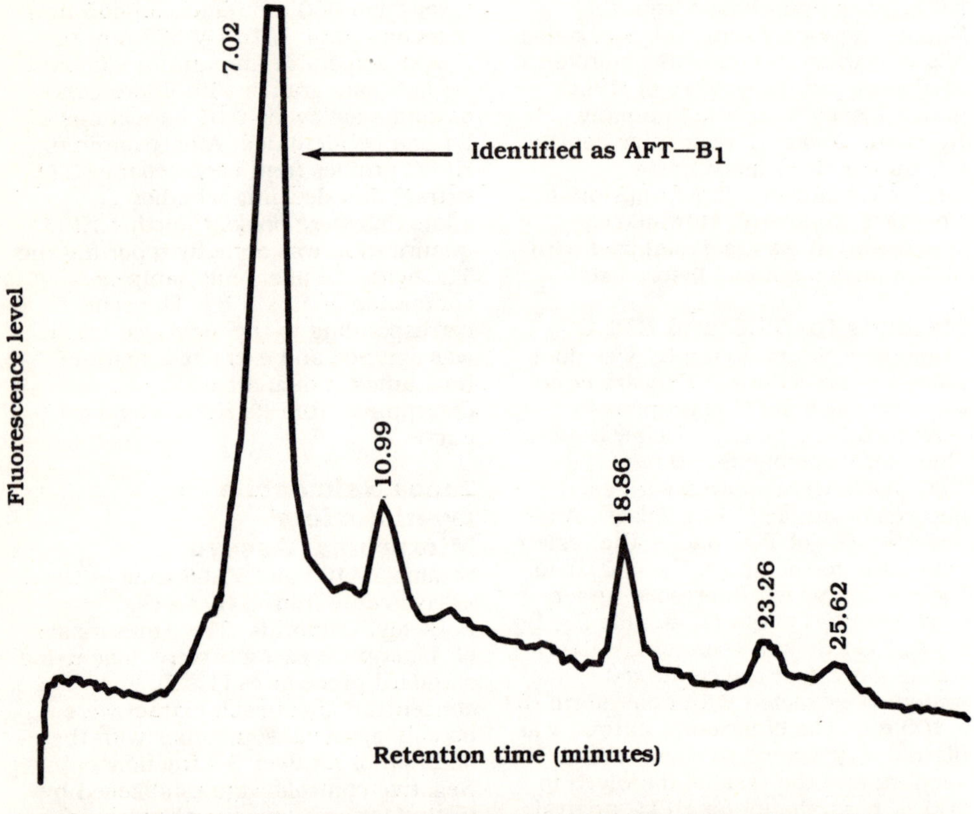

Figure 1. HPLC profiles for the TLC/*Aspergillus* mix containing 1.08 ppm of aflatoxin B_1, using a solvent system of methanol:water 65:35 (v/v) with 1 ml/minute flow rate through an Altex reversed phase Ultrasphere 5 ODS column; HPLC with fluorescence detection (excitation 360 nm, emission 455 nm). The peak with retention time of 7.02 minutes was identified as aflatoxin B_1; the other peaks with retention times of 10.99, 18.86, 23.26 and 25.62 minutes were not identified.

containing 1.08 ppm of aflatoxin B_1 showed that the sample is positive (+ +) with activation (+ S-9) and the TA98 strain gives more net revertants/plate for 100 μl volume.

Table 1 summarizes both mutagenicity and chemical (TLC and HPLC) analysis results on the 12 allergenic mold samples tested in the laboratory.

Discussion

Of the 12 samples analyzed by a combination of TLC and HPLC, four samples contained aflatoxin B_1. The single sample containing 1.08 ppm of aflatoxin B_1 was found to be highly mutagenic (>1000 net revertants/100 μl) in TA100 and TA98 strains with activation. The amount of aflatoxin B_1 equal to 0.062 ppm corresponds to 6.2 nanograms/100 μl of the sample; this is too low a level to show any positive result in the Ames test. Thus, the other three samples with low levels of aflatoxin B_1 gave only ± on the Ames test. The HPLC technique using fluorescence detection, on the other hand, is sensitive enough to detect such low levels of aflatoxin B_1.

Since *Aspergillus* mold extracts are widely used for hyposensitization, these findings suggest that a careful screening of allergenic mold extracts from a large number of manufacturing sources is warranted. The potential risk of such iatrogenic aflatoxin-induced illness as Reye's syndrome (10) from hyposensitization immunotherapy needs to be determined through clinico-epidemiological studies of an exposed population.

Table 1. Results of Ames test and chemical analysis for aflatoxin presence in 12 allergenic samples

	Mutagenicity (Ames test)				Chemical analysis		
	Strain	TA98	Strain	TA100			
Sample	+S9	-S9	+S9	-S9	TLC[a/]		HPLC[b/]
						FTB_1	
1	- [c/]	-	+ [d/]	+	+		+ (0.062 ppm)
2	-	-	+	+	+		+ (0.054 ppm)
3	-	-	+ +	+ +	+		+ (1.08 ppm)
						AFTs [e/]	
4	-	-	-	-	-		-
5	-	-	-	-	-		-
6	NT [f/]	-	NT	-	-		-
7	-	-	-	-	-		-
8	-	-	-	-	-		-
9	-	-	-	-	-		-
10	-	-	-	-	-		-
						$AFTB_1$	
11	-	-	+	+	+		+ (0.096 ppm)
						AFTs	
12	-	NT	NT	NT	-		-

a/ TLC = thin-layer chromatography
b/ HPLC = high-pressure liquid chromatography
c/ - = tested with negative results
d/ + = tested with positive results
e/ AFTs = aflatoxins B_1, B_2, G_1 and G_2
f/ NT = not tested

Acknowledgements
The authors wish to acknowledge the McGregor Allergy Clinic, Houston, Texas, for financial support for this investigation.

References

1. Ames, B.N., F.D. Lee and W.E. Duston. 1973. An improved bacterial test system for the detection and classification of mutagens and carcinogens. Proceedings of the National Academy of Science (USA) 70:782-786.

2. Ames, B.N., J. McCann and E. Yamasaki. 1975. Methods for detecting carcinogens and mutagens with the *Salmonella*/ mammalian microsome mutagenicity test. Mutation Research 31:347-364.

3. Baer, H. 1978. Allergenic extracts. *In* Allergy: Principles and Practice, vol. 1, E. Middleton, Jr., C.E. Reed and E.F. Ellis, eds. C.V. Mosby, St. Louis, Missouri, USA. Pp. 217-230.

4. Beljaars, P.R., C.A.H. Verusdonk, W.E. Paulsch and D.H. Liem. 1973. Collaborative study of two-dimensional thin layer chromatographic analysis of aflatoxin B_1 in peanut butter extracts, using the antidiagonal spot application technique. Journal of the Association of Official Analytical Chemists 56:1444-1451.

5. Davis, N.D., and U.L. Diener. 1980. Confirmatory test for the high pressure liquid chromatographic determination of aflatoxin B_1. Journal of the Association of Official Analytical Chemists 63:107-109.

6. Gregory, III, J.F., and D.B. Manley. 1982. High performance liquid chromatographic quantitation of aflatoxin metabolite in animal tissues. Journal of the Association of Official Analytical Chemists 65:869-875.

7. Kaad, P.H., and P.A. Ostergaard. 1982. The hazard of mould hyposensitization in children with asthma. Clinical Allergy 12:317-320.

8. Legator, M.S., G. Kline, V.M. Sadagopa Ramanujam, B.R. Cunningham, J.B. Ward, Jr., and M.M. Gad-el-Karim. 1983. Aflatoxin B_1 in mould extracts used for desensitization. Lancet ii:915.

9. McCann, J.E., E. Choi, E. Yamasaki and B.N. Ames. 1975. Detection of carcinogens as mutagens in the *Salmonella*/microsome test: Assay of 300 chemicals. Proceedings of the National Academy of Science (USA) 72:5135-5139.

10. Pollack, J.D. 1979. Models of chemical and virus interaction and their relation to a multiple etiology of Reye's Syndrome. *In* Reye's Syndrome II, J.F.S. Crocker, *et al.* eds. Grune and Stratton, New York, N.Y., USA. Pp. 341-60.

Development of DNA Adduct Technology to Monitor Human Exposure to Cancer-Causing Mycotoxins

T.R. Irvin, Laboratory of Toxicology, Department of Anatomy, College of Veterinary Medicine, Texas A & M University, College Station, Texas, USA

Abstract

For epidemiological modeling of human myctoxin exposure, a study was done to characterize the capacity of placental tissue to metabolize and biotransform aflatoxin B_1 to DNA-binding moieties in mice. Formation of aflatoxin B_1-DNA adducts in mouse placental cells are described as functions of time and dosage. Mice treated on the seventeenth day of pregnancy were administered 5 to 20 mg/kg [^{3}H]aflatoxin B_1 and decapitated 2 to 24 hours later. The uterus and liver of each animal were isolated. Samples of placental tissue were pooled, frozen and later subjected to cesium chloride/cesium sulfate density centrifugation to yield purified placental DNA isolate free of protein and RNA contamination. The DNA-bound aflatoxin B_1 moieties were measured by liquid scintillation; placental DNA was spectrophotometrically quantitated by absorbance at 260 nm. At 5 and 10 mg aflatoxin B_1/kg doses, placental tissue contained approximately 30-40% of aflatoxin B_1-DNA adduct levels observed in maternal liver. At 15 to 20 mg/kg, aflatoxin B_1-DNA adduct levels within placental and maternal liver DNA were virtually equal. At 12 and 24 hours after dosing, aflatoxin B_1-DNA adduct levels in the placenta were greater than adduct levels in maternal liver DNA by 25% and 68%, respectively. The time course of aflatoxin B_1 DNA adducts in placental tissue suggests that placental tissue absorbs aflatoxin B_1 over an extended period of time after exposure. New methodologies to assay myctoxin-DNA interactions in certain tissues (such as the placenta) in humans undergoing chronic mycotoxin exposure are also discussed.

Resumen

Se efectuó un estudio con el fin de caracterizar la capacidad del tejido placentario para metabolizar y biotransformar la aflatoxina B_1 en moléculas ligadoras al ADN en ratas y se evaluó la utilidad que tienen los resultados en la creación de modelos epidemiológicos de la exposición del ser humano a las micotoxinas. La formación de aductores de aflatoxina B_1 y de ADN en las células placentarias de las ratas se describe como una función del tiempo y de la dosis. Se administraron de 5 a 20 mg/kg [^{3}H]aflatoxina B_1 a ratas que se encontraban en el decimoséptimo día de gestación y que fueron decapitadas 2 a 24 horas después. Se aislaron el hígado y el útero de cada animal. Asimismo, se combinaron y congelaron muestras del tejido placentario que, posteriormente, fueron sometidas a centrifugación por densidad con cloruro de cesio o sulfato de cesio para obtener un aislamiento puro de ADN placentario que estuviera libre de proteínas y de la contaminación con ARN. Las moléculas de aflatoxina B_1 ligadoras al ADN se midieron mediante el centelleo líquido; el ADN placentario se cuantificó espectrofotométricamente mediante la absorbancia a 260 nm. A dosis de 5 y 10 mg de aflatoxina B_1/kg, el tejido placentario contenía aproximadamente 30 a 40% de los niveles de aductores de aflatoxina B_1 y de ADN observados en el hígado materno. A dosis de 15 a 20 mg/kg, los niveles de *aductores de aflatoxina B_1 y de ADN en el ADN placentario y el materno fueron prácticamente iguales. Doce a veinticuatro horas después de la dosificación, los niveles de aductores de aflatoxina B_1 y de ADN en la placenta fueron mayores que los niveles de aductores de ADN del hígado materno en un 25% y 60%, respectivamente. El curso del tiempo de los aductores de aflatoxina B_1 y de ADN en el tejido placentario sugiere que éste absorbe aflatoxina B_1 durante un período prolongado después de la exposición. En el presente trabajo también se tratan métodos nuevos para evaluar las interacciones entre las micotoxinas y el ADN en ciertos tejidos (como la placenta) de seres humanos que están expuestos crónicamente a las micotoxinas.*

Noninvasive Methods for Epidemiological Modeling of Mycotoxin Exposure: Model Studies with Aflatoxin B_1

The aflatoxins are a family of heterocyclic, secondary mold metabolites produced by strains of *Aspergillus* spp. constituting a group of modified coumarins with a fused dihydrofurofuran moiety (17). Due to the wide geographical distribution of these molds, aflatoxins have been found to contaminate a variety of food and feed commodities (e.g., maize, peanuts, soybeans and cottonseed) produced in many parts of the world. The fact that food-spoilage molds can incorporate toxic substances such as the aflatoxins into food and feed products has come to be fully appreciated only within the last two decades. Scientific and regulatory attention has been directed towards the public health and economic implications of aflatoxin production for several reasons:

- Recognition of frequent animal feed contamination with aflatoxins;
- Frequent causal association of aflatoxins with animal diseases;
- Potential occurrence and transmission of aflatoxin residues in tissues of commercially important food animals; and
- Potential for human disease (including cancer) from chronic aflatoxin ingestion.

Aflatoxin B_1 is the major component of naturally occurring aflatoxin mixtures and is the aflatoxin of primary interest in public health (16). This compound is one of the most potent hepatotoxins and hepatocarcinogens known for laboratory and livestock animals. Aflatoxin B_1 induces massive liver cirrhosis as well as liver tumors in mammalian and avian species when administered by either intraperitoneal or dietary routes. Epidemiological studies have linked aflatoxin contamination of human food supplies in Asia and Africa with increased incidence of human liver cancer.

Like many xenobiotics, aflatoxin B_1 is metabolized by microsomal mixed-function oxidases primarily associated with the endoplasmic reticulum of organs, such as the liver, kidneys and lungs (6). *In vivo* studies have demonstrated the correlation of species sensitivity and target-organ susceptibility to the toxic and carcinogenic effects of aflatoxin B_1 with the metabolic competence of target tissues to activate aflatoxin B_1 to aflatoxin B_1-2,3-oxide (Figure 1). This electrophilic intermediate is believed to mediate many toxic effects of aflatoxin B_1 by binding to critical cellular macromolecules, such as DNA. Comparative studies have shown a direct proportion between species-specific hepatoxic and hepatocarcinogenic effects of aflatoxin B_1 and the sensitivity of hepatic DNA to aflatoxin B_1 covalent modification and damage.

Epidemiological studies seeking to relate aflatoxin B_1 exposure to reported instances of human liver cancer have focused attention on the need to document DNA-bound aflatoxin B_1 moieties in human tissues and fluids. Chromatographic analyses of urine from laboratory animals, as well as humans undergoing chronic aflatoxin B_1 ingestion, have correlated the excretion of aflatoxin B_1-N^7-guanine, the principal product of aflatoxin B_1-2,3-oxide interaction with DNA *in vivo* and *in vitro* with dose-dependent aflatoxin B_1-DNA adduct levels in liver (6). Similar studies employing polyclonal antibodies generated against protein-bound aflatoxin B_1 have permitted urinary aflatoxin B_1-DNA adduct detection to levels of 1 moiety in 10^7 nucleotides (1).

Investigations seeking noninvasive methods to document human exposure to a wide range of carcinogenic chemicals have focused attention on chemico-DNA adduct formation in the placenta. The placenta, like maternal organs, is rich in mixed-function oxidases and has been demonstrated *in vitro* to biotransform a number of xenobiotic compounds to DNA-binding, electrophilic intermediates (11).

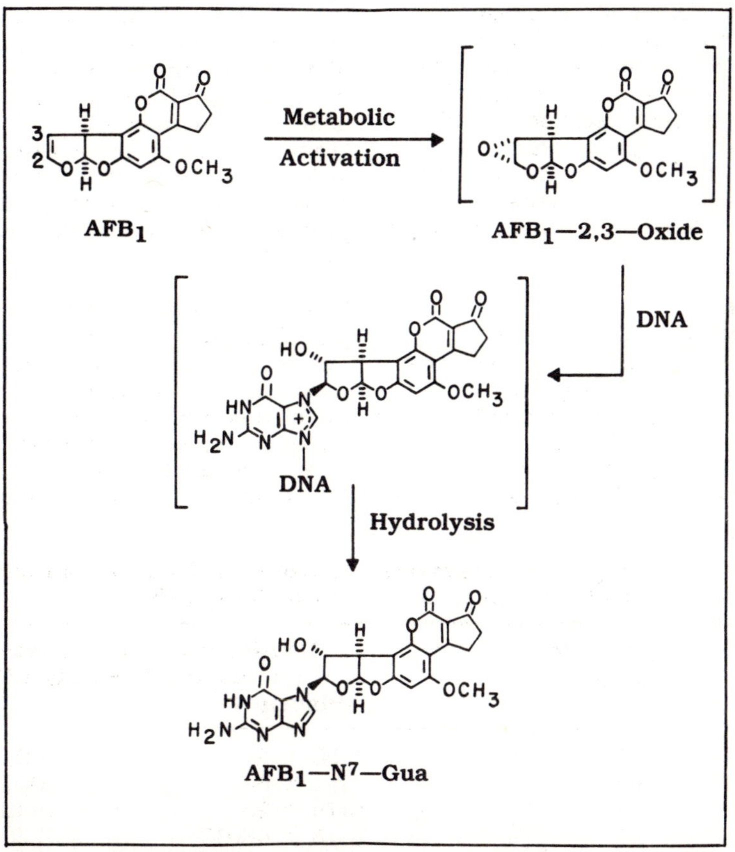

Figure 1. Metabolic activation of aflatoxin B_1 to its DNA-binding metabolite aflatoxin B_1-2,3-oxide

Compounds known to undergo metabolism in the placenta include polycyclic aromatic hydrocarbons, aromatic amines and steroidal drugs. However, while numerous studies have documented placental enzymatic activity for metabolizing environmental toxins, studies relating specific placental-borne DNA adducts to adduct levels in maternal or conceptus tissues have not been reported.

To evaluate the placenta as a site for epidemiological modeling of human mycotoxin exposure, the current study was conducted to characterize the capacity of placental tissue to metabolize and biotransform aflatoxin B_1 to DNA-binding moieties in mice. The initial section of this paper describes the formation of aflatoxin B_1-DNA adducts in mouse placental cells as functions of both time and dosage. The second section describes novel methodologies to assay myctoxin-DNA interactions in certain tissues, such as the placenta, in humans undergoing chronic mycotoxin exposure.

Methods

Tissue DNA was isolated from pregnant C57B6 mice, treated on the seventeenth day of pregnancy, via cesium salt density centrifugation (9,10). Pregnant mice were administered 5 to 20 mg/kg [^{3}H]aflatoxin B_1 and were decapitated 2 to 24 hours postdosing. The uterus and liver from each animal were isolated, and the placental disk tissue associated with each fetus was removed; placental tissue from each mouse was subsequently pooled and frozen. Samples of placental tissue were subjected to cesium chloride/cesium sulfate density centrifugation, yielding a purified placental DNA isolate free of protein and RNA contamination. The DNA-bound aflatoxin B_1 moieties were measured by liquid scintillation, and placental DNA was spectrophotometrically quantitated by absorbance at 260 nm.

Results

Initially, levels of aflatoxin B_1 adduction were compared as a function of dosage within placental and maternal liver DNA isolated from aflatoxin B_1-treated C57B6 mice (Table 1). Binding levels extended from 156 to 2618 picomoles aflatoxin B_1/mg of placental DNA and from 502 to 3226 picomoles aflatoxin B_1/mg maternal liver DNA.

The ratio of aflatoxin B_1 residues in placental versus maternal liver DNA indicated a dichotomy between lower and higher dosage levels. At 5 and 10 mg aflatoxin B_1/kg doses, placental tissue contained approximately 30 to 40% of the aflatoxin B_1-DNA adduct levels observed in maternal liver; at

Table 1. Dose-dependent formation of aflatoxin B_1-DNA adducts in placental and maternal liver tissue of C57B6 mice[a/]

AFB_1 dosage (mg/kg)	DNA-bound AFB_1 moieties[b/] (picomoles AFB_1/mg DNA) Placenta	Liver
5	156 ± 2	502 ± 62
10	354 ± 64	1007 ± 167
15	1475 ± 25	1542 ± 145
20	2618 ± 4601	3226 ± 1351

a/ Time of sacrifice: 2 hours postdosing
b/ Values represent the average ± SD of three experiments

higher doses (15 to 20 mg/kg), aflatoxin B_1-DNA adduct levels within placental and maternal liver DNA were virtually equal. One explanation for the dose-dependent alteration in placental versus liver aflatoxin B_1-DNA adduct profile is saturation of aflatoxin B_1 hepatic uptake with subsequent, increased blood-borne aflatoxin B_1 for placental absorption and biotransformation. However, the linearity of the dose-response curve for aflatoxin B_1 adduction of liver DNA would appear to argue against hepatic uptake saturation. An alternative hypothesis is saturation of conjugation enzymes in the placenta which, at low doses, detoxify aflatoxin B_1 metabolites to water-soluble derivatives; at higher doses, saturation of these enzymes would make greater quantities of aflatoxin B_1-2,3-oxide available for placental DNA binding and adduct formation.

Previous studies of aflatoxin B_1-DNA adduct formation in target and nontarget tissues for aflatoxin B_1 toxicity and carcinogenicity have demonstrated the presence of enzymatic activity capable of removing DNA-bound aflatoxin B_1 moieties (6,9,10). In the current work, removal of DNA-bound aflatoxin B_1 residues in placental and liver tissue of mice was investigated by monitoring the time course of aflatoxin B_1-DNA adduct levels within the tissues (Table 2). Two hours postdosing, maternal liver contained levels of DNA-bound aflatoxin B_1 moieties 2.5 times that of placenta tissue; by six hours postdosing, placental and maternal liver DNA levels were virtually equal.

At 12 and 24 hours postdosing, aflatoxin B_1-DNA adduct levels in the placenta were greater than adduct levels in maternal liver DNA by 25% and 68%, respectively. Examination of other maternal tissues, including the kidneys and lungs, indicated that at 24 hours postdosing placental tissue contained the greatest levels of DNA-bound aflatoxin B_1 adducts (13). The time course of aflatoxin B_1-DNA adducts within placental tissue suggests that, unlike the liver which reaches maximum aflatoxin B_1 levels within 30 minutes after i.p. dosing, placental tissue absorbs aflatoxin B_1 over an extended period of time after exposure.

Monitoring Mycotoxin-DNA Interactions in Humans: Development of DNA Postlabeling Methods for Sterigmatocystin

Investigations of mycotoxin-DNA interactions in laboratory animals typically employ radiolabeled toxins to detect and quantitate formation of covalent DNA adduct species. To permit extension of these laboratory

Table 2. Time course of aflatoxin B_1-DNA adduct formation in placental and maternal liver tissue of C57B6 mice[a/]

Time postdosing (hours)	DNA-bound AFB_1 moieties[b/] (picomoles AFB_1/mg DNA)	
	Placenta	Liver
2	353 ± 64	1007 ± 167
6	408 ± 121	385 ± 109
12	454 ± 27	350 ± 15
24	568 ± 25	324 ± 49

a/ Dosage administered: 10 mg/kg
b/ Values represent the average ± SD of three experiments

studies of genotoxic mycotoxins to humans and animals chronically exposed to mycotoxins, alternative techniques have been sought. Immunological assays employing monoclonal and polyclonal antibodies to mycotoxin-DNA adducts permit the detection of 1-mycotoxin residue in 10^6 to 10^7 nucleotides (1); however, the versatility of immunoassays is limited by the need to produce highly selective antibodies for each relevant mycotoxin.

A general approach to the detection of DNA damage and repair by genotoxic mycotoxins without the need for radiolabeled compounds has recently been described by Reddy *et al*. (14). Mycotoxin-adducted DNA nucleotides, generated by nuclease digestion of tissue DNA from mycotoxin-treated animals, are enzymatically labeled with ^{32}P and quantitated by thin-layer chromatography (TLC). Initially, DNA isolated from animals that had been given the carcinogenic mycotoxin sterigmatocystin (ST) was studied. This heterocyclic, food-borne mycotoxin is produced by *Aspergillus*, *Penicillum* and *Bipolaris* spp. and exhibits considerable hepatotoxic and hepatocarcinogenic potency in a number of laboratory animal species (15).

Interest in ST as a food-borne carcinogen is due principally to its structural similarity to aflatoxin B_1. Sterigmatocystin is a biosynthetic precursor to aflatoxin B_1 in molds grown under varying nutrient conditions and can be a co-contaminant with aflatoxin B_1 in mold-adulterated foods and feeds. Comparative toxicological studies employing aflatoxin B_1 and ST have found that ST exhibits one to two orders-of-magnitude decreased potency as a rat hepatocarcinogen and approximately one order less potency as an acutely lethal toxin. Assessment of the comparative mutagenicities of ST and aflatoxin B_1 in *Salmonella typhimurium* has shown that aflatoxin B_1 exhibits approximately ten times more mutagenic potency than ST (18).

Research to identify mechanisms underlying the carcinogenic properties of ST has focused on tissue-specific metabolic activation of the compound to chemically reactive intermediates. *In vitro*, ST has been shown to undergo microsomal-mediated activation (5) to an electrophilic intermediate, ST-1,2-oxide (Figure 2). The chemically reactive metabolite predominantly binds to DNA at the N^7 position of guanine to form ST-N^7-guanine as the major DNA-bound moiety. While the *in vivo* formation of this ST-DNA reaction product has not been documented, exposure of rat liver *in vitro* to ST via perfusion has documented the formation of ST-N^7-guanine as the major DNA adduct (7).

DNA postlabeling technology has been employed to describe the time course of ST modification of liver DNA from rats undergoing single, subacute exposures to the toxin. The approach allows ultrasensitive detection of ST-DNA interactions at long periods of time after administration. In addition, the results provide a tool for detection of ST-DNA reaction products in humans exposed to the toxin.

Methods

Male Fischer 344 rats were administered 0.33 to 9.0 mg ST/kg via intraperitoneal injection and sacrificed by cervical dislocation. Liver DNA was isolated by phenol:chloroform extraction and digested with a mixture of spleen phosphodiesterase and micrococcal nuclease creating 3'-carcinogen-modified nucleoside monohosphates along with mixtures of nucleoside monophosphates of nonadducted bases; preparative TLC of this enzymatic digest separates nonadducted nucleotides and creates an enriched fraction of ST-modified

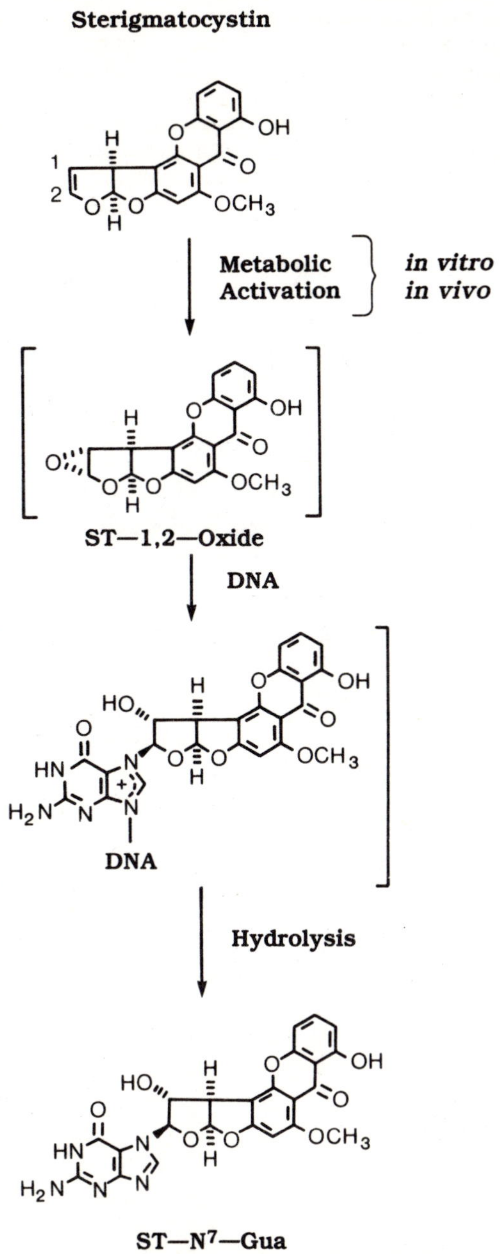

Figure 2. ***In vitro*** **metabolic activation of sterigmatocystin**

nucleotides (Figure 3). Adducted nucleotide fractions were subsequently incubated with gamma-[^{32}P]ATP and polynucleotide kinase isolated from T_4 bacteriophage in which the terminal phosphate of ATP is transferred to the 5'-position of the ST-modified nucleotide creating a 3',5'-bis-phosphate. Unreacted phosphorous-32 radioctivity is subsequently removed via a second round of preparative TLC, and the individual postlabeled, ST-modified nucleotides are separated by two-dimensional TLC, creating maps or "fingerprints" of ST-modified DNA samples.

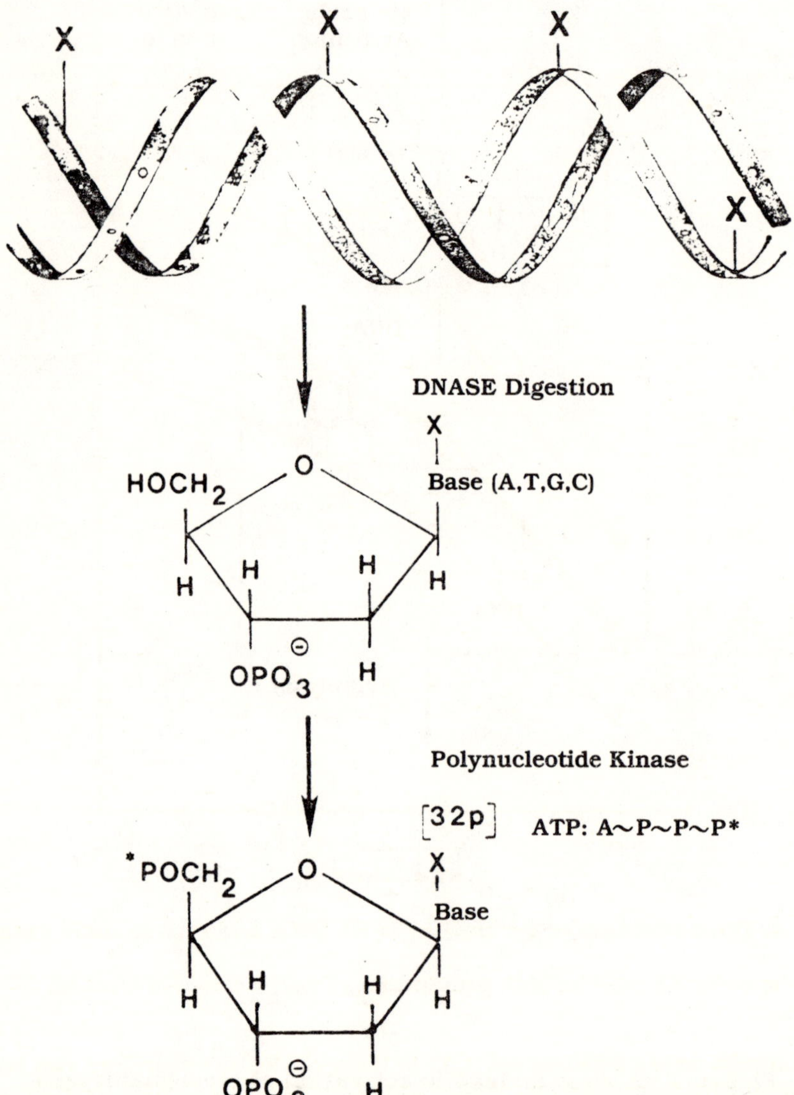

Figure 3. Diagrammatic summary of DNA postlabeling technology

Results

Rats were administered 0.33, 1.0, 3.0 and 9.0 mg ST/kg and sacrificed two hours later. The extent of ST-DNA covalent interaction in liver was an approximately linear function of dosage in the range of 1 to 9 mg ST/kg (Figure 4). A three-fold increase in dosage resulted in an increase in DNA modification of approximately 2.4 times. The level of ST-DNA adducts observed after a 0.33 mg ST/kg dose indicated a substantial departure from the linearity observed with higher doses; a three-fold decrease in dosage (from 1 to 0.33 mg ST/kg) was accompanied by a twenty-fold decrease in DNA modification. Explanations for

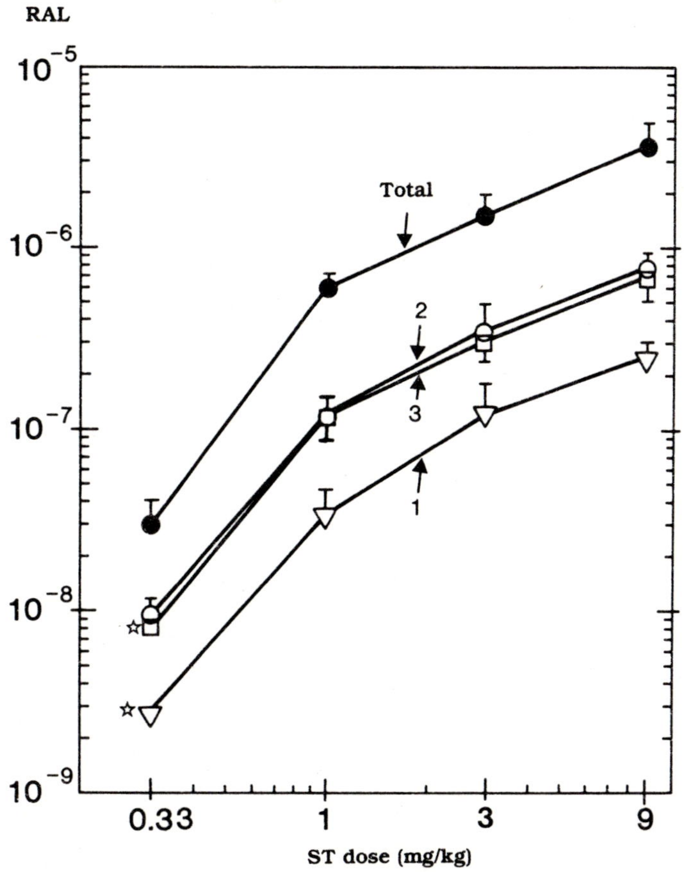

Figure 4. Dose-responsive formation of ST-DNA adducts in male rats

1,2,3 = specific ST-DNA adduct populations, Total = total DNA-bound ST residues

Note: Animals were administered 0.33 to 9 mg ST/kg and sacrificed two hours postdosing. DNA-bound adduct values are given in terms of RAL (relative adduct level), an expression of number of nonadducted nucleotides per ST-adducted nucleotide.

the nonlinearity of dose responses observed at higher dosages include a dose-dependent decrease in ST activation, a more efficient detoxification of ST at lower dosages or a greater rate of removal of ST-DNA adducts in tissues undergoing low-level, subtoxic ST exposure.

Average adduct levels from duplicate time-course experiments monitoring ST-liver DNA adducts were determined two hours to 105 days after administration of an i.p. 9 mg/kg ST dose (Figure 5). Maximum ST-DNA adduct levels, observed at the two-hour time point, were reduced by 89% at 24 hours. The rate of *in vivo* adduct removal decreased to 7% per day between day 1 and day 14 postdosing, and to 0.46% between day 14 and day 105 postdosing; 0.5% of the two-hour adduct levels still remained 105 days after toxin administration.

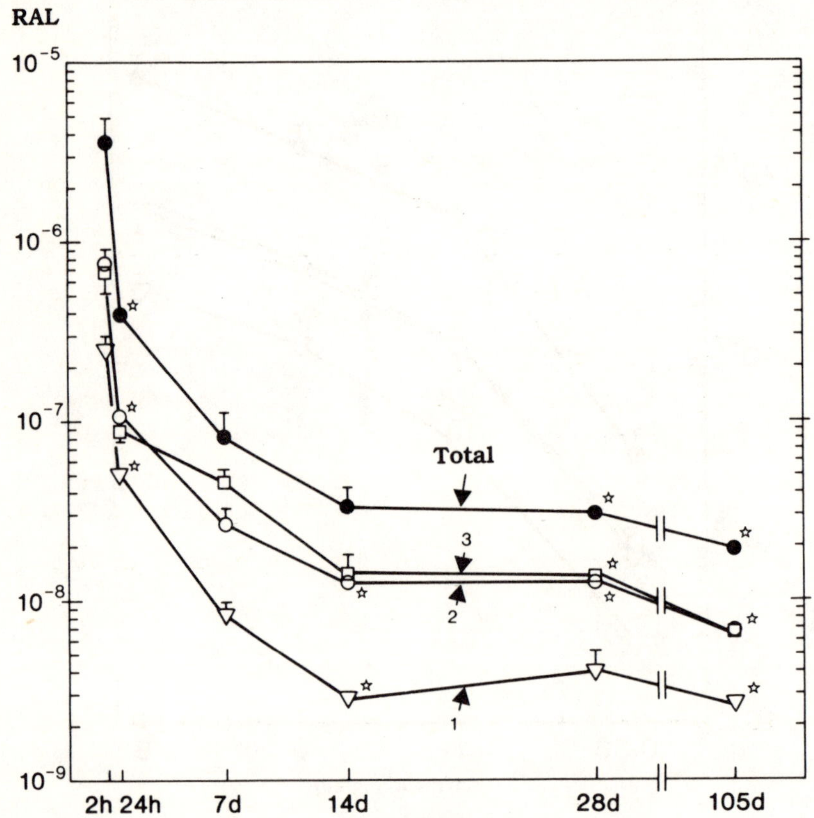

Figure 5. Time course of ST-DNA adducts in male rats

1,2,3 = specific ST-DNA adduct populations, Total = total DNA-bound ST residues

Note: Animals were administered 9 mg ST/kg and sacrificed two hours to 105 days postdosing. DNA-bound adduct values are given in terms of RAL (relative adduct level), an expression of number of nonadducted nucleotides per ST-adducted nucleotide.

Two features are notable from these findings. The values indicate a triphasic pharmacokinetic profile for DNA-bound ST moieties, i.e., rapid loss during the first 24 hours ($t_{1/2}$ = 12 hours), followed by a slower decline from 1 to 14 days postdosing ($t_{1/2}$ = 7 days) and an extremely slow decline from days 14 to 105 post treatment ($t_{1/2}$ = 109 days). Also, ST adduct levels in rat liver can be measured three months after a single, subacute exposure. Sensitivity of the procedure for ST-DNA provides a basis for adduct studies in humans undergoing chronic, low-level ST exposure.

Discussion

Recognition of the genetic and carcinogenic hazards posed by food- and feed-borne mycotoxins has fueled intense scientific research to design animal studies that probe cellular mechanisms responsible for mycotoxin-induced disease. Successful identification of the principal mycotoxin-DNA reaction products in laboratory animals, as well as in acutely exposed humans, has provided key information necessary to construct molecular models describing perturbations in DNA structure and function subsequent to mycotoxin damage. Rapid advances in the technology to monitor expression of cellular oncogenes responsible for tumor formation portend improved diagnosis of cellular transformation, as well as accelerated design of new drugs to inhibit tumor proliferation and metastasis.

The technology described in the current sutdy, which monitored mycotoxin exposure and effects via DNA adduct levels in target-organ tissue (liver) or sentinental tissue (placenta), provides a clear approach to extend biochemical studies of mycotoxicology to address critical human and veterinary health issues.

Future Plans

Reports of mycotoxin-contaminated grain being sold to Mexico for human consumption have led to serious concern about mycotoxin-induced disease in the country (3,12). As a result, wide-ranging research collaboration between scientists at Texas A & M University and the National Autonomous University of Mexico (UNAM) has recently been initiated to apply DNA adduct technologies to document human mycotoxin consumption and disease among certain Mexican populations.

The discovery of a diverse and complex mixture of mycotoxins in many areas of Mexico indicates an urgent need for an assessment of food-borne mycotoxins to:

- Determine which of the many mycotoxins to which rural populations are exposed are principal agents for human and animal mycotoxicoses;
- Assess whether mixtures of naturally occurring mycotoxins act synergistically to increase the toxicity of the individual mycotoxin components; and
- Suggest viable postharvest treatments to eliminate the toxic effects of mycotoxins while retaining the organoleptic properties of food products.

Initial joint studies have focused on an assessment of the multiple interactions of mycotoxins previously documented in Mexican foodstuffs (4,8), as well as on mutagenic and DNA-damaging components in current rural food samples (2). Current toxicokinetic studies of placental uptake and metabolism of these food-borne mycotoxins will permit a broader design and clinical application of DNA-postlabeling technology to routinely monitor human populations at risk from high-level mycotoxin exposure.

Acknowledgements

The author wishes to acknowledge the technical support of P. Hey, P.C. Mertes, C. Maynard, A. O'Neill, E.K. Stevens, S. Wadhwani and F. Wall in this research. The work was supported by USDA grants AH6658 and 6859, contracts 5218 and 5434 from Shell Development Company, Texas A & M CEMR grant 18845, TEES project 901664 and Project H6735 (predictive tests for mycotoxicity) by the Texas Agricultural Experiment Station.

References

1. Autrup, H., J. Wakhisi, A. Wasunna and C.C. Harris. 1985. Detection of 8.9-dihydro-(7'-guanyl)-9- hydroxyaflatoxin B_1 in human urine. Environmental Health Perspectives 62:105-108.

2. Carvajal, M., and T.R. Irvin. 1986. Identification and genotoxic potency of complex mycotoxin mixtures isolated from Mexican foodstuffs. (Submitted.)

3. Carvajal, M., and T.R. Irvin. 1986. Identification and toxicological potency of mycotoxin mixtures found as contaminants in Mexican foodstuffs. CRIS report for TAES project H6735. Texas Agricultural Experiment Station, College Station, Texas, USA.

4. Donnelly, K.C., T.R. Irvin and R.H. Bailey. 1986. Mutagenic potency of binary mixtures of aflatoxins and sterigmatocystins. Toxicologist 6(1):151.

5. Essigmann, J.M., L.J. Barker, M.A. Fowler, M.A. Francisco, V.N. Reinhold and G.N. Wogan. 1979. Sterigmatocystin-DNA interactions: Identification of a major adduct formed after metabolic activation *in vitro*. Proceedings of the National Academy of Science (USA) 76:179-183.

6. Essigmann, J.M., R.G. Croy, R.A. Bennett and G.N. Wogan. 1982. Metabolic activation of aflatoxin B_1: Patterns of DNA adduct formation, removal, and excretion in relation to carcinogenesis. Drug Metabolism Reviews 13(4):581-602.

7. Essigmann, J.M., D.L. Donahue, D.L. Story, G.N. Wogan and H. Brunengraber. 1980. Use of the isolated perfused rat liver to study carcinogen-DNA adduct formation from aflatoxin B_1 and sterigmatocystin. Cancer Research 40:4085-4091.

8. Irvin, T.R., and M. Carvajal. 1986. *In vitro* prenatal toxicity of complex mycotoxin mixtures isolated from mold-adulterated foods. (Submitted.)

9. Irvin, T.R., and G.N. Wogan. 1985. Quantitative and qualitative characterization of aflatoxin B_1 adducts formed *in vivo* within the ribosomal RNA genes of rat liver DNA. Cancer Research 45:3497-3502.

10. Irvin, T.R., and G.N. Wogan. 1984. Quantitation of aflatoxin B_1 adduction within the ribosomal RNA gene sequences of rat liver DNA. Proceedings of the National Academy of Sciences (USA) 81:664-668.

11. Juchau, M.R. 1973. Placental metabolism in relation to toxicology. CRC Critical Reviews in Toxicology 2(2):125-158.

12. Marshall, E. 1984. Yellow rain experts battle over cold mold. Science 221:526-529.

13. Post, D.J., and T.R. Irvin. Metabolic activation of aflatoxin B_1 by fetal and maternal tisues of the C57B6 mouse. Toxicologist (1):386.

14. Reddy, M.V., T.R. Irvin and K. Randerath. 1985. Formation and persistence of sterigmatocystin-DNA adducts in rat liver determined via ^{32}P-postlabeling analysis. Mutation Research 152:85-96.

15. Terao, K. Sterigmatocystin: A masked potent carcinogenic mycotoxin. Journal of Toxicology and Toxin Reviews 2:77-110.

16. Wogan, G.N. 1973. Aflatoxin carcinogenesis. Methods in Cancer Research 7:325-344.

17. Wogan, G.N. 1966. Chemical nature and biological effects of the aflatoxins. Bacteriological Reviews 30(2):460-470.

18. Wong, J.J., R. Singh and D.P.H. Hsieh. 1977. Mutagenicity of fungal metabolites related to aflatoxin biosynthesis. Mutation Research 44:447-450.

Sampling and Detection Techniques for Aflatoxin in Maize

J.W. Dickens, Agricultural Research Service, US Department of Agriculture, North Carolina State University, Raleigh, North Carolina, USA

Abstract

*Because a small percentage of the kernels in a lot of maize (*Zea mays *L.) may contain very high concentrations of aflatoxin, accurate determination of the concentration is difficult. The variance and skew increase as sample size and/or lot concentration decrease. Therefore, the distribution of sample concentrations may be different from that of lot concentrations when small samples are used. Equations to compute the variance associated with sampling, subsampling and analysis have been developed. Probe sampling may be used soon after the blending of a lot of maize by harvesting, handling or other operations; otherwise, stream sampling should be utilized. The relationship between the weight percent of bright greenish-yellow fluorescent (BGYF) kernels and the aflatoxin concentration in the sample may be expressed by the following equation:*

μg/kg in sample = 197 (weight percentage of BGYF kernels)

Because most samples of maize with more than 20 μg/kg aflatoxin have BGYF kernels, an effective screening program might require chemical analysis only for those samples with BGYF kernels.

Resumen

*Dado que es posible que un porcentaje pequeño de granos en una gran cantidad de maíz (*Zea mays *L.) contenga concentraciones muy altas de aflatoxina, resulta difícil determinar la concentración con exactitud. La varianza y la asimetría aumentan conforme disminuye el tamaño de la muestra y/o la concentración del lote. Por consiguiente, la distribución de las concentraciones en la muestra suele diferir de la de las concentraciones en los lotes cuando se emplean muestras pequeñas. Se han creado ecuaciones para calcular la varianza relacionada con el muestreo, submuestreo y análisis. Se puede usar un muestreo de prueba poco después de mezclar un lote de maíz al cosechar, manejar el maíz o efectuar otras operaciones; de otro modo, es preciso utilizar el muestreo de flujo. La siguiente ecuación se puede emplear para expresar la relación que existe entre el porcentaje en peso de granos con la fluorescencia característica color amarillo verdoso brillante (FAVB) y la concentración de aflatoxina en la muestra:*

μg/kg en la muestra = 197 (porcentaje en peso de granos FAVB)

Como la inmensa mayoría de las muestras de maíz que presentan más de 20 μg/kg de aflatoxina contienen granos FAVB, es posible que los programas eficaces de selección requieran un análisis químico de las muestras que presenten granos FAVB.

The objectives of sampling maize (*Zea mays* L.) for aflatoxin analyses may be to check for the presence, distribution or concentration of aflatoxin in a given lot or among a population of lots. The method of sampling should be appropriately designed. To check for the presence of aflatoxin or to determine the incidence of aflatoxin among different lots, the sampling procedure should be biased towards the inclusion of kernels that are more likely to contain aflatoxin. For example, kernels with visible mold growth or insect damage probably contain more aflatoxin than kernels that appear sound; the sampling procedure should be selective or biased to insure that the analytical sample contains such high-risk kernels. If the objective is to estimate the average aflatoxin concentration within a lot or to determine the distribution of a population of lots according to their aflatoxin concentrations, the sampling procedure must be unbiased, with every kernel within the sample collected in a completely random manner without regard to physical characteristics or location within the population of kernels that make up the lot. The objectives of most aflatoxin tests require unbiased sampling.

The aflatoxin concentration in a lot of maize is usually estimated from a sample drawn from the lot. A previous study has demonstrated that replicated aflatoxin tests on the same lot of shelled maize may be highly variable (22). Because the toxin concentration in individual kernels of a lot may range from 0 to over 500,000 μg/kg (12,19), a large sample of kernels is required to insure that the sample concentration is in reasonable agreement with the lot concentration. Subsampling error and analytical error may also cause significant differences between test results and lot concentrations. A thorough review of research related to sampling and testing maize for aflatoxin has been published by Dickens and Whitaker (4).

Aflatoxin in maize may be determined by the presence of kernels that exhibit bright greenish-yellow fluorescence (BGYF) under long-wave ultraviolet (UV) light (15). Through screening, an estimate may be made of the percentage of BGYF kernels in a sample from the lot (5). In this paper, sampling techniques for reducing errors in aflatoxin tests for maize are discussed, focusing on the efficacy of the BGYF method.

Sources of Error in Aflatoxin Tests

The distribution of replicated aflatoxin tests about their mean is an important consideration in the design and evaluation of aflatoxin testing programs. The distribution of tests may be skewed or symmetrical, depending upon the test conditions. The distribution is symmetrical when an equal number of tests are above and below the mean of all the tests (the median equals the mean). The distribution is positively skewed when more than half of the replicated tests are less than the mean of all the tests (the median is less than the mean). When small samples are used, the distribution of sample concentrations about the true aflatoxin concentration in a lot is positively skewed, so that more than half of the sample concentrations are less than the lot concentration (22). Consequently, there is more than a 50% probability that the aflatoxin concentration in a small sample is less than the concentration in the lot. As sample size increases, the skew of the distribution of sample concentrations will decrease, and according to the central limit theorem, will approach a normal distribution (14). For a given sample size, the skew also decreases as lot concentration increases.

The official first action method for maize, as specified by the Association of Official Analytical Chemists (2), does not designate sample size, but it requires that the entire sample of shelled maize be ground to pass through a no. 14 sieve, and that a 1- to 2-kg subsample of this material be ground to pass through a no. 20 sieve. A 50-g subsample of the latter material is then analyzed by the CB (AOAC) method. Whitaker *et al.* (22) developed the following equations for the variance (error) terms related to this test procedure:

1. $V = S + C + F + Q$
2. $S = 3.9539\ P/Ws$
3. $C = 0.1196\ P/Wc$
4. $F = 0.0125\ P/Wf$
5. $Q = 0.0699\ P^2/Nq$

The symbols V, S, C, F and Q are, respectively, the variance or the total error of the test, the error in sampling the shelled maize, the error in subsampling the coarsely ground material, the error in subsampling the finely ground material and the error in the quantification of aflatoxin in the chloroform extract from the finely ground subsample. The mass of the sample in kg is Ws, Wc is the mass of the coarsely ground subsample in kg, Wf is the mass of the finely ground subsample in kg, Nq is the number of times the aflatoxin in the solvent extract is quantified on a separate thin-layer chromatography (TLC) plate and P is the aflatoxin concentration (μg/kg) in the lot.

Equations 2, 3 and 4 are based on laboratory studies in which sampling and subsampling procedures were designed to eliminate bias. Only the inherent heterogeneity of aflatoxin concentration among the kernels and among the comminuted particles causes the indicated variances. When errors are introduced because of sampling and subsampling procedures, or when the particles in the comminuted samples are larger than in the laboratory study, the variances will be greater than those estimated by the equations.

Equation 5 reflects the variance among replicated measurements of the aflatoxin concentration in the chloroform extract from the finely ground maize according to the CB method. A densitometer was used to quantify the aflatoxin in spots on the TLC plates (6). A study by Whitaker and Dickens (21) suggests that even when a densitometer is used, most of the error indicated by Equation 5 is due to errors in quantification of aflatoxin on TLC plates. High-performance liquid chromatography or other analytical procedures may reduce this error, but comparable data on these analytical procedures are not presently available.

Error reduction

Equations 2, 3 and 4 specify mass rather than the number of maize kernels or particles in the sample or the subsamples, because mass is directly correlated with the number of kernels or particles and is a more convenient measurement. However, the number of kernels or particles in a sample or subsample is the important criteria. Therefore, for a given mass of subsample, the amount of subsampling error is reduced by grinding the sample more finely and thus increasing the number of particles in the subsample.

Changes that can improve the precision of aflatoxin tests are increases in sample size, increases in the size of the subsample used for aflatoxin analysis and an increase in the number of analyses. Different costs are associated with each change, and careful study is required to determine the testing program that will provide the most precision for a given cost. The optimum balance in the sample size, degree of comminution, subsample size and number of analyses will vary according to such factors as the cost of the sample to be comminuted, the cost of sample comminution and subsampling and the cost of analysis. In general, more precise aflatoxin testing programs will be more costly.

Sampling Procedures

Samples may be taken from maize growing in the field, during handling, during storage, and at other points in the production, marketing and processing system. Each type of sampling presents a different situation regarding distribution of the toxin and accessibility of test material. When feasible, samples should be taken after the lot has been reduced to a smaller particulate size. For example, it is better to sample shelled maize rather than ears, and ground maize is best.

A sample is more unbiased when a lot of maize has been recently blended by such operations as harvesting, loading or unloading and turning. This is because mold growth, which may have occurred in spots, will have been more evenly distributed. For example, moisture condensation or leaks during storage may have caused a portion of the lot to mold with a resulting high concentration of aflatoxin (18). It is impossible to predict where to probe the contents of a storage bin in order to obtain a sample with the same aflatoxin concentration as that of the entire lot.

Stream sampling
The most effective sampling method is to take small portions from a moving stream at intervals and to combine these portions into a sample; crosscut samplers are commercially available that automatically cut through the stream at predetermined intervals. When an automatic crosscut sampler is not available, a cup may be passed through the stream at periodic intervals, thus collecting a sample. The stream should be sampled frequently, but the amount taken at each interval should be small. The samples must be taken from the stream during the entire time that the lot is being moved.

Probe sampling
Probe sampling is probably adequate for shelled maize that has recently been blended by harvesting or handling operations. Recommended methods for taking probe samples are published by the American Oil Chemists' Society (1).

Field sampling
When samples are taken before harvest, representative sampling of the field is difficult as *Aspergillus flavus* may be erratically distributed among ears in the field. Also, no kernels or any number of kernels on an ear may be aflatoxin-contaminated. The answer is to collect a large number of widely distributed ears to obtain a representative sample for shelling. Since maize is shelled during most harvesting operations, sampling should be coordinated with harvesting so that a large number of ears is represented in the sample (3).

Sample preparation
Because it is not feasible to solvent extract the aflatoxin from a large sample of maize, the sample must be comminuted in order to extract the aflatoxin from a small subsample of finely comminuted material. Subsampling error is inversely proportional to the number of particles in the subsample, so a larger subsample is required for coarsely ground material than for finely ground material. The Association of Official Analytical Chemists (2) recommends a two-stage grinding operation, but in aflatoxin laboratories of the North Carolina Department of Agriculture and North Carolina State University a hammer mill is used to grind an entire sample so that it will pass through a screen with 1-mm openings (9).

Sampling for Survey Studies

A positively skewed distribution of sample means about the lot mean indicates that the distribution of sample concentrations determined in a survey will probably be different than the distribution of the true concentrations of the sampled lots. When small samples are used, most of the samples will contain a lower aflatoxin concentration than the sampled lot, and a few samples will contain much higher concentrations than the sampled lot. Expected agreement increases when a larger sample is used.

It is difficult to make a general recommendation about the sample size that should be used in surveys. Larger samples will increase accuracy, but such costs as maize, transportation and sample preparation may be limiting. If the purpose of the survey is limited to a determination of the average aflatoxin concentration in all lots sampled, relatively small samples are required. When an accurate estimate of the distribution of lots according to aflatoxin concentration is desired, larger samples are required. A 4.54-kg sample of shelled maize is probably sufficient for most survey purposes. Davis *et al.* (3) have suggested protocols for mycotoxin surveys and discussed problems related to sampling.

Sampling for Quality Control

For regulatory purposes, the North Carolina Department of Agriculture (9) analyzes 4.54-kg samples of shelled maize using the procedure outlined by the Association of Official Analytical Chemists (2). Many commercial operations employ the same procedure for their aflatoxin control programs. Errors in aflatoxin tests may cause good lots to test as bad (false rejects), and bad lots to test as good (false accepts).

Most procedures for dealing with aflatoxin-contaminated products are expensive; therefore, it is important for the owners of the product that the risks of false rejects or false accepts be minimized. Since the accuracy of a properly designed aflatoxin testing program is generally directly proportional to the cost of the program, it is not usually economically feasible to employ a testing program that eliminates all risk for the owner. A program that maximizes the benefit/cost ratio would be the reasonable compromise. The benefit is the savings to the owner in the rejection of the bad lot before additional investment is made for processing. The costs include the cost of the testing program, the cost for rejecting lots and the cost of accepting bad lots for further processing. The relative size of these costs and savings depends on many factors, including the value of the commodity, its intended use, the cost of removing aflatoxin contamination, the cost of diverting contaminated lots to nonfood uses, the cost of the aflatoxin tests and the incidence of bad lots. As a result, different types of testing programs may be desirable.

At many points in the marketing and processing system, the benefits and costs can be treated as economic values rather than as questions of human health. Evaluation of health risk can be deferred until the last stage in the marketing/processing system. At this stage, sampling error is often reduced because of comminution and blending during processing operations; i.e., aflatoxin tests using cornmeal are usually more accurate than those using shelled maize.

Several rapid analytical methods, generally referred to as minicolumn methods, have been developed to detect aflatoxin in maize (10,20). These methods are rapid enough to keep pace with the marketing operation, do not require expensive laboratory space and equipment and are economical. The accuracy of these methods is often limited because samples are too small and because inadequate comminution and subsampling procedures are used. The same sampling and subsampling procedures should be used for minicolumn methods as for other methods, but unfortunately this greatly increases the time and labor requirements.

Using Bright Greenish-Yellow Fluorescence to Detect Aflatoxin Contamination

A bright greenish-yellow fluorescence (BGYF) under longwave (365 nm) ultraviolet light has been associated with the presence of aflatoxin in maize, cottonseed and pistachio nuts (7,8,13). Examination of maize for BGYF has been proposed as a rapid screening method to detect aflatoxin-contaminated lots at the time of marketing. Previous studies indicate that the aflatoxin content of samples with BGYF kernels may range from none to very high concentrations (16,17). When there are no BGYF particles in 4.54-kg samples of cracked maize, the probability is very low that the sample contains aflatoxin (15). Therefore, an effective aflatoxin screening program may consist of accepting lots with no BGYF in the samples, and only analyzing those with BGYF kernels.

Marketing tolerances for aflatoxin concentrations of 20 μg/kg or more in maize, depending on its intended use, have been accepted in the USA. For a BGYF screening method to be practical, it must provide a dependable, quantitative estimate of aflatoxin concentrations ranging from 20 to 100 μg/kg in commercial lots. However, studies on white maize produced and stored on farms in Missouri in 1971 indicated that the weight percentage of BGYF particles did not provide a satisfactory quantitative estimate for aflatoxin (11).

Dickens and Whitaker (6) conducted a study to determine the correlation between the weight percentage of BGYF kernels and the aflatoxin concentration in samples from lots of yellow maize marketed in eastern North Carolina during the 1977 and 1978 marketing season. Maize kernels that exhibited BGYF under long-wave ultraviolet light were hand-picked from the samples. The BGYF kernels from 113 4-kg samples contained an average of 8665 μg/kg compared to an average of 46 μg/kg aflatoxin in the non-BGYF kernels. A regression analysis between the aflatoxin concentration and the weight percentage of BGYF kernels in 2304 4.5-kg samples produced the regression equation:

μg/kg in sample = 197 (weight percentage of BGYF kernels)

The correlation coefficient for the regression analysis was 0.90.

Testing programs to reduce aflatoxin concentrations in purchased lots of maize, based on either the BGYF method or the AOAC chemical assay method (2), were compared by Dickens and Whitaker (6). The average aflatoxin concentration in lots accepted by the AOAC method was 4, 10 or 18 μg/kg when an acceptance level of 20, 50 or 100 μg/kg, respectively, was used. For the BGYF method, the average aflatoxin concentration in accepted lots was 10, 16 or 22 μg/kg when an acceptance level of $<0.10\%$, $<0.25\%$ or $<0.50\%$ BGYF, respectively, was used. Approximately the same percentage of lots were accepted by both methods when either the low, medium or high acceptance level was used. These results indicate that the AOAC chemical assay method is more discriminating than the BGYF method, but the percentage difference in the average aflatoxin concentrations of the lots accepted by the two methods diminishes when the acceptance level is increased. Both the cost and efficacy of the two methods should be considered when choosing between them.

References

1. American Oil Chemists' Society. 1973. Official and Tentative Methods of Analysis. Champaign, Illinois, USA.

2. Association of Official Analytical Chemists. 1980. Official Methods of Analysis, 13th ed. Washington, D.C., USA. Pp. 229-434.

3. Davis, N.D., J.W. Dickens, R.L. Freie, P.B. Hamilton, O.L. Shotwell, T.D. Wyllie and J.F. Fulkerson. 1980. Protocols for surveys, sampling, post-collection handling and analysis of grain samples involved in mycotoxin problems. Journal of the Association of Official Analytical Chemists 63:95-102.

4. Dickens, J.W., and T.B. Whitaker. 1983. Aflatoxin testing procedures for corn, pp. 35-37. *In* Aflatoxin and *Aspergillus flavus* in Corn, U.L. Diener, R.L. Asquith and J.W. Dickens, eds. Southern Cooperative Services Bulletin 279. Alabama Agricultural Experiment Station, Auburn, Alabama, USA.

5. Dickens, J.W., and T.B. Whitaker. 1981. Bright greenish-yellow fluorescence and aflatoxin in recently harvested yellow corn marketed in North Carolina. Journal of the American Oil Chemists' Society 58:973A-975A.

6. Dickens, J.W., W.F. McClure and T.B. Whitaker. 1980. Densitometric equipment for rapid quantitation of aflatoxin on thin layer chromatograms. Journal of the American Oil Chemists' Society 57:205-208.

7. Dickens, J.W., and R.E. Welty. 1975. Fluorescence in pistachio nuts contaminated with aflatoxin. Journal of the American Oil Chemists' Society 52:448-450.

8. Fennell, D.S., R.J. Bothast, E.B. Lillehoj and R.E. Peterson. 1973. Bright greenish-yellow fluorescence and associated fungi in white corn naturally contaminated with aflatoxin. Cereal Chemistry 50:404-414.

9. Food and Drug Protection Division. 1978. Training manual on screening methods for aflatoxin. North Carolina Department of Agriculture, Raleigh, North Carolina, USA.

10. Holaday, C.E., and J. Landsden. 1975. Rapid screening method for aflatoxin in a number of products. Journal of Agricultural and Food Chemistry 23:1134-1136.

11. Kwolek, W.F., and O.L. Shotwell. 1979. Aflatoxin in white corn under loan. V. Aflatoxin prediction from weight percent of bright greenish-yellow fluorescent particles. Cereal Chemistry 56:342-345.

12. Lee, L.S., E.B. Lillehoj and W.F. Kwolek. 1980. Aflatoxin distribution in individual corn kernels from intact ears. Cereal Chemistry 57:340-343.

13. Marsh, P.B., M.E. Simpson, R.J. Ferreti, G.V. Merola, J. Donoso, G.O. Craig, M.W. Trucksess and P.S. Work. 1969. Mechanism of formation of a fluorescence in cotton fiber associated with aflatoxins in the seeds at harvest. Journal of Agricultural and Food Chemistry 17:468-472.

14. Remington, R.D., and M.A. Shork. 1970. Statistics with Application to the Biological and Health Sciences. Prentice-Hall, Englewood Cliffs, New Jersey, USA. P. 418.

15. Shotwell, O.L. 1983. Aflatoxin detection and determination in corn, pp. 38-43. *In* Aflatoxin and *Aspergillug flavus* in Corn, U.L. Diener, R.L. Asquith and J.W. Dickens, eds. Southern Cooperative Series Bulletin 279. Alabama Agricultural Experiment Station, Auburn, Alabama, USA.

16. Shotwell, O.L., and C.W. Hesseltine. 1981. Use of bright greenish-yellow fluorescence as a presumptive test for aflatoxins in corn. Cereal Chemistry 58:124-127.

17. Shotwell, O.L., M.L. Goulden, A.M. Jepson, W.F. Kwolek and C.W. Hesseltine. 1975. Aflatoxin occurrence in some white corn under loan, 1971. III. Association with bright greenish-yellow fluorescence in corn. Cereal Chemistry 52:670-677.

18. Shotwell, O.L., M.L. Goulden, R.J. Bothast and C.W. Hesseltine. 1975. Mycotoxins in hot spots in grains. I. Aflatoxin and zearalenone occurrence in stored corn. Cereal Chemistry 52:687-697.

19. Shotwell, O.L., M.L. Goulden, E.B. Lillehoj, W.F. Kwolek and C.W. Hesseltine. 1977. Aflatoxin occurrence in 1973 corn at harvest. III. Aflatoxin distribution in contaminated, insect-damaged corn. Cereal Chemistry 54:620-626.

20. Velasco, J. 1975. Fluometric measurement of aflatoxin adsorbed on florisil in minicolumns. Journal of the Association of Official Analytical Chemists 58:757-763.

21. Whitaker, T.B., and J.W. Dickens. 1981. Errors in aflatoxin analyses of raw peanuts by thin layer chromatography. Peanut Science 8:89-92.

22. Whitaker, T.B., J.W. Dickens and R.J. Monroe. 1979. Variability associated with testing corn for aflatoxin. Journal of the American Oil Chemists' Society 56:789-794.

Detection and Determination of Aflatoxins in Maize

D.M. Wilson, Department of Plant Pathology, Coastal Plain Experiment Station, University of Georgia, Tifton, Georgia, USA

Abstract
Many different methods are available for aflatoxin analysis. Maize is often contaminated with a combination of aflatoxins B_1 and B_2 and less often with a combination of B_1, B_2, G_1 and G_2. Safe handling of analytical samples in laboratories working with the aflatoxins or with the aflatoxigenic species of the Aspergillus *group is essential. Common methods for screening maize for aflatoxin contamination include the BGYF (bright greenish-yellow fluorescence) presumptive test and any one of several minicolumn methods. These methods should not be used to measure quantities of aflatoxin, but are effective for identifying lots of maize requiring further testing. Quantitative procedures for aflatoxins B_1, B_2, G_1 and G_2 include several thin-layer chromatographic (TLC) and high-performance liquid chromatographic (HPLC) methods. The present enzyme-linked immunosorbent assay (ELISA) methods are not sufficiently developed to be used for quantification, but they can be effectively calibrated to give positive results above a predetermined amount of aflatoxin B_1. Current ELISA methods do not assess amounts of aflatoxin B_2, G_1 and G_2.*

Resumen
Se cuenta con diversos métodos para analizar las alfatoxinas. Con frecuencia, el maíz es contaminado por una combinación de aflatoxina B_1 y B_2 y, menos frecuentemente, con una combinación de aflatoxina B_1, B_2, G_1 y G_2. Es de vital importancia tomar medidas de seguridad al manejar muestras analíticas en los laboratorios que trabajan con aflatoxinas o con especies aflatoxigénicas del grupo Aspergillus. *Los métodos más comunes de seleccionar el maíz para determinar la presencia de contaminación por aflatoxinas incluyen la prueba de presunción de FAVB (fluorescencia color amarillo verdoso brillante) y diversos métodos de microcolumnas. Estos métodos no se deben utilizar para medir cantidades de aflatoxina, pero resultan eficaces para identificar lotes de maíz que requieran más pruebas. Los procedimientos cuantitativos para las aflatoxinas B_1, B_2, G_1 y G_2 incluyen diversos métodos de cromatografía de capa fina y de cromatografía líquida de alta presión. Los métodos actuales del ensayo inmunosorbente conjugado con una enzima (ELISA) no se han desarrollado lo suficiente para emplearlos en la cuantificación, pero se pueden calibrar eficazmente para dar resultados positivos por encima de una cantidad de aflatoxina B_1 determinada con anterioridad. Los métodos ELISA de los que se dispone hoy día no sirven para evaluar cantidades de aflatoxina B_2, G_1 y G_2.*

Aflatoxin contamination of food and feed is important in human and animal health, because the aflatoxins are toxic and carcinogenic. The toxic and carcinogenic properties of the aflatoxins make experimental safety a very important issue in all work with aflatoxins or the fungi, *Aspergillus flavus* Link ex Fries and *A. parasiticus* Speare, that produce aflatoxins.

Safety

Chemical safety

Guidelines for mycotoxin safety are given at the beginning of chapter 26 in the Association of Official Analytical Chemists (AOAC) *Official Methods of Analysis* manual (2). The International Agency for Research on Cancer (IARC) also has a good publication on safety precautions (22). These safety guidelines are appropriate for both crude and pure aflatoxin preparations. The chemicals should only be handled with gloves and used only in properly ventilated hoods or glove boxes. Since dry aflatoxins can disperse in the laboratory, most analytical laboratories should buy commercially prepared standards. Spills and contaminated laboratory surfaces should be treated with 1% sodium hypochlorite bleach for ten minutes, followed by 5% aqueous acetone. Glassware should be rinsed, ideally before washing, in methanol and soaked in 1% NAOCl bleach for ten minutes, after which 5% acetone should be added for 30 minutes.

Biological safety

Spores and other viable propagules of *A. flavus*, *A. parasiticus* and other fungi can cause three types of disease in humans: allergy, poisoning and infection (19). Airborne spores and dust containing propagules of the *A. flavus* group may cause allergies in some people and the inhaled particles may contain aflatoxins (47). Two thin-layer chromatographic (TLC) methods have been developed to measure aflatoxins in maize and grain dust (14,47). *Aspergillus flavus* infection in humans is uncommon but possible.

Hill and co-workers (19) found between 10^4 and 10^9 viable fungal propagules per m^3 of air containing maize dust. The majority of the *A. flavus* propagules in air samples were deposited on the stages of the Andersen sampler corresponding to the trachea, primary bronchi and secondary bronchi in the human respiratory system (19). *Aspergillus flavus* spores and propagules in maize dust associated with inoculation, shelling, grinding and extraction procedures are sufficiently hazardous to require the use of gloves, masks, protective clothing and efficient dust collection mechanisms.

Sampling

Samples taken for aflatoxin analysis are subject to large sampling errors because aflatoxin is irregularly distributed. The first consideration in any experiment should be how to take the sample, and then how to prepare the sample for analysis. Protocols have been published on field sampling techniques (9), test plot inoculation and sampling techniques (59,60), and lot sampling techniques (22,58).

Aflatoxin contamination in maize is generally less variable in single fields, single test plots or single lots than in some other crops. A 4.54 kg sample is usually sufficient for maize, especially when several analytical samples are averaged to approximate the true mean (58). However, in mixed lots from different sources, a larger initial sample should be taken. The total sample should be ground to pass through a 0.85 mm sieve, thoroughly mixed or divided and properly

Asperguillus flavus **spores must not be inhaled; masks and other safety precautions are essential**

subsampled before analytical samples are taken. Sampling protocols for test plots must be part of the experimental design and need to be appropriate for the experimental objectives.

Standards

Criteria for mycotoxin standards (38) and procedures for checking the concentration and purity of aflatoxin standards can be found in chapter 26 of the AOAC manual (2). The use of calibrated standards in all analytical laboratories is essential. Prepared standards are available from several commercial companies at reasonable prices, and analytical laboratories should routinely use these standards. Velasco (57) found that cyclohexane, heptane and toluene could be substititued for benzene in standards if the solutions were not exposed to light, but aflatoxins in these solvents degraded rapidly upon exposure to light. Chang-Yeng *et al.* (5) found that solvent composition affects aflatoxin fluorescence, which should be taken into consideration when standards are prepared for high-performance liquid chromatography (HPLC).

Aflatoxin Testing Methods

Presumptive and screening methods

Aflatoxin analysis can be approached in many ways. Maize is more often contaminated with aflatoxin B_1 and B_2 than with a combination of B_1, B_2, G_1 and G_2. Some applications require only presumptive or screening tests, while others require the quantitation of B_1 alone or several of the aflatoxins. A black light test to detect the bright greenish-yellow fluorescence (BGYF) (16,49) indicative of potential aflatoxin contamination of maize should only be used to identify lots for further chemical analyses. Bright greenish-yellow fluorescence should never be used to set maize prices in the market place.

Another commonly used screening technique is the application of one of several minicolumns to detect aflatoxin contamination above a predetermined level (20,39,44). Shannon and Shotwell conducted a collaborative study of two minicolumn methods and found that a combination method using the Holaday extraction and the Velasco minicolumn was the most satisfactory (44). It is important to understand that a minicolumn technique should not be used for quantitative purposes. Madhyastha and Bhat (29) recently developed a minicolumn confirmation method for aflatoxins. They confirmed the identity of aflatoxins on the developed minicolumn by applying 20% H_2SO_4, 20% HCl or trifluoroacetic acid (TFA) in 20% HNO_3. All acids changed the fluorescence from blue to yellow, with the TFA in 25% HNO_3 having the lowest detection limit.

Quantitative methods

Many of the methods adopted by scientific groups and government agencies are based on TLC detection and quantitation procedures that have been evaluated in one or more collaborative studies. The only reason that HPLC methods are not recommended more often is that few collaborative studies on HPLC methods have been conducted. The *Official Methods of Analysis* (2) published by the AOAC is probably the most widely consulted manual for aflatoxin analysis; it recommends the CB method for aflatoxin analysis in maize. Other societies and agencies that recommend or publish methods include: American Industrial Hygiene Association, American Oil Chemists' Society, International Union of Pure and Applied Chemistry, Tropical Products Institute, American Association of Cereal Chemists and the International Agency for Research on Cancer. Schuller *et al.* (40) published an excellent review of sampling plans and collaboratively studied methods for aflatoxin analysis. This valuable paper should be consulted to help devise or review analytical protocols for aflatoxin analysis. Nesheim (32) and Shotwell (46) also have excellent reviews on aflatoxin analysis.

Analytical laboratories can and should participate in one or more check sample programs. The American Oil Chemists' Society conducts the Smalley Mycotoxin Check Sample Program (30). Information on this program can be obtained from Dr. J.D. McKinney, Ranchers Cotton Oil, P.O. Box 2596, Fresno, California 93745. The International Agency for Research on Cancer conducts the International Mycotoxin Check Sample Program (17). Information on this program may be obtained from Dr. M.D. Friesen, International Agency for Research on Cancer, 150 cours Albert Thomas, 69372 Lyon Cedex 2, France.

Thin-layer chromatography—The official AOAC CB method (2) is the standard by which other methods are judged. A good discussion of this method can be found in the IARC mycotoxin manual (22). Shotwell and Goulden (48) compared the AOAC BF peanut method and the AOAC cottonseed method with the CB method in maize. The BF method uses a methanol:water (55:45) extraction solution. Neither of these solvents extracted aflatoxins from maize as efficiently as the chloroform:water (250:15) extraction of the CB method. Lee and Catalano (28) developed a scaled-down cleanup column as a solvent-saving modification of the CB method. Alexander and Baur (1) described one of many two-dimensional TLC procedures which have been developed for use in samples with interfering substances. Laboratories that use fluorodensitometry for quantitative measurements must avoid fading of aflatoxin spots on TLC plates. Nesheim (31) found that fading could be delayed by covering the layer on the TLC plate with another glass plate.

The CB method is an excellent TLC method, but it has two major disadvantages; it is expensive because it uses large amounts of solvents (thus creating a disposal problem), and the major solvent is $CHCl_3$, which may be a hazard to workers. In many research applications, alternative methods may satisfy the experimental objective in a less expensive and safer manner. Dantzman and Stoloff (8) developed a screening method omitting the column chromatography step of the CB method and directly spotting the residual oil from maize extracted with $CHCl_3$-water. Spilman (50) modified this method by adding benzene:CH_3CN (98:2) to the residual oil and measuring the volume to obtain quantitative TLC results.

Many other methods have also been published. Kamimura *et al.* (25) recently published a simple, rapid HPTLC method which compared favorably with the CB method. Davis *et al.* (11) devised an approach using the fluorescence of the iodine derivative of

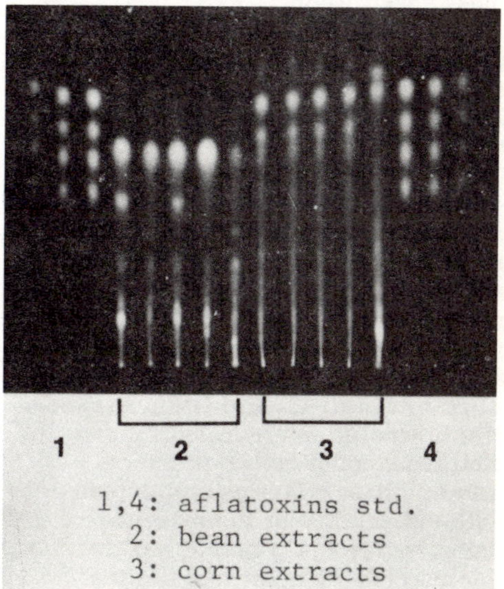

Thin-layer chromoatography plate showing separation of aflatoxins B_1, B_2, G_1 and G_2 standards (1 and 4), bean extracts (2) and maize extracts (3)

aflatoxin B_1 for quantitation and TLC confirmation. Shannon and Shotwell (43) developed a method for determination of aflatoxin in roasted maize, and Bagley (3) referred to a method especially adapted for maize detoxified by ammonia treatment. Josefsson and Möller (24) developed a multimycotoxin screening method for the detection of aflatoxins, ochratoxin, patulin, sterigmatocystin and zearalenone in maize; Seitz and Mohr (42) and Thomas *et al*. (55) developed methods for aflatoxin and zearalenone determination in maize.

Aflatoxin identity needs to be confirmed regardless of the TLC method used. Nesheim and Brumley (33) have written a review of confirmation methods. The AOAC method is based on the trifluoroacetic acid (TFA) reaction with B_1, G_1 or M_1 (37). The TFA procedure or direct acetylation (4) can be carried out on a TLC plate before development. Trucksess *et al*. (56) recently published a rapid TLC method using a disposable silica gel column for cleanup and confirmation by gas chromatography/ mass spectroscopy.

High-performance liquid chromatography—Aflatoxin analysis using HPLC for separation and detection is quite similar to TLC because many of the same types of sampling and extraction procedures are used. The major advantages of HPLC over TLC are speed, automation, accuracy and precision. Both normal-phase and reverse-phase HPLC separations have been developed for aflatoxin analyses. Early experimental work by Seitz (41) and Garner (18) on HPLC separations revealed that aflatoxins could be separated on normal-phase columns and detected with either a UV or fluorescence detector. Seitz (41) noted that the fluorescence detector had limited usefulness for aflatoxin B_1 and B_2 with normal-phase separations. Panalaks and Scott (34) developed a silica-gel packed flow cell for fluorometric detection of B_1 and B_2 with normal-phase separations. A silica-gel packed cell was used by Pons (36) and Thean *et al*. (54) in two different HPLC methods for determination of aflatoxins in maize. The major disadvantage of the packed cell is instability. The cell needs to be repacked often and the detector signal weakens with time. The advantages are that no derivative is necessary for detection, and the mobile phase can be recycled.

Reverse-phase HPLC separations of aflatoxins are more widely used than normal-phase separations. However, the fluorescence intensities of B_1 and G_1 are diminished in reverse-phase solvent mixtures, so their derivatives are generally prepared before injection. Analysts should be aware that the B_{2a} and G_{2a} derivatives are not stable in methanol, which should therefore be used with caution, especially in the injection solvent. Acetonitrile-water mixtures do not degrade B_{2a} and G_{2a} rapidly and are preferable to methanol-water mobile phases.

Several reverse-phase methods have been published (7,51), including three with comparisons to the CB method (12,21,53). Stubblefield and Shotwell (51) found that M_1 and M_2, as well as B_1, B_2, G_1 and G_2, could be resolved and detected with a UV detector at 350 nm using reverse-phase chromatography. The methods developed by Hutchins and Hagler (21), DeVries and Chang (12) and Tarter *et al*. (53) all use TFA derivitization and apparently compare favorably with other methods. Diebold *et al*. (13) used laser fluorometry to detect aflatoxin B_{2a} after reverse-phase chromatography.

Davis and Diener (10) found that the iodine derivative of B_1 could be used for confirmation and developed a reverse-phase method with fluorescence detection for this derivative. This work led to the development of a post-column iodination method to enhance B_1 and G_1 fluorescence after reverse-phase chromatography. Shepard and Gilbert (45) investigated the conditions needed for the post-column iodination reaction to enhance fluorescence of aflatoxin B_1 and G_1. The detection limit for B_1 was about 20 pg at a signal to noise ratio of 3.

ELISA and RIA

Aflatoxin B_1 in maize can be determined using solid phase radio-immunoassay (RIA) (26,52) or enzyme-linked immunosorbent assay (ELISA) techniques (6,15,27,35). Aflatoxin M_1 can also be determined using ELISA (23). The ELISA techniques are more suited to field use than RIA techniques and probably will be extensively developed and utilized. The major advantages of ELISA include speed, ease of sample preparation, ease of use and a potentially low cost per analysis. The disadvantages include specificity for only B_1 and cross reactivity with other aflatoxins. At present, ELISA procedures are qualitative or semiquantitative at best, and their major application is screening for aflatoxin B_1 below a predetermined concentration. The color developed by the enzyme-mediated reaction gives an indication of the amount of B_1 present. To be useful for applications where quantitation is critical, ELISA techniques need much more refinement, and methods need to be developed that will distinguish between B_1, B_2, G_1 and G_2 individually or collectively.

Selection of Analytical Approach

Regulatory and experimental methods for aflatoxin analysis do not always need to be the same. Regulatory applications need to be quantitative and legally acceptable, but acceptable methods may vary within a country or between countries.

Aflatoxin analysis in experimental work must be tailored to the objectives, and a part of the experimental design should be a careful selection of method. Inexpensive BGYF and minicolumn data may be sufficient for some experimental purposes, or quantitative data on B_1, B_2, G_1 and G_2 may be required. Costs and data requirements can be at odds when quantitative data is necessary. In the long run, TLC and ELISA are not necessarily less expensive, because HPLC requires a single large initial investment, whereas TLC and ELISA both use disposable plates. Large analytical laboratories are probably more suited for HPLC, while laboratories where only a few samples will be tested may prefer TLC. With further development, ELISA methods will possibly become more versatile.

References

1. Alexander, R.J., and M.C. Baur. 1977. Note on a two-dimensional TLC procedure for determining aflatoxins in corn. Cereal Chemistry 54:699-704.

2. Association of Official Analytical Chemists. 1984. Official Methods of Analysis, 14th ed., Sidney Williams, ed. Arlington, Virginia, USA.

3. Bagley, E.B. 1979. Decontamination of corn containing aflatoxin by treatment with ammonia. Journal of the American Oil Chemists' Society 56:808-811.

4. Cauderay, P. 1979. Rapid chemical confirmation method for aflatoxins B_1 and G_1 by direct acetylation on a thin layer plate before chromatography. Journal of the Association of Official Analytical Chemists 62:197.

5. Chang-Yen, I., V.A. Stoute and J.B. Felmine. 1984. Effect of solvent composition on aflatoxin fluorescence. Journal of the Association of Official Analytical Chemists 67:306-308.

6. Chu, F.S., and I. Ueno. 1977. Production of antibody against aflatoxin B_1. Applied Environmental Microbiology 33:1125-1128.

7. Cohen, H., and M. Lapointe. 1981. High pressure liquid chromatographic determination and fluorescence detection of aflatoxins in corn and dairy feeds. Journal of the Association of Official Analytical Chemists 64:1372-1376.

8. Dantzman, J., and L. Stoloff. 1972. Screening method for aflatoxin in corn and various corn products. Journal of the Association of Official Analytical Chemists 55:139-141.

9. Davis, N.D., J.W. Dickens, R.L. Freie, P.B. Hamilton, O.L. Shotwell and T.D. Wyllie. 1980. Protocols for surveys, sampling, post-collection handling, and analysis of grain samples involved in mycotoxin problems. Journal of the Association of Official Analytical Chemists 63:95-102.

10. Davis, N.D., and U.L. Diener. 1980. Confirmatory test for the high pressure liquid chromatographic determination of aflatoxin B_1. Journal of the Association of Official Analytical Chemists 63:107-109.

11. Davis, N.D., M.L. Guy and U.L. Diener. 1981. Improved fluorometric-iodine method for determination of aflatoxin in corn. Journal of the Association of Official Analytical Chemists 64:1074-1076.

12. DeVries, J.W., and H.L. Chang. 1982. Comparison of rapid high pressure liquid chromatographic and CB methods for determination of aflatoxins in corn and peanuts. Journal of the Association of Official Analytical Chemists 65:206-209.

13. Diebold, G.J., N. Karny, R.N. Zare, and L.M. Seitz. 1979. Laser fluorometric determination of aflatoxin B_1 in corn. Journal of the Association of Official Analytical Chemists 62:564-569.

14. Ehrlich, K.C., and L.S. Lee. 1984. Mycotoxins in grain dust: Method for analysis of aflatoxins, ochratoxin A, zearalenone, vomitoxin, and secalonic acid. Journal of the Association of Official Analytical Chemists 67:963-967.

15. El-NaKib, O., J.J. Pestka and F.S. Chu. 1981. Determination of aflatoxin B_1 in corn, wheat, and peanut butter by enzyme-linked immunosorbent assay and solid phase radioimmunoassay. Journal of the Association of Official Analytical Chemists 64:1077-1082.

16. Fennell, D.I., R.J. Bothast, E.B. Lillehoj and R.E. Peterson. 1973. Bright greenish-yellow fluorescence and associated fungi in white corn naturally contaminated with aflatoxin. Cereal Chemistry 50:404-414.

17. Friesen, M.D., E.A. Walker and M. Castegnaro. 1980. International mycotoxin check sample program. I. Report on the performance of participating laboratories. Journal of the Association of Official Analytical Chemists 63:1057-1066.

18. Garner, R.C. 1975. Aflatoxin separation by high-pressure liquid chromatography. Journal of Chromatography 103:186-188.

19. Hill, R.A., D.M. Wilson, W.R. Burg and O.L. Shotwell. 1984. Viable fungi in corn dust. Applied Environmental Microbiology 47:84-87.

20. Holaday, C.E. 1981. Minicolumn chromatography: State of the art. Journal of the American Oil Chemists' Society 58:931A-934A.

21. Hutchins, J.E., and W.M. Hagler. 1983. Rapid liquid chromatographic determination of aflatoxins in heavily contaminated corn. Journal of the Association of Official Analytical Chemists 66:1458-1465.

22. International Agency for Research on Cancer. 1982. Some mycotoxins. *In* Environmental Carcinogens: Selected Methods of Analysis, vol. 5. IARC Scientific Publication 44. Lyon, France.

23. Jackman, R. 1985. Determination of aflatoxins by enzyme-linked immunosorbent assay with special reference to aflatoxin M_1 in raw milk. Journal of the Science of Food and Agriculture 36:685-698.

24. Josefsson, B.G.E., and T.E. Möller. 1977. Screening method for the detection of aflatoxins, ochratoxin, patulin, sterigmatocystin, and zearalenone in cereals. Journal of the Association of Official Analytical Chemists 60:1369-1371.

25. Kamimura, H., M. Nishijima, K. Yasuda, H. Ushiyama, S. Tabata, S. Matsumoto and T. Nishima. 1985. Simple, rapid cleanup method for analysis of aflatoxins and comparison with various methods. Journal of the Association of Official Analytical Chemists 68:458-461.

26. Langone, J.J., and H. Van Vunakis. 1976. Aflatoxin B_1: Specific antibodies and their use in radioimmunoassay. Journal of the National Cancer Institute 56:591-595.

27. Lawellin, D.W., D.W. Grant and B.K. Joyce. 1977. Enzyme-linked immunosorbent analysis for aflatoxin B_1. Applied Environmental Microbiology 34:94-96.

28. Lee, L.S., and E.A. Catalano. 1981. Pons scaled-down clean-up column adapted for use in solvent-saving modification of the CB method for aflatoxin. Journal of the American Oil Chemists' Society 58:949A-951A.

29. Madhyastha, M.S., and R.V. Bhat. 1984. Application of TLC chemical confirmatory tests to minicolumn chromatography of aflatoxins. Journal of the American Oil Chemists' Society 61:907-908.

30. McKinney, J.D., and G.C. Cavanagh. 1977. The American Oil Chemists' Society's Smalley mycotoxin check sample program: An evaluation. Annales de la Nutrition et de l'Alimentation 31:519-529.

31. Nesheim, S. 1971. Fading of aflatoxin spots on TLC plates during fluorescence densitometry. Journal of the Association of Official Analytical Chemists 54:1444-1445.

32. Nesheim, S. 1979. Method of aflatoxin analysis. *In* Trace Organic Analysis: A New Frontier in Analytical Chemistry. Proceedings of the Ninth Materials Research Symposium. National Bureau of Standards Special Publication 519. Pp. 355-372.

33. Nesheim, S., and W.C. Brumley. 1981. Confirmation of identity of aflatoxins. Journal of the American Oil Chemists' Society 58:945A-948A.

34. Panalaks, T., and P.M. Scott. 1977. Sensitive silica gel-packed flow cell for fluorometric detection of aflatoxins by high pressure liquid chromatography. Journal of the Association of Official Analytical Chemists 60:583-589.

35. Pestka, J.J., P.K. Gaur and F.S. Chu. 1980. Quantitation of aflatoxin B_1 and B_1 antibody by an enzyme-linked immunosorbent microassay. Applied Environmental Microbiology 40:1027-1031.

36. Pons, W.A., Jr. 1979. High pressure liquid chromatographic determination of aflatoxins in corn. Journal of the Association of Official Analytical Chemists 62:586-594.

37. Przybylski, W. 1975. Formation of aflatoxin derivatives on thin layer chromatographic plates. Journal of the Association of Official Analytical Chemists 58:163-164.

38. Rodricks, J.V. 1973. Criteria for mycotoxin standards. Journal of the Association of Official Analytical Chemists 56:1290-1291.

39. Romer, T.R., N. Ghouri and T.M. Boling. 1979. Minicolumn screening methods for detecting aflatoxin: State of the art. Journal of the American Oil Chemists' Society 56:795-797.

40. Shuller, P.L., W. Horowitz and L. Stoloff. 1976. A review of sampling plans and collaboratively studied methods of analysis for aflatoxins. Journal of the Association of Official Analytical Chemists 59:1315-1343.

41. Seitz, L.M. 1975. Comparison of methods for aflatoxin analysis by high pressure liquid chromatography. Journal of Chromatography 104:81-89.

42. Seitz, L.M., and H.E. Mohr. 1976. Simple method for simultaneous detection of aflatoxin and zearalenone in corn. Journal of the Association of Official Analytical Chemists 59:106-109.

43. Shannon, G.M., and O.L. Shotwell. 1975. A quantitative method for determination of aflatoxin B_1 in roasted corn. Journal of the Association of Official Analytical Chemists 58:743-745.

44. Shannon, G.M., and O.L. Shotwell. 1979. Minicolumn detection for aflatoxin in yellow corn: Collaborative study. Journal of the Association of Official Analytical Chemists 62:1070-1075.

45. Shepard, M.J., and J. Gilbert. 1984. An investigation of HPLC post-column iodination conditions for the enhancement of aflatoxin B_1 fluorescence. Food Additives and Contamination 1:325-335.

46. Shotwell, O.L. 1983. Aflatoxin detection and determination in corn. *In* Aflatoxin and *Aspergillus flavus* in Corn, U.L. Diener, R.L. Asquith and J.W. Dickens, eds. Southern Cooperative Series Bulletin 279. Alabama Experiment Station, Auburn, Alabama, USA. Pp. 38-45.

47. Shotwell, O.L., W.R. Burg and T. Diller. 1981. Thin layer chromatographic determination of aflatoxin in corn dust. Journal of the Association of Official Analytical Chemists 64:1060-1063.

48. Shotwell, O.L., and M.L. Goulden. 1977. Aflatoxin: Comparison of analyses of corn by various methods. Journal of the Association of Official Analytical Chemists 60:83-88.

49. Shotwell, O.L., and C.W. Hesseltine. 1980. Use of bright greenish yellow fluorescence as a presumptive test for aflatoxin in corn. Cereal Chemistry 58(2):124-127.

50. Spilman, J.R., Jr. 1985. Modification of the rapid screening method for aflatoxin in corn for quantitative use. Journal of the Association of Official Analytical Chemists 68:453-458.

51. Stubblefield, R.D., and O.L. Shotwell. 1977. Reverse phase analytical and preparative high pressure liquid chromatography of aflatoxins. Journal of the Association of Official Analytical Chemists 60:784-790.

52. Sun, P.S., and F.S. Chu. 1978. A simple solid-phase radioimmunoassay for aflatoxin B_1. Journal of Food Safety 1:67-75.

53. Tarter, E.J., J.P. Hanchay and P.M. Scott. 1984. Improved liquid chromatographic method for determination of aflatoxins in peanut butter and other commodities. Journal of the Association of Official Analytical Chemists 67:597-600.

54. Thean, J.E., D.R. Lorenz, D.M. Wilson, K. Rodgers and R.C. Gueldner. 1980. Extraction, cleanup, and quantitative determination of aflatoxins in corn. Journal of the Association of Official Analytical Chemists 63:631-633.

55. Thomas, F., R.M. Eppley and M.W. Trucksess. 1975. Rapid screening method for aflatoxins and zearalenone in corn. Journal of the Association of Official Analytical Chemists 58:114-116.

56. Trucksess, M.W., W.C. Brumley and S. Nesheim. 1984. Rapid quantitation and confirmation of aflatoxins in corn and peanut butter, using a disposable silica gel column, thin layer chromatography, and gas chromatography/mass spectrometry. Journal of the Association of Official Analytical Chemists 67:973-975.

57. Velasco, J. 1981. Replacement of benzene as a solvent for aflatoxin standards. Journal of the American Oil Chemists' Society 58:938A-939A.

58. Whitaker, T.B., and J.W. Dickens. 1983. Evaluation of a testing program for aflatoxin in corn. Journal of the Association of Official Analytical Chemists: 56(5):1055-1058.

59. Widstrom, N.W., D.M. Wilson and W.W. McMillian. 1981. Aflatoxin contamination of preharvest corn as influenced by timing and method of inoculation. Applied Environmental Microbiology 42:249-251.

60. Widstrom, N.W., D.M. Wilson, and W.W. McMillian. 1982. Evaluation of sampling methods for detecting aflatoxin contamination in small test plots of maize inoculated with *Aspergillius flavus*. Journal of Environmental Quality 11:655-657.

Forecasting Vegetative Stress Via Remote Sensing Techniques

C.J. Johannsen, Agricultural Data Network, and K.J. Ranson, Laboratory for Applications of Remote Sensing, Purdue University, West Lafayette, Indiana, USA

Abstract
The ability to detect plant stress from space-borne platforms has become more promising because of increasing knowledge about stress (chlorophyll and temperature changes) and the development of better instruments. The recent Landsat satellites with thematic mappers (TM) and the Spot satellite with improved resolution promise better detection of stress, although these intruments still require a cloud-free day for acquiring data. Future satellite instruments, such as those proposed in the Earth Observation System (EOS), which will be placed on the Space Station in about 1994, offer further possibilities for detecting stress. Models that use climatological as well as remote sensing data are encouraging for predicting areas of stress. It should not be necessary to wait until plants are strongly affected before knowing about drought conditions.

Resumen
La capacidad de detectar (mediante los cambios en la clorofila y la temperatura) desde plataformas espaciales los efectos que las condiciones ambientales adversas tienen en las plantas se ha convertido en una posibilidad real a causa de los mayores conocimientos que se tienen acerca de estos efectos y la creación de mejores instrumentos. Los nuevos satélites Landsat, provistos de planímetros temáticos (PT) y el satélite Spot, con una mejor resolución de imagen, prometen una mejor detección, aunque todavía se requieren días despejados para que estos instrumentos puedan obtener información. Los instrumentos de satélite del futuro, como los propuestos en el Sistema de Observación de la Tierra (SOT), que se colocarán en la Estación Espacial alrededor de 1994, ofrecen nuevas posibilidades de detección. Los modelos que emplean información climatológica y de detección a distancia ofrecen muchas posibilidades para predecir las áreas donde las plantas pueden estar experimentando los efectos de las condiciones adversas. No debería ser necesario esperar a que las plantas den muestras de estar seriamente afectadas para determinar que hay sequía.

Recent research in remote sensing shows that much has been learned about crop identification and the delineation of maturity stages, cultural practices and stress (2). These achievements have come about because of the design and construction of better remote sensing instruments, an improved understanding of the interaction of soil and vegetation, and the refinement of analysis techniques. This paper will consider remote sensing, use of data bases and geographic information systems, as well as the extent to which vegetation (especially maize) and vegetation stress can be identified. Finally, the promise of future work on modeling for stress detection will be evaluated.

Remote Sensing

Remote sensing is the science and art of acquiring information about material objects from measurements made at a distance and without physical contact; included in this definition are photography, scanning images, radar, sonar and similar data-gathering techniques.

An important extension of the definition is data extraction or analysis to obtain useful information. Data acquired by remote sensing are measurements of variations in the electromagnetic energy that may reveal spectral, spatial and temporal variations in the scene (15). Researchers need to think seriously about these variations before planning to acquire or use a remote sensing product. An agricultural scene, for example, can be identified by the color of light emanating from it (spectral variations), by the relatively uniform shapes of local crop fields (spatial variations), by the way in which the scene changes during the growing season (temporal variations), or by a combination of these factors.

Frequent reference is made in remote sensing articles to the electromagnetic spectrum (Figure 1). The optical

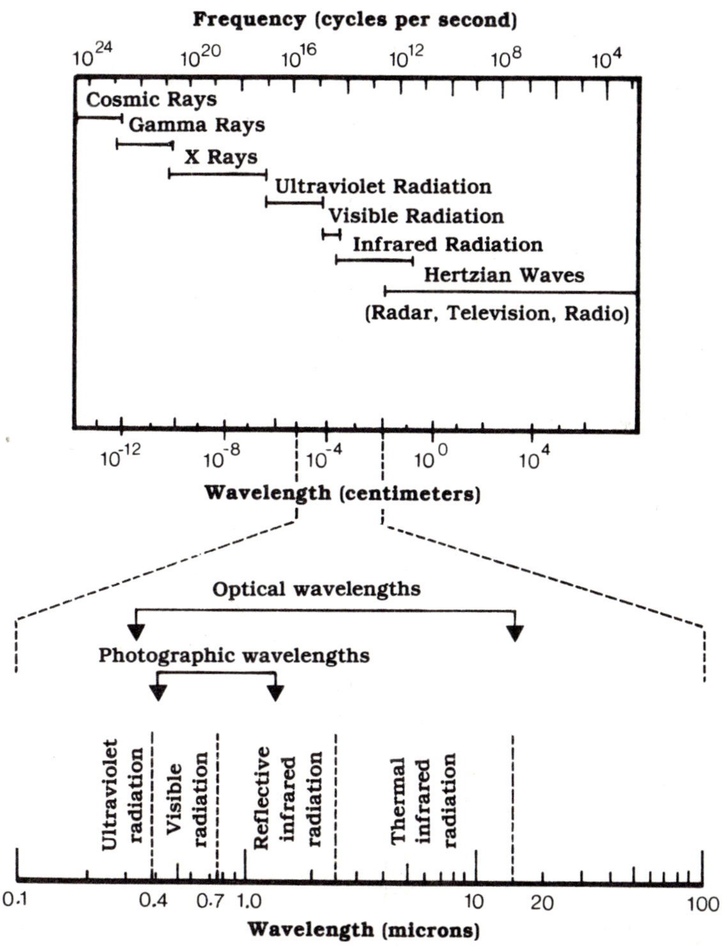

Figure 1. The electromagnetic spectrum; the lower part emphasizes the regions of primary importance to most remote sensing users

Note the relatively small range of wavelengths to which the eye is sensitive

wavelength portion of this spectrum covers the range of 0.3 to 15.0 micrometers. The visible portion (0.4 to 0.7 micrometers) is the most familiar to us since this is the range to which the human eye is sensitive. Wavelengths below 0.4 micrometers are ultraviolet readings and have little value to land surface remote sensing, since much of the energy in these wavelengths is absorbed by the atmosphere. Those wavelengths from 0.7 to approximately 3.0 micrometers are called the reflective infrared; the region from 3.0 to 15.0 micrometers is the emmisive or thermal infrared region. Wavelengths above 1 cm are in the microwave region. This region, where data are collected by passive microwave and radar sensors, has become more important in recent years because of improved design and collection capabilities, providing the opportunity to collect data on cloudy days, since measurements from optical wavelengths are limited to cloud-free days.

Remote sensing in agriculture began when the US National Aeronautics and Space Administration (NASA) provided funds to the US Department of Agriculture (USDA), and research was initiated at the University of California at Berkley; the University of Michigan; the Agricultural Research Service station at Wesleco, Texas; and the Laboratory for Agricultural Remote Sensing (LARS) at Purdue University, West Lafayette, Indiana. Measurements were made of plants and soils in the laboratory and field using spectrometers and radiometers. The first aircraft scanner data over the Purdue Agronomy Farm were obtained with the University of Michigan aircraft in 1965. Other US universities initiated remote sensing work in the late 1960s and early 1970s.

The first satellite data were actually obtained from the Apollo flights in 1964. A scene over the Imperial Valley in California was digitized and analyzed by Anuta *et al.* (1). Landsat data were first analyzed in 1972 with 80-meter (0.64 ha) resolution followed by thematic mapper (TM) data of 30-meter (0.09 ha) resolution from Landsats 4 and 5, which are currently operating. The French launched the Spot Satellite in February, 1986, and it will provide 10-meter (0.01 ha) resolution data. The National Oceanographic and Atmospheric Administration (NOAA) GOES satellites with their advanced very high resolution radiometer (AVHRR) became available in 1980, giving scientists the opportunity to map vegetation on a global basis with spatial resolutions of 1 and 4 km.

Data Bases and Geographic Information Systems

The techniques available for analyzing remotely sensed data have been reviewed by Reeves (19), Swain and Davis (20) and Bauer (2). In particular, ancillary data, such as surface observations, soil maps and weather information can be combined with remotely sensed data. When this information is correlated in an orderly format (for example, geographically arrayed by a computer) it is referred to as a data base.

An example of the use of a data base would be the combination of elevation data with Landsat data (8). In mountainous terrain, certain tree species exist within certain elevation ranges. Therefore, digital, geographically oriented topographic data can be merged with Landsat data to separate species that appear spectrally similar.

Data bases also permit more flexibility in the use of remotely sensed data as well as ancillary data. Weismiller and colleagues (25) spatially registered Landsat data at a scale of 1:24,000 and overlaid this with digitized township, watershed and physiographic boundaries. This technique allowed separation of soil associations by three landscape positions; the data base can also delineate by categories the hectares of soil and vegetation by slope group and by watersheds. The accuracy of runoff estimates in watershed analysis is greatly increased by this approach (12).

With the addition of temporal remotely sensed data and ancillary data, land cover can be determined by soil type, soil interpretations can be provided, erosion hazard areas determined, land use changes charted and a variety of other applications done. Some of these data base applications can be obtained without using remote sensing data directly.

Developing data bases ultimately leads to the need for geographic information systems (GIS). A GIS is a formal process for gathering, storing, analyzing and disseminating information about natural resources and socioeconomic data (8). Many resource scientists have found that such a system provides a cost-effective procedure for planning, developing and organizing natural resources research.

Identification of vegetation

From its beginnings in the 1960s, remote sensing in agricultural research has concentrated on crop identification. Early work, extensively reviewed by Colwell (4), concentrated on physiological studies of individual plant leaves. The interaction of electromagnetic energy with individual leaves becomes increasingly complex in an assemblage of leaves in a crop canopy. Efforts have centered on parameters to determine the reflectance of a vegetative canopy, including:

- Transmittance of leaves;
- Number and arrangement of leaves;
- Characteristics of other components of the vegetation canopy (stalks, trunks, limbs);
- Characteristics of the background, such as reflectance of soil and residues;
- Solar zenith angle;
- Look angle; and
- Azimuth angle

A series of regional and global projects was conducted to provide a focus on agricultural remote sensing and to improve the technology. These programs were the Corn Blight Watch Experiment (18), the Crop Identification Technology Assessment of Remote Sensing (CITARS) project (7), the Large Area Crop Inventory Experiment (LACIE) (17) and the AgRISTARS program (6). Through these programs, researchers developed and refined analysis techniques and began use of multitemporal data with a national focus on remote sensing by a number of federal agencies, including NASA, NOAA, the USDA and the US Department of State.

This research determined that it was difficult to quantify reflectance for a specific crop because of dynamic changes due to growth, development stages, stress and varying cultural practices. Therefore, instead of focusing on a specific crop, remote sensing scientists devoted their attention to such research factors as leaf area index (LAI), percent soil cover and leaf angle distribution (LAD) (2).

Identification of stress

Vegetative stress may be described as an adverse condition imposed on the plant from biological or environmental factors. Crop growth and yield are influenced by light, carbon dioxide supply, temperature, water supply and nutrients interacting with the genetically determined biochemical and physiological systems of the plant. When changes in any one of these factors exceed the ability of the plant to compensate, growth and yield are reduced and the limiting factor constitutes stress (14).

The effects of stress on a plant can be manifested in a variety of ways that may be detected by remote sensing technology. Desiccation of plant tissue causes changes in cellular composition and structure which affect the reflectance of sunlight (14). For example, Figure 2 shows that reflectance measured from maize leaves increases as leaf moisture content decreases (9). These changes in reflectance are accompanied by changes in the plant canopy, such as the wilting of soybean leaves or rolling of maize leaves. Changes in plant architecture, while conserving water in the plant, also produce changes in composite (plant + soil) reflectance that can be detected from aircraft and satellite-borne instruments.

Nutrient deficiency is another form of stress that produces characteristic reflectance patterns. Plants deficient in nitrogen tend to have reduced chlorophyll density and consequently reduced absorption of red light (0.68 μm) (16). In addition, nitrogen-stressed plants will have less foliage than normal plants; the result is higher canopy reflectance in the red band and lower reflectance in the near- and middle-infrared regions. The composite effect of reduced nitrogen levels on reflectance of maize is illustrated in Figure 3.

A useful indicator of stress in plants is the relative change in the amount of green leaves or phytomass. Transformations of spectral data that utilize chlorophyll absorption (Red) and near-infraded (NIR) wavelengths have been shown to be sensitive to leaf area and phytomass. The ratio of NIR to Red (NIR/Red) spectral data and the normalized difference (NIR - Red)/(NIR + Red) are two commonly used vegetation indices (18). Figure 4 presents the relationships of NIR/Red with maize canopy LAI and phytomass from data acquired at the Purdue

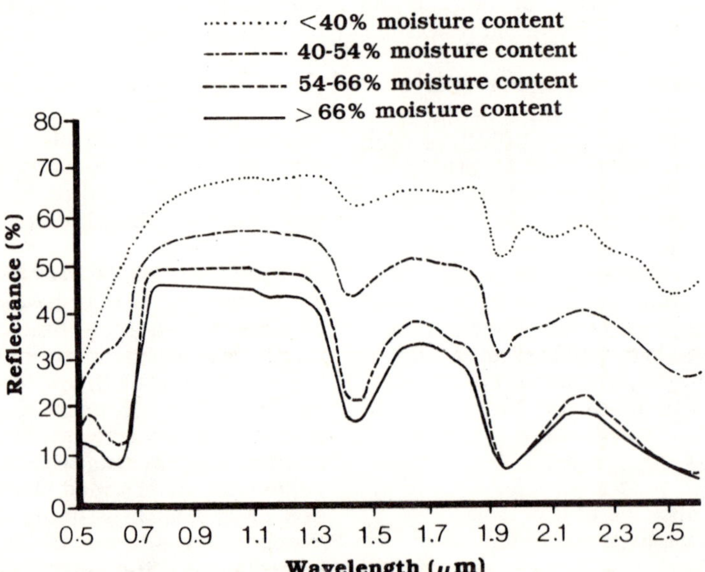

Figure 2. Spectral reflectance of maize leaves with different moisture contents

Source: Hoffer and Johannsen (9).

Agronomy Farm. Tucker and co-workers have demonstrated the feasibility of monitoring global changes in green phytomass using AVHRR data transformed to normalized differences (19,20). The effects of sun angle, look angle, atmosphere, and canopy structure are known to affect the usefulness of vegetation indices. Research to correct for these effects continues in NASA, NOAA and the USDA.

Plant temperature is also known to vary with stress. If transpiration is reduced by a deficit of water, damage from disease or insects to conducting vessels, or by excess soil water salinity, then the net result is an increase in plant temperature (10). Instruments sensitive to thermal infrared portions of the spectrum (8 to 14 μm) can be used to detect crop temperatures. For example, experiments using thermal-infrared scanners and ground observations have demonstrated that recently irrigated crops were up to 20°C cooler than nonirrigated portions of the same field (24). Jackson and colleagues have developed a crop water stress index based on a linear relationship between the difference in air temperature, remotely measured canopy temperature and air vapor pressure deficit (11). These relationships suggest that remotely sensed temperatures of crops, coupled with meteorological parameters, may be used to effectively monitor stress over large areas. These techniques may be limited in areas where meteorological ground stations are scarce.

Modeling for Stress Detection

The task of detecting and recognizing stress in crops is made more difficult by the natural variability in the scene

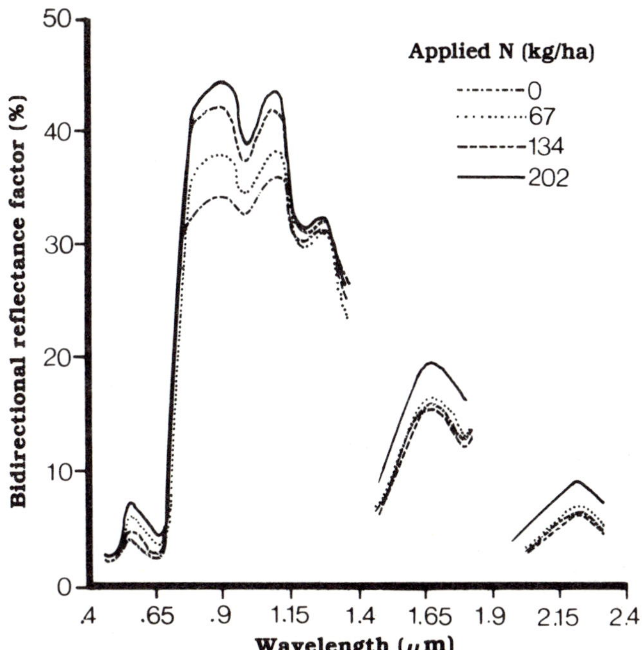

Figure 3. Spectral reflectance of maize grown with different levels of applied nitrogen

Source: Walburg *et al.* (23).

which may mask subtle stress-induced changes in crop reflectance or temperature. In the most promising techniques for monitoring stress, models are used that indicate the stress potential for a given crop and region. The Energy Crop Growth (5) and Crop Water Stress Index (11) models have been used to assess crop stress and production. The early warning and crop condition assessment project under the NASA AgRISTARS program developed crop stress indicator models that combine satellite observations with daily precipitation, maximum and minimum

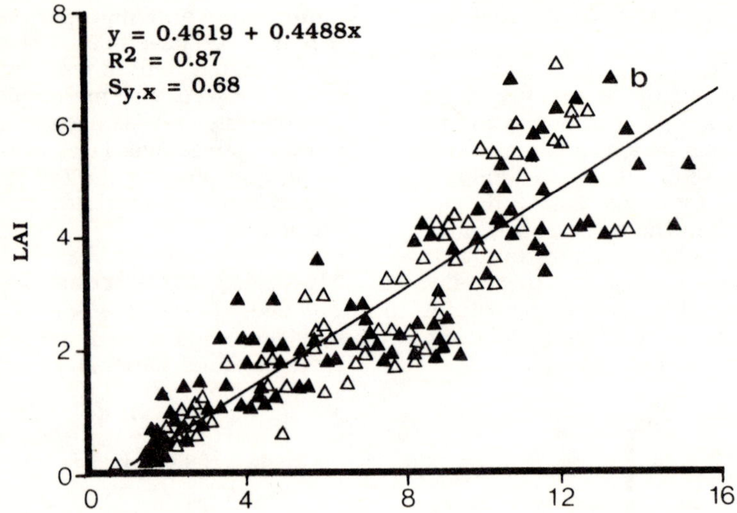

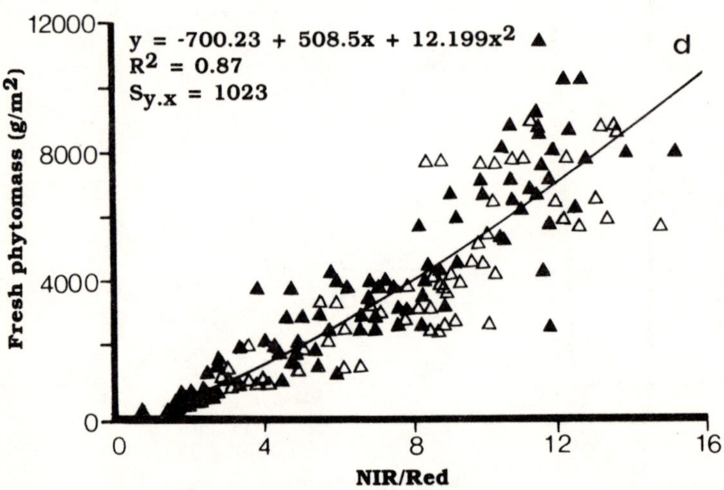

Figure 4. Relationship of maize leaf area index (LAI) and fresh phytomass to the ratio of near-infrared to red (NIR/Red) reflectance

temperatures, evapotranspiration and solar radiation (3). In addition, a satellite-derived stress index was developed using day and night temperatures obtained from NOAA weather satellites and ground-measured air temperatures. This index approximates the ratio of actual to potential evapotranspiration, which is related to crop water stress (25). The application of this technique has been limited to areas with a good network of ground meteorological stations; however, the concept has global implications.

One of the most significant developments of crop assessment models is the derived capability to combine a referenced data base with a geographic information system. This makes it possible to produce computer-generated maps that pinpoint areas where the potential for crop stress may be high.

References

1. Anuta, P.E., S.J. Kristof, D.W. Levandowski, T.L. Phillips and R.B. MacDonald. 1971. Crop, soil and geological mapping from digitized spectral satellite photography. *In* Seventh International Symposium on Remote Sensing of the Environment, Ann Arbor, Michigan, USA. Pp. 1983-2016.

2. Bauer, M.E. 1985. Spectro inputs to crop identification and condition assessment. IEEE Proceedings 73:1071-1085.

3. Boatwright, G.O., and V.S. Whitelead. 1986. Early warning and crop condition assessment research. IEEE Transactions of Geoscientific Remote Sensors GE-24:54-64.

4. Colwell, R.N., 1983. Manual of Remote Sensing. American Society of Photogrammetry, Falls Church, Virginia, USA.

5. Daughtry, C.S.T., K.P. Gallo, L.L. Biehl, E.T. Kanemasu and G. Asrar. 1984. Spectral estimates of agronomic characteristics of crops. *In* Proceedings of the Machine Processing Remotely Sensed Data Symposium, Purdue University, West Lafayette, Indiana, USA. Pp. 248-256.

6. Hall, F.G. 1984. Remote sensing of vegetation at regional scales. *In* Proceedings of the SPIE (Remote Sensing) 475:70-80.

7. Hall, F.G., M.E. Bauer and W.A. Malila. 1974. First results from the crop identification by remote sensing technology assessment project. *In* Proceedings of the Ninth International Symposium on Remote Sensing of the Environment. Pp. 1171-1192.

8. Hoffer, R.M. 1975. Natural resources mapping in mountainous terrain by computer analysis of ERTS-1 satellite data. Laboratory for Applications of Remote Sensing Research Bulletin 919. Purdue University, West Lafayette, Indiana, USA.

9. Hoffer, R.M., and C.J. Johannsen. 1969. Ecological potentials in spectral signature analysis. *In* Remote Sensing in Ecology, P.L. Johnson, ed. University of Georgia Press, Athens, Georgia, USA.

10. Jackson, R.D. 1984. Remote sensing of vegetation: Characteristics for farm management. Paper 475-10 presented at the Society of Photooptical Instrumentation Engineers. Arlington, Viriginia, USA, April 29 - May 4, 1984.

11. Jackson, R.D., S.B. Idso, R.J. Reginatto and P.J. Pinter. 1981. Canopy temperature as a crop water stress indicator. Water Resources Research 17:1133-1138.

12. Johannsen, C.J., and T.W. Barney. 1981. Remote sensing applications for resource management. Journal of Soil and Water Conservation 36:128-133

13. Justice, C.O., J.R.G. Townshend, B.N. Holben and C.J. Tucker. 1985. Analysis of the phenology of global vegetation using meteorological satellite data. International Journal of Remote Sensing 6:1271-1318.

14. Knipling, E.B. 1967. Physical and physiological basis for difference in reflectance of healthy and diseased plants. Paper presented as the Workshop on Infrared Color Photography in the Plant Sciences. Florida Department of Agriculture, Winter Haven, Florida, USA, March 2-3, 1967.

15. Landgrebe, D.A. 1976. Computer-based remote sensing technology: A look into the future. Remote Sensing of the Environment 5:229-246.

16. Lawlor, D.W. 1979. Effects of water and heat stress on carbon metabolism of plants with C3 and C4 photosynthesis. *In* Stress Physiology in Crop Plants, H. Mussell and R. Staples, eds. John Wiley and Sons, New York, N.Y., USA.

17. MacDonald, R.B., and F.G. Hall. 1980. Global crop forecasting. Science 208:670-679.

18. MacDonald, R.B., M.E. Bauer, R.D. Allen, J.W. Clifton, J.D. Erickson and D.A. Landgrebe. 1972. Results of the 1972 corn blight watch experiment. *In* Proceedings of the Eighth International Symposium on Remote Sensing of the Environment. Pp. 157-190.

19. Reeves, R.G., ed. 1975. Manual of Remote Sensing. American Society of Photogrammetry, Falls Church, Virginia, USA.

20. Swain, T.H., and S.M. Davis, eds. 1978. Remote Sensing: The Quantitative Approach. McGraw-Hill, New York, N.Y., USA.

21. Tucker, C.J. 1979. Red and photographic infrared linear combinations for monitoring vegetation. Remote Sensing of the Environment 8:127-150.

22. Tucker, C.J., J.R.G. Townshend and T.E. Goff. 1985. African land-cover classification using satellite data. Science 227:369-374.

23. Walburg, G., M.E. Bauer, C.S.T. Daughtry and T.L. Housley. 1982. Effects of nitrogen nutrition on the growth, yield and reflectance characteristics of corn. Agronomy Journal 74:677-683.

24. Wiegand, C.L., M.D. Heilman and A.H. Cerbermann. 1969. Detailed plant and soil thermal regime in agronomy. *In* Proceedings of the Fifth International Symposium on Remote Sensing of the Environment, University of Michigan, Ann Arbor, Michigan, USA. Pp. 325-342.

25. Weismiller, R.A., I.D. Persinger and O.L. Montgomery. 1977. Soil inventory from digital analysis of satellite sensor and topographic data. Soil Science Society of America Journal 41:161-170.

Aspergillus flavus Infection of Maize: Silks and Kernels

G.A. Payne, Department of Plant Pathology, North Carolina State University, Raleigh, North Carolina, USA

Abstract
Preharvest contamination of maize with aflatoxin is a serious problem in the southeastern USA. Understanding the infection process of Aspergillus flavus *in maize and the factors that influence this process is critical for developing control strategies. The purpose of this review is to summarize what is known about the infection process and to discuss the infection process in relation to the epidemiology of aflatoxin contamination.* Aspergillus flavus *readily colonizes maize silks that are external to the husk and rapidly moves down the ear, first colonizing internal silks, then the glumes and kernel surfaces, but rarely penetrating into the cob pith. The fungus resides on the silks, glumes and kernel surfaces until late in the development of the maize kernel. Once the kernel approaches physiological maturity (moisture content approximately 32%) the fungus enters the kernels, predominantly through the pedicel region.* Aspergillus flavus *is capable of direct penetration into kernels in the absence of insects. Surface colonization of kernels, however, appears to be more prevalent than internal infection; this may play a major role in the epidemiology of the disease. The presence of the fungus on kernel surfaces allows it to rapidly invade kernels that are subsequently injured by insect feeding.*

Resumen
La contaminación del maíz con aflatoxinas antes de la cosecha es un problema grave en el sureste de Estados Unidos. El comprender el proceso de infección de Aspergillus flavus *en maíz y los factores que lo afectan es esencial para crear estrategias de control. El objeto de esta revisión es resumir lo que se conoce acerca del proceso de infección y tratar el tema a la luz de la epidemiología de la contaminación con aflatoxinas.* Aspergillus flavus *coloniza con facilidad los estigmas externos a las brácteas y luego desciende rápidamente por la mazorca, invadiendo primero los estigmas interiores, luego las glumas y la superficie de los granos, pero rara vez penetra la médula del olote. El hongo vive en los estigmas, glumas y superficies de los granos hasta la etapa tardía del desarrollo del grano de maíz. Cuando el grano se aproxima a la madurez (su contenido de humedad es de cerca de 32%), el hongo lo penetra principalmente a través de la región del pedicelo.* Aspergillus flavus *es capaz de introducirse en los granos directamente cuando no hay insectos. Sin embargo, la colonización de la superficie al parecer es más frecuente que la infección interna, hecho que quizá desempeñe una función primordial en la epidemiología de la enfermedad. La presencia del hongo en las superficies de los granos le permite invadir con rapidez los granos que son dañados por los insectos al alimentarse.*

Understanding of the aflatoxin contamination process in maize was transformed just over a decade ago when it was learned that the crop could become contaminated with aflatoxin in the field (2,18,21). Prior to 1975, *Aspergillus flavus* and *A. parasiticus*, the two fungi that produce aflatoxin, were considered storage molds, or fungi capable of colonizing only damaged kernels in storage. The presence of *A. flavus* in preharvest maize raised the possibility that these fungi might have limited parasitic abilities, a concept that still seems heretical to many. Recent studies (16,24,25,27), however, have demonstrated that *A. flavus* has the limited parasitic ability to colonize silks and invade developing maize kernels. The objective of this paper is to summarize the evidence for silk and kernel infection by *A. flavus* and to characterize the infection process. In addition, the role of silk and kernel infection in the preharvest aflatoxin problem of maize is discussed. Since the last review of this subject area (27), much more has been learned about the infection process in maize, but many aspects of the process and the factors governing infection are not known.

Early Field Studies

Aspergillus flavus has been observed on preharvest maize for many years. Since aflatoxin was unknown until 1963, the fungus was considered a problem only as an ear rotter (4,5,7,33). Taubenhaus (33) first reported field infection by *A. flavus* in Texas in 1920, when he studied both the yellow and black ear molds, *A. flavus* and *A. niger*. He observed that *A. flavus* was found most commonly in the top one-third of the ear and almost always on varieties that had erect ears which collected water. Furthermore, he reported that infection was always associated with insect damage. Although he equated infection with sporulation, it has subsequently been established that kernels can be infected and show no visible sporulation (16,21,31,37). Taubenhaus probably underestimated the number of infected kernels. He tried several inoculation techniques and concluded that infection by *A. flavus* required kernel damage. Taubenhaus should be given credit for his early contributions, but since he may not have accurately determined the number of infected kernels, and since he did not attempt to establish infection without damage, he should not be cited as showing that insects are required for infection by *A. flavus*. Field infection by *A. flavus* was also reported in Missouri in 1927 (4), in Florida in 1930 (7) and in New South Wales, Australia, in 1947 (5).

Early reports of *A. flavus* infection in the field in the midwestern USA were uncommon. In Indiana, Tuite (35) found an average of only 0.6, 0.09 and 0.02% infection by *A. flavus* in 1956, 1957 and 1958, respectively. In 1971, 95 fields from 20 counties of western Indiana were sampled, and infection by *A. flavus* ranged from 0.2% in the northern counties to 1.2% in the southern counties. In a similar study, Rambo *et al.* (30) surveyed the northern, central and southern counties of both Indiana and Kentucky. In 1971, one sample from a northern county in Indiana contained 5.0% infected kernels. In 1972, *A. flavus* was found in 3.6% of the samples from southern Indiana, but the highest percentage kernel infection of any sample was 1.0%. No *A. flavus* was found in the northern counties. Anderson *et al.* (2) sampled all of the major maize-growing regions of the USA in 1971 and found the highest incidence of aflatoxin in the warmer, more humid growing areas. An extensive study conducted in central Illinois in 1974 (9) failed to recover any *A. flavus*.

Although *A. flavus* and aflatoxin contamination are present in the Corn Belt of the USA in some years, the incidence of both is greater in the southeastern USA (10,21). In a survey of maize grown in South Carolina in 1973, Hesseltine *et al.* (10) found 184 of 305 samples with one or more kernels infected with *A. flavus*. Some kernels had insect damage, but 221 of 297 samples had no visible injury and contained an average of 58% surface-colonized kernels and 4.4% infected kernels. A field study conducted in North Carolina in 1978 (15) revealed that under natural infection, 4.4% of the undamaged kernels were infected. The presence of *A. flavus* in kernels free of damage was an early indication that the fungus could directly invade kernels.

Attempts to obtain infection by *A. flavus* in the absence of damage have been tried by several investigators. In the Philippines, Ilag (12) sprayed ears of maize with a spore suspension of *A. flavus* or *A. parasiticus* and obtained an average infection of 10.5, 13.5 and 1.0% for *A. flavus, A. parasiticus* and an uninoculated control, respectively. Rambo *et al.* (31), however, had little success obtaining kernel infection by silk-inoculating maize in Indiana. Of 12 ears inoculated, no visible *A. flavus* was present, but upon plating the kernels, it was found that 1% of the kernels from three ears were infected with *A. flavus*. Jones *et al.* (16) were the first to extensively study silk infection as a mode of entry for the fungus. They examined silk inoculation of ears in the field and in controlled-environment studies. In a field study, silks were inoculated with *A. flavus* and the ears enclosed in plastic bags. An average kernel infection of 15.9% and a mean aflatoxin contamination of 83.5 $\mu g/kg^{-1}$ were obtained. Aflatoxin accumulation was positively correlated with the number of infected kernels. In a controlled-environment study, with no insects, silk-inoculated ears had an average kernel infection rate of 9.8% and a mean infection concentration of 866 $\mu g/kg^{-1}$. The correlation between number of infected kernels per ear and aflatoxin B_1 concentration was $r = 0.96$. These data, together with earlier studies, show that *A. flavus* can colonize maize silks and invade developing kernels.

Infection Process

Colonization of external silks

Aspergillus flavus has been isolated from silks in the field (16,25) and it has been reported to readily colonize both attached and detached silks (16,24,25). Silk condition greatly influences how well the fungus develops on the external silks and moves down the silk and infects kernels. In field studies, Jones *et al.* (16) found colonization most likely one week after silking. Silks four weeks after silking served as a substrate only if they were covered with a plastic bag after inoculation. Marsh and Payne (24) extensively studied silk colonization by *A. flavus* using scanning electron microscopy (SEM). They reported silk condition to be a better indicator of silk susceptibility than chronological age since many factors influence the rate of silk senescence in the field. In a comparison of three silk stages (green unpollinated, yellow-brown and brown) they found yellow-brown silks to be the most susceptible. Silks at this stage have begun to senesce but are still succulent. In four to eight hours, conidia of *A. flavus* germinated on these silks, first nearest the pollen grains. Then the fungus spread rapidly across the silk, producing extensive growth and lateral branching. By 48 hours conidiophores and conidia were present on pollen grains. In contrast, few conidia germinated on unpollinated silks by 24 hours and those that did failed to establish significant mycelial growth. Brown silks also supported little growth of the fungus, and growth was concentrated around pollen grains.

Aspergillus flavus penetrated yellow-brown silks both directly and indirectly through cracks and intercellular gaps (24). Internal colonization of the silks was restricted to the parenchymatous tissue, and growth was oriented parallel to the silk axis.

Internal silk and kernel colonization

Aspergillus flavus grows down silks very rapidly. In a controlled-environment chamber with a 34°C day and a 30°C night, a color mutant of *A. flavus* was recovered from the tip of the ear two days after inoculation and from the base four days after inoculation (25). The fungus was found on the glumes of the kernels and adjacent silks six days after inoculation, but not on the seed pericarp. Growth of the fungus from incubated silk segments was first observed from the cut ends, indicating that the fungus may move down the silks internally. Such directed growth down the silks could explain the rapidity by which the fungus reaches the base of the ear.

Colonization of external and internal silks in the field follows a similar pattern. In a field study in 1981 (25), *Aspergillus flavus* was found on external silks of 30% of the uninoculated ears 18 days after 50% silk emergence. *Aspergillus flavus* was present in the tip, mid- and base regions of 100, 67 and 33% of the silk-colonized ears, respectively. By kernel dough stage (nine days later) 72% of the ears had externally colonized silks and the fungus was present on the surface of kernels in 47% of the ears. By August 14, 32 days after silking, 75% of the ears had colonized silks and 50% had kernels that were surface colonized. Little internal infection of kernels was present at this date. More recent data (next section) indicate that the moisture content of kernels at this date was probably too high for extensive internal infection. Extensive surface colonization of kernels has also been reported in Georgia. Wilson *et al.* (38) found that 91 to 100% of ears sampled from inoculated and uninoculated plots in 1975 and 1976 contained *A. flavus* 56 days after silking. They concluded, however, that the fungus present at 56 days did not come from inoculum sprayed on silks.

To follow the growth of *A. flavus* on ears in the field, Marsh and Payne (25) inoculated silks with a color mutant (5T) of *A. flavus*. The rate and path of colonization by the mutant inoculated onto silks were similar to those observed for *A. flavus* under natural infection. The frequency of colonization was greatest in the silks, followed by the kernels and then the cob pith. Colonization of the pith of the cob, however, was uncommon. The color mutant was found on 39% of the kernels 53 days after inoculation. These results, in contrast to those of Wilson *et al.* (38), indicate that the fungus present on kernels late in the season came from inoculum sprayed on the silks. Kernels from these ears were removed and examined by both dissecting microscope and SEM. Colonies of the mutant originated from the pericarp on the silk scar and/or sides of the kernel where the glumes were in contact with the kernel. Colonization was mycelial and no conidiophores were present. A good correlation (r = 0.65) was found between surface-contaminated kernels and the presence of *A. flavus* in the silk or silk bundle adjacent to those kernels (25). Colonization of a particular kernel, however, did not require that it be adjacent to a colonized silk and the fungus appeared to be spread across the glumes and kernel surfaces. Rambo *et al.* (31) reported some superficial spreading of *A. flavus* mycelium among kernels, especially near the tip. Koehler (17) reported growth of *Cephalosporium acremonium* down the ear by way of the glumes. Lillehoj *et al.* (19) detected bright greenish-yellow (BGY) fluorescence of glumes, which is considered indirect evidence for the presence of *A. flavus*.

The distribution of surface-colonized kernels is not uniform or consistent throughout the season. In general, more colonized kernels are present at the ear tip, and the number of colonized kernels fluctuates. Jones *et al.* (16) found infected kernels to be banded in the middle of the ear in a field study, but to be evenly distributed across the ear in a phytotron study.

Kernel infection

The colonization of silks shortly after pollination and the rapid growth of *A. flavus* down the silks suggest that the fungus may be following the same path as does the pollen tube, i.e., the stylar canal. Such a route has been proposed for fungi by Wolf *et al.* (39) and for *A. flavus* by Jones (14). Also, Anderson *et al.* (2) reported BGY fluorescence in the crown area of kernels near the silk scar. Evidence presented by Marsh and Payne (24), however, suggests that the stylar canal is not the major point of entry by *A. flavus*. Of 80 plated kernel halves from ears in the milk, dough and dent stages, crown infection was found in one, seven and three kernels respectively, whereas tip infection was found in 4, 15 and 27 kernels, respectively. Only five kernels had sporulation on the silk scar. When the matching halves of these kernels were examined by scanning electron microscopy, *A. flavus* was found only in the tip cap region. Entry of *A. flavus* in the tip cap region was also found by Tsurata *et al.* (34) in scanning electron micrographs of mature maize kernels removed from the ear and inoculated with *A. flavus*.

The timing of silk colonization probably precludes frequent infection through the stylar canal. Silks are most susceptible to colonization when they are yellow brown and the greatest number of kernels become infected when silks of this age are inoculated. By this time, however, an abscission layer has formed at the silk attachment site, and many silks have become detached. Thus, silk detachment has probably occurred before the fungus has advanced to the silk-attachment site.

Strong evidence that the tip cap is the entry point of the fungus is supported by earlier studies that consistently show the fungus in the tip region of kernels and little fungal growth elsewhere. Fennell *et al.* (8) sectioned whole kernels that did not show BGY fluorescence. In kernels where they did not find obvious development of *A. flavus*, they often found mycelium associated with or external to the hilar layer. Rambo *et al.* (31) also observed extensive growth of *A. flavus* around the pedicel region of the kernel. The germ also appeared to be frequently colonized by the fungus.

Fennell *et al.* (8) plated surface-sterilized, undamaged or slightly cracked kernels and observed *A. flavus* most often in the germ (68%), next in the tip cap (50%) and least in the endosperm (12%). Rambo *et al.* (31) surface sterilized, split and plated undamaged kernels; the frequency of *A. flavus* sporulation was 54% in the germ, 38% in the endosperm and 8% in both germ and endosperm. Jones *et al.* (16) also reported extensive growth of the fungus over the germ with growth in the endosperm present only in severely infected kernels.

The tip cap and germ regions of maize kernels are common areas of colonization by some other fungi. Manns and Adams (23) found mycelium of *Fusarium* and *Cephalosporium* spp. most often in the tip cap. Branstetter (4) also commonly found infection in the tip of the kernel and stated that if the kernel is contaminated in any region it will also be contaminated in the tip. In studies on the infection process of *F. moniliforme, C. acremonium, Gibberella zeae* and *Penicillium* spp. Koehler (17) found the fungi most commonly in the tip cap, followed by the germ, floury endosperm and the horny endosperm. He reported that these fungi rarely entered in any way except via the tip cap.

The route of colonization by *A. flavus* from the tip cap into the maize kernel is not known. *Diplodia zeae* can penentrate the suberized membrane of the testa at its thinner spots, over the embryo and at its junction with the closing layer in the kernel tip (13). Johann (13) has reported that the closing layer is resistant to fungal penetration but suggests that hyphae can transverse the hilum before the closing layer is formed, or the hyphae may pass around the ends of the closing layer when there is a delayed or incomplete junction of the closing layer with the suberized membrane of the testa. Salama and Mishricky (32) suggested that *F. moniliforme* may infect immature kernels through this same area, which they refer to as the placento-chalazal region.

Internal infection of kernels by *A. flavus* appears to occur late in the development of the kernel. In controlled environment studies (Payne, Thompson and Lillehoj, unpublished) little infection occurred until kernel moisture was below 32%. Internal infection was 6, 18 and 27% at 30, 27 and 23% moisture, respectively. These findings correlate well with those of Koehler (17), who found that infection of maize kernels by *F. moniliforme* occurred late in development. One month after pollination, when the kernels were about to dent, only 7% of the ears were infected, and only 3.2% of the kernels had surface contamination. In contrast, 40 days later, when grain moisture was 24.9%, 71% of the kernels that had surface contamination, and 28.3% had internal infection.

Factors Influencing the Infection Process

Several factors are important in silk and kernel infection by *A. flavus*. Inoculum levels, drought, temperature and silk age can affect the infection process. Factors that have been shown to affect infection are temperature, drought and inoculum levels.

Inoculum levels

One of the least understood areas in the epidemiology of *A. flavus* is the source of inoculum. Conidia of the fungus appear to be present each year, but their airborne population fluctuates. Bothast *et al.* (3) and Ilag (12) found low levels of conidia in fields. Ilag (12) found higher levels of conidia in warehouses and adjacent areas. Holtmeyer and Wallin (11) collected airborne conidia of *A. flavus* during the growing season each year from 1976 through 1978 at several locations in Missouri. Conidia were detected in the air over the fields at least 17% of the days in each year during the three-year study. In 1976 and 1977, conidia were collected on more than 70% of the days at two locations, but less than 40% of the days at another location. In one location in 1978, conidia were collected on 88% of the sampling dates.

Jones *et al.* (15) were able to correlate levels of airborne conidia with environmental factors by daily collection of *A. flavus* conidia in irrigated and unirrigated plots. In 1978, 465 colonies of spores were obtained from the irrigated plots and 941 from the unirrigated plots. In 1979, 274 colonies were collected from the irrigated and 322 from the unirrigated plots. Higher levels of airborne conidia in the unirrigated plots correlated well with an increased number of infected kernels in those plots. Also, in unirrigated plots where conidia levels were high, hybrids that pollinated during weeks with high levels of airborne conidia contained a higher percentage of ears with visible growth of *A. flavus*. These data indicate that inoculum may be limiting for maximum aflatoxin contamination in some locations and in certain years. The premise that inoculum may be limiting is also supported by studies that have shown increases in aflatoxin levels in response to inoculation with conidia of the fungus (20,22,28).

Future work should be directed toward understanding the factors that influence inoculum levels of the fungus. Wicklow has shown that *A. flavus* can produce sporogenic sclerotia (36), and these sclerotia are present in infected maize seed (37). The role of these sclerotia in the epidemiology of the fungus is not known; however, the population of *A. flavus* rises dramatically in the soil following harvest (37).

Drought stress

Jones *et al.* (15) reported that plants exposed to drought stress in the field had more infected kernels than samples from irrigated plots. In 1978, they found 1.6% of the kernels from irrigated plots were infected with *A. flavus*, whereas 7.3% from unirrigated plots were infected. In 1979, the number of infected kernels was lower, but more infected kernels were present in the unirrigated plots. They attributed the increased levels of infection in unirrigated plots to higher levels of inoculum, and to reduced leaf area making the silks more accessible to conidia of the fungus. Irrigation also reduced kernel infection when silks were inoculated and covered with a bag, indicating that drought stress may directly affect the infection process (28).

Temperature

Temperature has one of the most striking influences on silk infection. Jones *et al.* (16) demonstrated that kernel infection was much greater at 32° to 38°C than at 21° to 26°C. Payne *et al.* (27,29) reported low kernel infection (2%) at the day/night temperature regime of 26°/22°C, but high infection (49%) at the day/night regime of 34°/30°C. These findings are consistent with reports by several investigators (26,40) that correlate serious aflatoxin contamination with high temperatures during development. Low temperatures at the time of inoculation may explain why Rambo *et al.* (31) were unable to obtain a high percentage of infected kernels in their silk inoculation studies in Indiana.

Temperature may influence infection through its effects on the plant, the fungus or both. Temperatures above 30°C are not favorable for maize (1) and may serve to predispose the plant to infection. On the other hand, high temperatures should be favorable for the growth of *A. flavus*, since the fungus has a high optimum temperature (36° to 38°C) for growth (6).

Discussion

Aspergillus flavus can colonize external silks, grow down the silks and invade developing kernels. Infection by this route can occur in the absence of insects, and it can lead to high levels of aflatoxin (16). Scanning electron microscopy studies have shown that the fungus readily colonizes yellow-brown silks and grows down the surface and interior of silks into the ear. Once in the ear, the fungus colonizes the glumes of the kernels, then the kernel surfaces and finally the tip of the kernel. Contamination of the kernel surfaces and the glumes is predominantly mycelial, but it persists throughout the season. Surface colonization is more prevalent than internal infection, and once *A. flavus* has colonized surfaces, it is in the position to invade kernels damaged by insect feeding. Therefore, under conditions of high insect activity, surface colonization of kernels may be the most important aspect of the infection process.

Internal kernel infection by *A. flavus* does not occur until late in the maturity of the kernel. The number of infected kernels greatly increases as kernel moisture falls below 32%. This is around the time of kernel maturity and coincides with the formation of the kernel abscission layer. The physiological activity in the kernel is lower at this time, and increased infection may reflect decreased resistance in the kernel. Alternatively, the moisture content *per se* may be important in stimulating the fungus, since *A. flavus* is a good competitor at low water activity levels. Whatever the mechanism, infection late in the maturity of the seed is common for many seed-infecting fungi.

The consistent association of insects with high levels of aflatoxin (26) indicates that insects are important in field contamination of maize. However, studies reviewed in this paper indicate that insects are not required to bring the fungus into the ear or to move it around within the ear. Furthermore, *A. flavus* is capable of infecting kernels and producing aflatoxin in the absence of insects (16,27). Insects are very important in inflicting damage and increasing the number of infected kernels. The numerous failures to correlate insect damage with aflatoxin contamination are probably due to the scattered distribution of the fungus within the ear. If, for example, insects enter the ear and feed on kernels that do not have surface or internal infection by *A. flavus*, there may be no correlation between insect damage and aflatoxin contamination.

The incidence and degree of maize kernel infection by *A. flavus* and the subsequent production of aflatoxin is a complex process. No single factor can be identified as responsible for the inconsistent results obtained in studies with the fungus. A major factor, however, is the weak parasitic ability of *A. flavus*. It is not a highly virulent pathogen such as *Biopolaris maydis*, but is predominantly a saprophyte with limited parasitic abilities. For infection and aflatoxin contamination to occur, many conditions must be met. If one important factor is missing, little disease occurs. In some years inoculum may be limiting; in other years temperature or insects may be limiting. To understand and solve this complex problem, the biology of the fungus must be understood as completely as possible, and the temptation to look for a rapid solution to the problem must be avoided.

References

1. Aldrich, S.R., W.O. Scott and E.R. Lenz. 1975. Modern Corn Production. A & L Publications, Champaign, Illinois, USA.

2. Anderson, H.W., E.W. Nehring and W.R. Wichser. 1975. Aflatoxin contamination of corn in the field. Journal of Agricultural and Food Chemistry 23:775-782.

3. Bothast, R.J., L.R. Beuchat, B.S. Emswiller, M.G. Johnson and M.D. Pierson. 1978. Incidence of airborne *Aspergillus flavus* spores in corn fields of five states. Applied Environmental Microbiology 35:627-628.

4. Branstetter, B.B. 1927. Corn root rot studies. Missouri Agricultural Experiment Station Research Bulletin 113:1-80.

5. Butler, F.C. 1947. Ear, cob, and grain rots of maize. Agricultural Gazette (New South Wales) 58:144-151.

6. Davis, N.D., and U.L. Diener. 1983. Biology of *A. flavus* and *A. parasiticus*. *In* Aflatoxin and *Aspergillus flavus* in Corn, U.L. Diener, R.A. Asquith and J.W. Dickens, eds. Southern Cooperative Service Bulletin 279. Alabama Agricultural Experiment Station, Auburn, Alabama, USA. Pp. 1-5.

7. Eddins, A.H. 1930. Corn diseases in Florida. Florida Agricultural Experiment Station Bulletin 210:1-35.

8. Fennell, D.I., R.J. Bothast, E.B. Lillehoj and R.E. Peterson. 1973. Bright greenish-yellow fluorescence and associated fungi in white corn naturally contaminated with aflatoxin. Cereal Chemistry 50:404-414.

9. Hesseltine, C.W., and R.J. Bothast. 1977. Mold development in ears of corn from tasseling to harvest. Mycologia 69:328-340.

10. Hesseltine, C.W., O.L. Shotwell, W.F. Kwolek, E.B. Lillehoj, W.K. Jackson and R.J. Bothast. 1976. Aflatoxin occurrence in 1973 corn at harvest. II. Mycological studies. Mycologia 68:341-353.

11. Holtmeyer, M.G., and J.R. Wallin. 1980. Identification of aflatoxin-producing atmospheric isolates of *Aspergillus flavus*. Phytopathology 70:325-327.

12. Ilag, L.L. 1975. *Aspergillus flavus* infection of pre-harvest corn, drying corn and stored corn in the Philippines. Philippine Phytopathology 9:37-41.

13. Johann, H. 1935. Histology of the caryopsis of yellow dent corn, with reference to resistance and susceptibility to kernel rots. Journal of Agricultural Research 51:855-883.

14. Jones, R.K. 1979. The epidemiology and management of aflatoxins and other mycotoxins. *In* Plant Disease, An Advanced Treatise, vol. 4: How Pathogens Induce Disease, J.G. Horsfall and E.B. Cowling, eds. Academic Press, New York, N.Y., USA. Pp. 381-392.

15. Jones, R.K., H.E. Duncan and P.B. Hamilton. 1981. Planting date, harvest date, and irrigation effects on infection and aflatoxin production by *Aspergillus flavus* in field corn. Phytopathology 71:810-816.

16. Jones, R.K., H.E. Duncan, G.A. Payne and K.J. Leonard. 1980. Factors influencing infection by *Aspergillus flavus* in silk-inoculated corn. Plant Disease 64:859-863.

17. Koehler, B. 1942. Natural mode of entrance of fungi into corn ears and some symptoms that indicate infection. Journal of Agricultural Research 64:421-442.

18. Lillehoj, E.B., D.I. Fennell and W.F. Kwolek. 1976. *Aspergillus flavus* occurrence in 1975 corn at harvest from a limited region of Iowa. Cereal Chemistry 54:366-372.

19. Lillehoj, E.B., W.F. Kwolek, D.I. Fennell and M.S. Milburn. 1975. Aflatoxin incidence and association with bright greenish-yellow fluorescence and insect damage in a limited survey of freshly harvested high-moisture corn. Cereal Chemistry 52:403-412.

20. Lillehoj, E.B., W.F. Kwolek, E.S. Horner, N.W. Widstrom, L.M. Josephson, A.O. Franz and E.A. Catalano. 1980. Aflatoxin contamination of preharvest corn: Role of *Aspergillus flavus* inoculum and insect damage. Cereal Chemistry 57:255-257.

21. Lillehoj, E.B., W.F. Kwolek, G.M. Shannon, O.L. Shotwell and C.W. Hesseltine. 1975. Aflatoxin occurrence in 1973 corn at harvest. I. A limited survey in the southeastern U.S. Cereal Chemistry 51:603-610.

22. Lillehoj, E.B., W.F. Kwolek, M.S. Zuber, A.J. Bockholdt, O.H. Calvert, W.R. Findley, W.D. Guthrie, E.S. Horner, L.M. Josephson, S. King, A. Manwiller, D.B. Sauer, D. Thompson, M. Turner and N.W. Widstrom. 1980. Aflatoxin in corn before harvest: Interactions of hybrids and locations. Crop Science 20:731-734.

23. Manns, T.F., and J.F. Adams. 1923. Parasitic fungi interval of seed corn. Journal of Agricultural Research 23:495-524.

24. Marsh, S.F., and G.A. Payne. 1984. Scanning EM studies on the colonization of dent corn by *Aspergillus flavus*. Phytopathology 74:557-561.

25. Marsh, S.F., and G.A. Payne. 1984. Preharvest infection of corn silks and kernels by *Aspergillus flavus*. Phytopathology 74:1284-1289.

26. McMillian, W.W., D.M. Wilson and N.W. Widstrom. 1985. Aflatoxin contamination of preharvest corn in Georgia: A six-year study of insect damage and visible *Aspergillus flavus*. Journal of Environmental Quality 14:200-202.

27. Payne, G.A. 1983. Nature of field infection of corn by *Aspergillus flavus*. *In* Aflatoxin and *Aspergillus flavus* in Corn, U.L. Diener, R.A. Asquith and J.W. Dickens, eds. Southern Cooperative Series Bulletin 279. Alabama Agricultural Experiment Station, Auburn, Alabama, USA. Pp. 16-19.

28. Payne, G.A., D.K. Cassel and C.R. Adkins. 1986. Reduction of aflatoxin contamination in corn due to irrigation and tillage. Phytopathology 76. (In press.)

29. Payne, G.A., D.L. Thompson and E.B. Lillehoj. 1981. Effect of silk age and incubation temperature on kernel infection by *Aspergillus flavus* in silk inoculated corn. Phytopathology 71:898-899. (Abstract.)

30. Rambo, G.W., J. Tuite and R.W. Caldwell. 1974. *Aspergillus flavus* and aflatoxin in preharvest corn from Indiana in 1971 and 1972. Cereal Chemistry 51:848-853.

31. Rambo, G.W., J. Tuite and P. Crane. 1974. Preharvest inoculation and infection of dent corn with *Aspergillus flavus* and *Aspergillus parasiticus*. Phytopathology 64:797-800.

32. Salama, A.M., and A.G. Mishricky. 1973. Seed transmission of maize wilt fungi with special reference to *Fusarium moniliforme* Sheld. Phytopathologische Zeitschrift 77:356-362.

33. Taubenhaus, J.J. 1920. A study of the black and yellow molds of ear corn. Texas Agricultural Experiment Station Bulletin 290:3-38. College Station, Texas, USA.

34. Tsuruta, O., S. Gohara and M. Saito. 1981. Scanning electron microscopic observations of a fungal invasion of corn kernels. Transactions of the Mycological Society (Japan) 22:121-126.

35. Tuite, J. 1961. Fungi isolated from unstored corn seed in Indiana in 1956-1958. Plant Disease Reporter 45:212-215.

36. Wicklow, D.T., and J.E. Donahue. 1984. Sporogenic germination of sclerotia in *Aspergillus flavus* and *A. parasiticus*. Transactions of the British Mycological Society 82:621-624.

37. Wicklow, D.T., and B.W. Horn. 1984. Sclerotium dispersal of *Aspergillus flavus* and *Eupenicillium ochrosalmoneum* from maize during harvest. Transactions of the British Mycological Society 83:299-303.

38. Wilson, D.M., N.W. Widstrom, L.R. Marti and B.D. Davis. 1981. *Aspergillus flavus* group, aflatoxin, and bright greenish-yellow fluorescence in insect-damaged corn in Georgia. Cereal Chemistry 58:40-42.

39. Wolf, M.J., C.L. Buzan, M.M. MacMasters and C.E. Rist. 1952. Structure of the mature corn kernel. II. Microscopic structure of pericarp, seed coat, and hilar layer of dent corn. Cereal Chemistry 29:234-237.

40. Zuber, M.S., and E.B. Lillehoj. 1979. Status of the aflatoxin problem in corn. Journal of Environmental Quality 8:1-5.

Maize Yields and the Incidence and Levels of Aflatoxin in Preharvest Maize

J.R. Wallin, Agricultural Research Service, US Department of Agriculture, and Department of Plant Pathology, H. Minor, Agronomy Department, and G.E. Rottinghaus, Department of Veterinary Biomedical Sciences, all of the University of Missouri, Columbia, Missouri, USA

Abstract
High levels of aflatoxin B_1 have long been suspected of being associated with the low maize yields caused by drought stress. In 1982, the average aflatoxin levels of four hybrids at six locations in the Missouri yield trials ranged from 0 ppb to a trace. In 1983, six hybrids were tested at seven locations; the average yield ranged from 439 to 7595 kg/ha and the aflatoxin B_1 levels were highest in maize from locations where the yield was 4896 kg/ha or less. In 1984, 12 hybrids were analyzed at seven locations. Average yields ranged from 3013 to 9479 kg/ha. Yields were low (3013 and 3515 kg/ha) at two locations where aflatoxin B_1 levels were the highest (37 and 38 ppb).

Resumen
Desde hace mucho tiempo se ha sospechado que los altos niveles de aflatoxinas B_1 se relacionan con los bajos rendimientos de maíz debidos a la sequía. En 1982, los niveles medios de aflatoxinas de cuatro híbridos en seis localidades en los ensayos de rendimiento de Missouri variaron de O ppb a trazos. En 1983, se sometieron a prueba seis híbridos en siete localidades; el rendimiento medio varió de 439 a 7596 kg/ha y los niveles de alfatoxina B_1 fueron mayores en el maíz proveniente de las localidades que presentaban un rendimiento de 4897 kg/ha o menos. En 1984, se analizaron 12 híbridos en siete localidades. Los rendimientos medios variaron de 3013 a 9479 kg/ha. Los rendimientos fueron bajos (3013 y 3515 kg/ha) en dos localidades donde los niveles de aflatoxina B_1 fueron mayores (37 y 38 ppb).

Drought stress at varied temperatures during different stages of development was shown to change aflatoxin formation and accumulation in a phytotron study by Thompson *et al.* (7). The results showed that both high temperature and inoculation at later developmental stages are important for producing high toxin levels. Drought stress observed by Manwiller and Fortnum (2) suggested that water stress somehow rendered the plant more susceptible to *Aspergillus flavus* infection and toxin production (1). Zuber and Lillehoj (8) proposed that good management practices include avoidance of drought during ear development. However, it may be impossible to avoid water stress when irrigation is not available.

In Missouri, 1982, 1983 and 1984 were excellent years to compare the influence of rainfall on maize yield and aflatoxin development in preharvest maize kernels because of the sharp contrasts between years. In 1982, temperatures during the growing season were relatively cool, and rainfall was adequate. The 1983 and 1984 growing seasons were hot and dry. Since the relation of maize yield at specific locations to aflatoxin levels had not appeared in the literature, this became the objective of the current study.

Materials and Methods

In 1982, samples of four maize hybrids grown in the Missouri Corn Yield Test and commonly grown in the state (DeKalb XL 72AA, Golden Harvest H2680, PAG S x 98 and US-13) (Figure 1) were brought to Columbia, Missouri, dried for six days at 60°C, shelled and observed for bright greenish-yellow fluorescence (BGYF) under long-wave ultraviolet light for the number of glowing particles per weight of sample. Subsequently, the kernels were ground for aflatoxin analyses and samples of the ground materials were taken to the University Veterinary Diagnostic Laboratory for aflatoxin analyses by the thin-layer chromatography (TLC) method (6).

In 1983, six other hybrids commonly grown in the state (Golden Harvest 2680, Funk 4507, Zimmerman 14, Pioneer Brand 3382A, Pioneer Brand 3382B and Pioneer Brand 3183), along with the check hybrid US-13, were sampled.

In 1984, 12 hybrids were involved in the trials: Cargill 967, DeKalb Pfizer XL-72AA, Funk G-4522, Golden

Figure 1. Location of maize yield test sites, Missouri, USA

Harvest H-2500, MFA 6708, Paymaster 8201, Pioneer Brand 3183, Pioneer Brand 3184, Pioneer Brand 3358, Pioneer Brand 3377, Sturdy Grow 935W and US-13.

The maize yield test data were obtained from the Missouri Crop Performance Reports (3,4,5). The weekly maximum and minimum temperatures and weekly total rainfall from June 1 to August 29 were plotted from data obtained from the Atmospheric Science Department, University of Missouri. Data from some locations were not available, so data from the closest weather recording station were substituted; in some cases, the closest station was as much as 25 miles from the plots. The aflatoxin analyses were related to meteorological data and yield.

Test Results

The aflatoxin content of preharvest maize in 1982 ranged from none to a trace at seven locations in the Missouri State Yield Trial. The summer rainfall conditions were localized and the Weather Service data were not representative of the maize plot locations, where local thunderstorms influenced yields; these data were not recorded. With the exception of one location (Novelty), weekly rainfall was adequate in August at every location. However, 3 inches of rain were recorded at Novelty in July and more than 1 inch in August. The weekly maximum temperature at all locations was below 35°C. Therefore, no heat or drought stress occurred at any location, and aflatoxin B_1 was rarely detected (Table 1).

The situation was drastically different in 1983. Aflatoxin B_1 was detected in preharvest maize at all locations at levels of 18 to 70 ppb (Table 2). Zimmerman 14 had the greatest toxin content, with four locations providing kernels exceeding 100 ppb. The aflatoxin B_1 levels of Pioneer Brand 3183, sampled at five locations, was 0, including two locations where levels in Zimmerman 14 were greater than 100 ppb. Toxin levels in Pioneer Brand 3382A kernels were less than 20 ppb. Funk's hybrid 4507 was also consistently low in aflatoxin B_1 levels, having less than 20 ppb at five locations and 20 to 40 ppb at one location. Aflatoxin B_1 levels were lowest (20 ppb on the average) for the six hybrids sampled at Marshall. More hybrids at Spickard had aflatoxin levels exceeding 100 ppb than at any other location.

The weekly temperature-rainfall data for each location differed drastically in 1982 and in 1983. In 1983, from Fairfax in northwestern Missouri to Portageville in the southeast, the July and August weekly maximum temperatures were greater than or equal to 35°C. At Fairfax, almost no rainfall was recorded from July 1 to August 29, and aflatoxin at levels of 20 to 80 ppb was detected (Table 3). At Portageville, almost no precipitation was recorded from July 15 to August 29, and extreme drought stress and aflatoxin of from 20 to greater than 100 ppb was noted. The same drought conditions apparently prevailed at all locations in July and August, and at every location high aflatoxin content was observed in maize grain. Comparative data for the two years portray the influence of heat and drought stress on the production of aflatoxin B_1 in preharvest maize grain.

In 1984, drought and high temperatures prevailed at three of the locations; yields were low (Table 4) and aflatoxin B_1 levels high (Table 2). This situation again demonstrated that drought and temperature stress lead to low yields, resulting in high levels of aflatoxin B_1.

Table 1. Maize yield (kg/ha) at nine locations, Missouri Maize Yield Test, 1982

	Location a/								
Variety	**FFX**	**NMC**	**GMC**	**MARSH**	**CAPE**	**ARC-dry**	**DRC-dry**	**ARC-irr**	**DRC-irr**
DeKalb XL72A	8,789	7,093	9,354	10,735	7,973	6,089	5,775	11,425	11,111
Golden Harvest H-2680	10,484	8,035	8,914	10,420	8,475	6,591	6,340	9,228	10,484
US-13	7,093	7,533	7,596	8,160	5,964	4,959	4,143	7,156	7,470
PAG SX98	9,859	8,224	9,354	10,358	9,605	8,098	6,089	8,726	10,484

a/ FFX = Fairfax, NMC = Spickard, GMC = Novelty, MARSH = Marshall, CAPE = Cape Girardeau, ARC = Bradford, DRC = Portageville, dry = unirrigated, irr = irrigated

Table 2. Maize yields and aflatoxin B_1 levels, Missouri Maize Yield Test, 1982-1984

	1982		1983		1984	
Location	**Yield (kg/ha)**	**Afl. B_1 (ppb)**	**Yield (kg/ha)**	**Afl. B_1 (ppb)**	**Yield (kg/ha)**	**Afl. B_1 (ppb)**
Cape Girardeau	8030	0	3451	70	7151	0
Columbia	6397	TRa/	4893	35	5012	18
Fairfax	9033	0	3764	48	7652	6
Marshall	9912	0	8218	18	9492	11
Novelty	8781	0	440	28	3011	37
O'Fallon	-	-	3512	45	9786	6
Spickard	7714	0	753	43	3512	38

a/ TR = trace

Table 3. Maize yield (kg/ha) at nine locations, Missouri Maize Yield Test, 1983

	Location[a/]								
Variety	**FFX**	**NMC**	**GMC**	**MARSH**	**CAPE**	**ARC-dry**	**ARC-irr**	**DRC-irr**	**O'FAL**
Funk G-4507	3,708	565	502	10,044	3,202	4,896	8,663	8,349	2,511
US-13	1,507	816	502	6,340	3,390	3,766	5,964	7,470	2,950
Golden Harvest H-2680	3,892	439	439	9,290	4,457	4,771	8,224	7,533	4,018
Pioneer 3382	6,089	1,318	377	7,093	4,018	5,713	10,358	8,412	3,264
Zimmerman Z14W	2,762	691	314	6,340	1,946	3,829	7,533	7,910	4,519
Pioneer 3183	4,708	565	377	10,044	3,578	6,089	9,165	8,035	3,829

a/ FFX = Fairfax, NMC = Spickard, GMC = Novelty, MARSH = Marshall, CAPE = Cape Girardeau, ARC = Bradford, DRC = Portageville, O'FAL = O'Fallon, irr = irrigated, dry = unirrigated

Table 4. Maize yield (kg/ha) at nine locations, Missouri Maize Yield Test, 1984

	Location[a/]								
Variety	**FFX**	**NMC**	**GMC**	**MARSH**	**CAPE**	**ARC-dry**	**ARC-irr**	**DRC-irr**	**O'FAL**
Golden Harvest H-2500	8,224	3,708	3,390	9,165	4,018	4,394	11,739	8,726	9,228
Funk G-4522	7,973	3,578	3,641	8,789	4,959	5,085	11,362	12,932	10,107
DeKalb-Pfizer XL-72AA	8,726	2,825	3,641	9,919	8,160	5,148	9,416	10,232	9,228
Paymaster 8201	8,538	3,139	3,766	10,358	7,345	4,645	11,425	10,735	10,295
Cargill 967	7,659	3,264	2,825	9,290	7,784	4,708	12,179	10,923	9,416
MFA 6708	7,470	3,829	3,390	9,228	7,721	5,022	11,111	12,555	10,170
Sturdy Grow 934W	5,901	2,448	2,072						
US-13	5,462	3,013	2,072	6,654	6,026	4,332	7,659	8,789	7,784
Pioneer 3377	8,475	4,645	3,453	10,358	8,538	5,462	12,304	12,367	9,605
Pioneer 3358	8,035	3,641	2,448	9,354	8,286	5,713	11,614	12,492	10,860
Pioneer 3183	8,538	4,206	2,699	11,739	8,098	5,587	11,802	14,062	10,797
Pioneer 3184	7,219	3,704	3,076	9,730	7,784	5,085	12,555	14,941	9,981

a/ FFX = Fairfax, NMC = Spickard, GMC = Novelty, MARSH = Marshall, CAPE = Cape Girardeau, ARC = Bradford, DRC = Portageville, O'FAL = O'Fallon, irr = irrigated, dry = unirrigated

References

1. Aflatoxin Committee. 1977. Information on aflatoxins. North Carolina Agricultural Extension Service Bulletin AG100. Raleigh, North Carolina, USA.

2. Manwiller, A., and B. Fortnum. 1979. A comparison of aflatoxin levels in preharvest corn from the South Carolina Coastal Plain for 1977-1978. South Carolina Agricultural Experiment Station Agronomy and Soils Research Service Bulletin 101. South Carolina, USA.

3. Minor, H.C., C.G. Morris, D. Knerr, E. Lawman and V. Sparks. 1984. Corn: 1984 Missouri Crop Performance. Special Report 316. Agricultural Experiment Station, University of Missouri. Columbia, Missouri, USA.

4. Minor, H.C., C.G. Morris, D. Knerr, E. Lawman and V. Sparks. 1983. Corn: 1983 Missouri Crop Performance. Special Report 304. Agricultural Experiment Station, University of Missouri. Columbia, Missouri, USA.

5. Minor, H.C, C.G. Morris, D. Knerr, E. Lawman and V. Sparks. 1982. Corn: 1982 Missouri Crop Performance. Special Report 289. Agricultural Experiment Station, University of Missouri. Columbia, Missouri, USA.

6. Rottinghaus, G.E., B. Olesen and A.D. Gweiler. 1982. Rapid screening methods for aflatoxin B_1, zearalenone, ochratoxin A, T-2 toxin, diacetoxyscirpenol, and vomitoxin. *In* Proceedings of the 25th Annual Meeting of the American Association of Veterinary Diagnosticians. Pp. 447-484.

7. Thompson, D.L., E.B. Lillehoj, K.J. Leonard, W.F. Kwolek and M.S. Zuber. 1980. Aflatoxin concentration in corn as influenced by kernel development stage and postinoculation temperature in controlled environments. Crop Science 20:609-612.

8. Zuber, M.S., and E.B. Lillehoj. 1979. Status of the aflatoxin problem in corn. Journal of Environmental Quality 8:1-5.

The Influence of Cultural Practices on Minimizing the Development of Aflatoxin in Field Maize

R.K. Jones, Agricultural Extension Service, Texas A&M University, College Station, Texas, USA

Abstract

The implications of preharvest aflatoxin contamination in maize have led to the evaluation of cultural practices to reduce the impact of the problem for maize producers. This paper summarizes practical considerations for growers to minimize potential losses at various stages during the production season. The influence of several factors is discussed: the choice of fields and hybrids; soil fertility; planting density; planting date; irrigation; harvest date and methodology; tillage; and crop rotation. Improved techniques for planting date and rate, fertility practices and modifications in harvesting methods are the most economical and useful techniques for reducing the risk of marketing contaminated grain. Irrigation is also extremely effective where practical. The potential benefits of tillage and crop rotations are also discussed, although the utility of these practices must be determined by further research prior to adoption.

Resumen

Los problemas que implica la contaminación del maíz con aflatoxinas en la fase de precosecha han dado como resultado que se evalúen las prácticas de cultivo a fin de reducir el efecto de este problema sobre los productores de maíz. En este artículo se presentan algunas recomendaciones prácticas para que los productores reduzcan al mínimo las pérdidas potenciales en diferentes etapas de la temporada de producción. Se analiza la influencia de diversos factores; la elección de campos e híbridos; fertilidad del suelo; densidad de siembra; fecha de siembra; irrigación; fecha y metodología de la cosecha; método de labranza y rotación de cultivos. Las técnicas más económicas y útiles para reducir el riesgo de comercializar grano contaminado son técnicas mejoradas en cuanto a la fecha y la densidad de la siembra, prácticas de fertilización y modificación de los métodos de cosecha. En aquellos casos en que resulta práctica, la irrigación es también sumamente eficaz. Asimismo, se analizan los beneficios potenciales de los métodos de labranza y la rotación de cultivos, aunque es necesario determinar, mediante nuevas investigaciones, la utilidad de estas prácticas antes de adoptarlas.

Understanding of the occurrence and development of aflatoxin in preharvest maize has increased greatly in the last decade, and methods of coping with the aflatoxin problem are beginning to emerge. The objective of these methods can be simply stated as an attempt to prevent aflatoxin from entering the food chain. Tactics that have evolved to achieve this objective include:

- Reducing the potential for aflatoxin production before harvest;
- Reducing the harvest and sale of contaminated maize;
- Finding ways to use contaminated maize; and
- Educating producers, buyers and consumers regarding aflatoxin.

This paper will deal exclusively with production practices that may reduce the potential for aflatoxin contamination of maize before harvest.

Infection by *Aspergillus flavus* and the subsequent synthesis of aflatoxin in maize kernels occurs in such a way that a classic disease cycle may develop. Wicklow and Donahue (26) recently proposed a rudimentary cycle that reflects the possible role of sporogenic sclerotia as primary inoculum in maize. Although many steps in this process are not yet fully understood, future investigations will likely elucidate the finer points.

Seasonal, climatic and biotic factors play important roles in the development of aflatoxin in the field. Some of these factors can be influenced directly by the maize producer, but among the methods used to control aflatoxin development, some practices are economical and others are too expensive for some producers to adopt, given the sporadic nature of the problem.

Producers have a vested interest in understanding the disease cycle of *Aspergillus flavus* in maize, peanuts and other crops, as such an understanding will be instrumental in economic evaluations of cultural control practices intended to minimize losses from the disease. In addition, the concept of a "disease triangle" involving host, pathogen and environment is a useful concept to apply when evaluating control tactics for aflatoxin in field maize. Many cultural practices that are effective in reducing the incidence of aflatoxin in maize can be directly assessed for their influence on the disease cycle and the disease triangle. In the following discussion, the decisions of 10 producers are assessed for their role in minimizing the aflatoxin problem in maize. The producers' practices are discussed in the sequence of the production year, and some effort is made to subjectively evaluate the ability to manipulate these factors economically to minimize the threat of aflatoxin.

Field Selection

Field selection is occasionally useful in minimizing the development of aflatoxin in maize, particularly on farms with mixed soil types, where diverse crops can be grown. In Texas, many farmers can effectively produce grain sorghum (*Sorghum vulgare*), guar (*Cyamopsis tetragonoloba*) or other more drought-tolerant crops in fields with sandy or shallow soil profiles. For most producers, however, other cultural manipulations are more useful than field selection when individual farms are evaluated. Soil profiles and their inherent moisture-holding capacities are more variable between farms at diverse locations than within farms. Economic reasons or consumption motivate growers to plant maize instead of another crop. However, in certain geographic areas some soils are not conducive to maize production, and aflatoxin contamination is a potential result of this unsuitability.

Hybrid Selection (Maturity Group)

Investigations in the United States during the last ten years have extensively examined the question of maturity group with respect to the severity of the aflatoxin problem. From 1977 to 1979 the influence of hybrid maturity group was evaluated at several locations in North Carolina (5,12,13). It was concluded that aflatoxin was influenced by a complex of factors that increased plant stress during pollination and grain-filling. Hybrids of short, medium and late maturity exhibited differing quantities of aflatoxin at harvest. However, these differences appeared to interact with other factors, including planting date, location, and within-year climatic factors (i.e., drought periods). Fortnum and Manwiller (6) observed similar results in evaluating 15 commercial hybrids from each of three maturity groups in South Carolina in 1979 and 1980. In other tests where the influence of maturity group has been examined (15,17,18,30), little or no consistency exists that can be attributed solely to maturity group.

Certain factors that can be regionally associated with reduced aflatoxin levels (i.e., husk cover, pericarp thickness, kernel hardness) may be useful sources of polygenic resistance to aflatoxin contamination, but most of these can be negated or overridden in some years. Thompson *et al.* (24) suggest that aflatoxin synthesis in maize inbreds is heritable, but that these differences may be relatively small and may not be sufficient to serve as a practical source for genetic control of aflatoxin contamination.

In Texas, producers do not consider aflatoxin in varietal selection. Unfortunately, most hybrids are selected on the basis of highest potential yields in optimum production seasons. Seed producers in the USA are guilty of performance testing under maximum production conditions despite the fact that in most regions (excluding the Corn Belt) these yield levels are neither the objective nor reality of the average yield per unit area harvested. In Texas, producers are encouraged to select hybrids from the results of on-farm testing. Educational programs stress the utility of regionally adapted hybrids without regard to maturity group but with particular attention to their ability to tolerate early planting (cold tolerance).

Fertilizer Programs

Higher levels of aflatoxins have been found to be associated with lower rates of soil-applied nitrogen (1,12). In these studies, the nitrogen content of foliage and grain was also lower in plots having more aflatoxin. Mineralization of nitrogen in highly organic soils may account for the reduced levels of aflatoxin associated with maize produced on these soils as compared to sandy soils. Drought stress may affect the uptake and translocation of nitrogen in maize and may also influence the physiological status of kernels as substrates for aflatoxin synthesis (10).

In general, balanced fertility programs are suggested to minimize aflatoxin contamination of field maize. In addition to an at-plant application of nitrogen, phosphorus and potassium at rates targeted for production goals, minor elements need to be added according to local recommendations to prevent infertility from being a limiting factor. Split nitrogen applications are suggested for sandy soils, in which extensive leaching may remove significant amounts of preplant nitrogen. Nitrification inhibitors may be useful in preserving the applied nitrogen; however, their ability to preserve plant nitrogen in sandy soils may not justify their cost. Gypsum application was found to reduce the incidence of seed colonization by *A. flavus* in peanuts (21), but this was probably a function of the role of calcium in pod formation and pod integrity and would not likely provide functional reductions of *A. flavus* infection of maize.

Intraspecific Competition (Planting Density)

Plant population is one factor that remains exclusively a decision of the producer. It is also one factor that could potentially have a dramatic effect on the epidemiology of *A. flavus* infection and subsequent aflatoxin production, but this question has not been intensively studied. Plant densities are a function of in-row numbers of plants per linear measure of row, as well as between-row spacings. In the USA, row spacings for maize usually vary from 60 to 100 cm;

the number of plants per row meter varies from 3 to 8. Consequently, plant densities can vary from 25,000 to as much as 80,000 plants per hectare.

Plant density could potentially reduce the incidence of infection by *A. flavus* by reducing exposure of the maize ears and silks (8,13,14,19,20) to airborne spores due to a canopy effect (10). Elevated plant population initiates increased aflatoxin contamination through intraspecific competition for moisture, nutrients and sunlight.

Optimum planting density varies from region to region and from field to field according to yield goals and potentials. In general, producers need to plant at densities that optimize yields for average years, particularly when irrigation is not available. In regions where soil moisture at planting is a major influence on total water available during the season, adjustments in seeding rate can be made to account for available soil moisture. Although this strategy has proved useful for some crops, such as wheat, it generally has little utility for maize because of the tremendous evapotranspirational (in-season) losses associated with maize production.

Planting Date

The influence of planting date has been investigated at diverse locations, and intensively within some geographical regions. The effect of planting date on the incidence of *A. flavus* infection and subsequent aflatoxin levels in preharvest maize is confounded by the influence of several other factors affecting the disease cycle and disease triangle. In North Carolina for both 1978 and 1979, maize planted in April had less aflatoxin at harvest than maize planted in May at four locations with hybrids consisting of early-, mid- and late-season maturity groups (12,13). In 1976 experiments in Florida and Georgia, Lillehoj *et al.* (16) reported a greater incidence of aflatoxin B_1 in maize planted in April and May than in maize planted in June. In 1974 experiments in Georgia, Widstrom *et al.* (30) reported an increase in the percent of aflatoxin-contaminated samples in maize planted on April 19 as opposed to maize planted on May 2.

As was shown in North Carolina, the influence of planting date on aflatoxin concentration at harvest is affected by location, within-year climatic factors (i.e., drought periods) and maturity factors of the hybrids tested (15,17,18,25). In most regions, spring-planted maize exhibits its optimum yield when delayed just past the frost-free date. In Texas, maize planting is not recommended until soil temperatures exceed 10°C at an 18-cm depth for three consecutive days, but producers are encouraged to plant as soon as these conditions are met.

Late planting shifts the ear development phase of most cultivars in southern Texas from April/May to May/June. This shift exposes the reproductive phase of the crop to higher temperatures, higher inoculum loads and increased insect activity. In some regions, planting dates are selected to take advantage of periods of higher natural rainfall that occur with statistical frequency at some particular time. The use of 100-year average rainfall charts and other data bases to predict planting dates that "on the average give the best chance of getting rain for optimum . . . [maize] grain fill" may be useful in nine out of ten years or so. Such an area would probably have an aflatoxin problem only one year in ten. In Texas and other regions of the southern USA, 100-year rainfall averages look fairly consistent, but experience has shown that averages hide a tremendous amount of year-to-year variations.

If planting dates are delayed by wet fields at planting time, or if drought stress occurs during pollination and grain filling, then producers must be conscious of other cultural practices that could lower the potential for aflatoxin development in their crop and take steps to prepare for them. Producers occasionally mention that planting delays result from the need to tend to more important crops; these attitudes show the need for further information about aflatoxin development.

Irrigation

The effects of irrigation on the aflatoxin content of maize have been evaluated in North Carolina (13), South Carolina (6) and elsewhere (3). Irrigation effects appear to be similar to adequate natural rainfall in reducing the incidence of *A. flavus* and aflatoxin. Fortnum and Manwiller (6) observed that irrigation suppressed aflatoxin levels in 14 or 15 hybrids (five from each of the three maturity groups) grown in South Carolina in 1979. The effect of irrigation on reducing aflatoxin was amplified in the presence of simulated corn earworm [*Heliothis zea* (Boddie)] damage.

The influence of irrigation was evaluated in North Carolina in 1978 and 1979. The effect of irrigation was more pronounced in 1978, when drought stress (as measured by leaf xylem water potential) occurred during the silking to late dough stages of grain development. The results of the 1979 tests were similar to those observed by Fortnum and Manwiller in their 1979 studies, which were conducted 200 km to the south. In the North Carolina study, irrigation affected airborne inoculum loads, subsequent kernel infection, aflatoxin concentrations at harvest and yields.

Producers generally use irrigation to optimize maize yields. In Texas, growers are encouraged to plant maize in fields where supplemental irrigation is available and sorghum where irrigation is not available. Maize is much more sensitive to drought than sorghum is, and yield economics support this strategy. However, the maize needs of the swine and poultry industries often dictate the need for maize planting in excess of available irrigated land.

Interspecific Competition (In-Season Biotic Stresses)

Aflatoxins have been shown to exist at higher levels in maize affected by other diseases (4). Competition from nematodes (5) and weeds (1) may also contribute to heightened plant stress. In Texas, major gene resistance and diligent screening and release programs have limited serious losses in recent years to the foliar pathogens of maize, including *Bipolaris maydis*, *Exerohilum turcicum* and *Puccinia polysora*. The selection of resistant cultivars has limited the potential contribution of these diseases to plant stress that may enhance aflatoxin contamination. Soil-borne diseases, including *Macrophomina phaseolina* (the incitant of charcoal rot) and plant parasitic nematodes, may reduce effective root surface areas and contribute to plant stress, although research on these interactions is not yet available.

Weed competition may also contribute to stress-induced aflatoxin problems, although as with many ectoparasitic nematode problems, significant competition usually occurs in years with average to above-average rainfall, and aflatoxins are usually rare to nonexistant in such crop seasons. Seasons in which above-average rainfall during the early vegetative stages of crop development is followed by drought during the reproductive stages may be considered seasons in which in-season biotic stresses from nematode and weed competition could have their greatest influence. These effects would likely be most evident in sandy soils that support high populations of nematodes and have limited moisture-holding capacities.

Most foliar diseases in commercial maize hybrids do not reduce yields sufficiently to suggest an overriding effect on aflatoxin levels. Exceptions do occur, as in the case of *B. maydis* on susceptible genotypes, but varietal selection usually will limit their severity. Nematode and weed control practices should follow general production recommendations, and when these factors are limiting, effective control measures should be used.

Insect damage may also contribute to aflatoxin in several ways. Root and stalk damage can alter the physiology of affected plants in much the same way as other biotic stresses, and control practices should be utilized according to local recommendations to limit damage.

Harvest Practices

Modifications in harvesting procedures can effectively reduce the risk of marketing contaminated grain. Producers who suspect that aflatoxins may be a problem can examine fields and collect ears for analysis two or three weeks before harvest (5,11,23). If *A. flavus* is visible on a high percentage ($>10\%$) of ears sampled, a grower may harvest early at high moisture (26 to 28%), and artifically dry the maize to a moisture content below 13%; this effectively halts aflatoxin build-up. This practice has been adopted in some areas where the natural dry-down of maize is slow, but has limited utility in regions with little late-season rainfall or where maturation occurs during hot periods of the year. For this practice to be adopted, the risk of further aflatoxin contamination must exceed the expense of early harvest and artificial drying.

Late-season rainfall has been shown to significantly contribute to aflatoxin content in maize left standing in the field (12,13) or in outdoor drying areas (9). Early warning systems for hurricanes and tropical storms along coastal areas should be used where practical to minimize exposure of maize for extended wet periods. Early harvest and artificial drying of suspect maize will likely prove economical in these situations.

During the actual harvest operation, reducing the combine header speed to minimize the harvest of fines and trash can lower aflatoxin content in maize. Kernels infected by *A. flavus* are extremely friable. In addition, insect-damaged kernels and kernel fragments can contain a high percentage of the aflatoxin in a given load. Combines that collect and harvest ears from fallen stalks should not be used in regions where aflatoxin is a problem. If hand harvesting is practiced in a region, workers should be trained to recognize visible sporulation of *A. flavus* and discard ears heavily infected with *A. flavus* or those in contact with the soil.

The incidence of aflatoxin contamination varies widely from field to field and also within fields. Producers have found some utility in separating maize from their best fields (fields with a history of high yields) from maize produced in traditionally low-yielding or late-planted fields. Maize from irrigated fields or high-yielding fields should be kept in tanks or truckloads separate from maize produced on late-planted fields or fields where irrigation problems (broken pumps, etc.) may have developed during the season. Marginal areas within a field may also be harvested separately to reduce the risk of aflatoxin contamination of clean maize.

After harvest, particularly of suspect, high-moisture maize, transit delays to the buying point or storage bin should be minimized. Aflatoxins have been shown to increase in truckloads of contaminated maize by as much as 6%

per hour of delay. Increased aflatoxin in storage from grain already contaminated in the field can have serious consequences for the producer.

Tillage

The effects of tillage can be divided into two categories, those that involve subsoiling to break up equipment hardpan layers that develop at various levels below the soil surface, and inversion tillage that can influence sclerotial inoculum in the upper soil profile. Documentation of the economics of these practices as they relate to aflatoxin in corn is limited. Subsoiling or chisel plowing can effectively disturb soil hardpans and allow root exploration of greater soil volumes. This likely has a double-edged effect on alleviating stress to maize plants at later stages of development by increasing the root/soil interface at lower soil depths. Also, root development is not restricted to upper soil zones where even modest drought periods deplete or eliminate soil moisture reserves due to the combined effects of plant uptake and surface evapotranspiration.

Inversion tillage (depending on depth) may disturb shallow equipment pans, but it might also have the benefit of burying sclerotia of *A. flavus* deeply enough to prevent sporogenic germination on the soil surface during the cropping season. Wicklow and Donahue (26) demonstrated the ability of diverse isolates to germinate and produce conidia on the surface of moist sand. They suggest that sporogenic sclerotial germination in *A. flavus* and *A. parasiticus* could be important in the dissemination of conidia as primary inoculum. Inversion tillage may be useful in limiting this source of inoculum, particularly in high-risk rotations (e.g., maize-maize or maize-peanut). The benefits to subsequent peanut crops may not be so evident, as aerial dispersion of primary inoculum may not be as important in that crop.

The role of sporogenic germination of sclerotia may also have important consequences in no-till or minimum-till systems. The economic evaluation of modifications in tillage practices requires further investigation.

Crop Rotations

The influence of crop rotation and minimum tillage on the populations of *A. flavus* conidia in peanut, maize and soybean rotations has been investigated recently (3). Populations of *A. flavus* were found to significantly increase in field plots of maize planted in 1975 and peanuts planted in 1976. Unfortunately, this work focused on conidial populations that were numerous but highly variable. Conidia of *A. flavus* are reported to have a very short life span when moistened (2) and would not likely overwinter. Wicklow and Horn (27) observed that a mixture of 10 sclerotium-forming isolates of *A. flavus* produced large numbers of sclerotia in wound-inoculated preharvest maize. Additional studies by this group quantified the sclerotial inoculum. (28,29). They suggested that sclerotia may represent an important source of inoculum in fields continuously cropped to maize. Pettit and Taber (22) showed that peanuts harvested from land planted to peanuts the previous year exhibited a higher incidence of *A. flavus* than peanuts grown on land planted to other crops in the previous year. It would be of interest to re-evaluate the rotations conducted by Cole *et al.* (3), with particular emphasis on quantification of sclerotial populations of *A. flavus* in association with the various maize, peanut, soybean and fallow cropping systems.

References

1. Anderson, H.W., E.W. Nehring and W.R. Wichser. 1975. Aflatoxin contamination of corn in the field. Journal of Agricultural and Food Chemistry 23:775-782.

2. Boller, R.A., and H.W. Schroeder. 1974. Production of aflatoxin by cultures derived from conidia stored in the laboratory. Mycologia 66:61-66.

3. Cole, R.J., R.A. Hill, P.D. Blankenship, T.H. Sanders and J.H. Gurren. 1982. Influence of irrigation and drought stress on invasion of *Aspergillus flavus* of corn kernels and peanut pods. Developments in Industrial Microbiology 23:229-236.

4. Doupnik, B., Jr. 1972. Maize seed predisposed to fungal invasion and aflatoxin contamination by *Helminthosporium maydis* ear rot. Phytopathology 62:1367-1368.

5. Duncan, H.E. 1979. Current information on aflatoxins in corn. North Carolina Agricultural Extension Service Bulletin AG-167. North Carolina State University, Raleigh, North Carolina, USA.

6. Fortnum, B.A., and A. Manwiller. 1985. Effects of irrigation and kernel injury on Aflatoxin B_1 production in selected maize hybrids. Plant Disease 69:262-265.

7. Griffin, G.J., K.H. Garren, and J.D. Taylor. 1981. Influence of crop rotation and minimum tillage on the population of *Aspergillus flavus* group in peanut field soil. Plant Disease 65:898-900.

8. Holtmeyer, M.G., and J.R. Wallin. 1981. Incidence and distribution of airborne spores of *Aspergillus flavus* in Missouri. Plant Disease 65:58-60.

9. Ilag, L.L. 1975. *Aspergillus flavus* infection of pre-harvest corn, drying corn and stored corn in the Philippines. Phytopathology 9:37-41.

10. Jones, R.K. 1979. The epidemiology and management of aflatoxins and other mycotoxins. Pp. 381-392. *In* Plant Disease: An Advanced Treatise, vol. 4, J.G. Horsfall and E.B. Cowling, eds. Academic Press, New York, N.Y., USA.

11. Jones, R.K. 1983. Minimizing the impact of corn aflatoxin. Plant Disease 67:1297-1298.

12. Jones, R.K., and H.E. Duncan. 1981. Effect of nitrogen fertilizer, planting date, and harvest date on aflatoxin production in corn inoculated with *Aspergillus flavus*. Plant Disease 65:741-744.

13. Jones, R.K., H.E. Duncan and P.B. Hamilton. 1981. Planting date, harvest date, and irrigation effects on infection and aflatoxin production by *Aspergillus flavus* in field corn. Phytopathology 71:810-816.

14. Jones, R.K., H.E. Duncan, G.A. Payne and K.J. Leonard. 1980. Factors influencing infection by *Aspergillus flavus* in silk-inoculated corn. Plant Disease 64:859-863.

15. LaPrade, J.C., and A. Manwiller. 1977. Relation of insect damage, vector, and hybrid reaction to aflatoxin B_1 recovery from field corn. Phytopathology 67:544-547.

16. Lillehoj, E.B., W.F. Kwolek, E.S. Horner, N.W. Widstrom, L.M. Josephson, A.O. Franz and E.A. Catalano. 1980. Aflatoxin contamination of preharvest corn: Role of *Aspergillus flavus* inoculum and insect damage. Cereal Chemistry 57:255-257.

17. Lillehoj, E.B., W.F. Kwolek, M.S. Zuber, A.J. Bockholt, O.H. Calvert, W.R. Findley, W.D. Guthrie, E.S. Horner, L.M. Josephson, S. King, A. Manwiller, D.B. Sauer, D. Thompson, M. Turner and N.W. Widstrom. 1980. Aflatoxin in corn before harvest: Interaction of hybrids and locations. Crop Science 20:731-734.

18. Lillehoj, E.B., W.F. Kwolek, W.F. Zuber, O.H. Calvert, E.S. Horner, N.W. Widstrom, W.D. Guthrie, G.E. Scott, D.L. Thompson, W.R. Findley and A.J. Bockholt. 1978. Aflatoxin contamination of field corn: Evaluation of regional test plots for early detection. Cereal Chemistry 55:1007-1013.

19. Marsh, S.F., and G.A. Payne. 1984. Scanning EM studies on the colonization of dent corn by *Aspergillus flavus*. Phytopathology 74:557-561.

20. Marsh, S.F., and G.A. Payne. 1984. Preharvest infection of corn silks and kernels by *Aspergillus flavus*. Phytopathology 74:1284-1289.

21. Mixon, A.C., D.K. Bell and D.M. Wilson. 1984. Effect of chemical and biological agents on the incidence of *Aspergillus flavus* and aflatoxin contamination of peanut seed. Phytopathology 74:1440-1444.

22. Pettit, R.E., and R.A. Taber. 1968. Factors influencing aflatoxin accumulation in peanut kernels and the associated mycoflora. Applied and Environmental Microbiology 16:1230-1234.

23. Thompson, D.L., G.A. Payne, E.B. Lillehoj and M.S. Zuber. 1983. Early appearance of aflatoxin in developing corn kernels after inoculation with *Aspergillus flavus*. Plant Disease 67:1321-1322.

24. Thompson, D.L., J.O. Rawlings, M.S. Zuber, G.A. Payne and E.B. Lillehoj. 1984. Aflatoxin accumulation in developing kernels of eight maize single crosses after inoculation with *Aspergillus flavus*. Plant Disease 68:465-467.

25. Tuite, J., R. Sensmeier, C. Koh-Knox and R. Noel. 1984. Preharvest aflatoxin contamination of dent corn in Indiana in 1983. Plant Disease 65:893-895.

26. Wicklow, D.T., and J.E. Donahue. 1984. Sporogenic germination of sclerotia in *Aspergillus flavus* and *A. parasiticus*. Transactions of the British Mycological Society 82:621-624.

27. Wicklow, D.T., and B.W. Horn. 1984. *Aspergillus flavus* sclerotia form in wound-inoculated preharvest corn. Mycologia 76:503-505.

28. Wicklow, D.T., B.W. Horn and R.J. Cole. 1982. Sclerotium dispersal of *Aspergillus flavus* and *Eupenicillium ochrosalmoneum* from maize during harvest. Transactions of the British Mycological Society 83:299-303.

29. Wicklow, D.T., B.W. Horn and R.J. Cole. 1982. Sclerotium production by *Aspergillus flavus* on corn kernels. Mycologia 74:398-403.

30. Widstrom, N.W., B.R. Wiseman, W.W. McMillian, W.F. Kwolek, E.B. Lillehoj, M.D. Jellum and J.H. Massey. 1978. Evaluation of commercial and experimental three-way corn hybrids for aflatoxin B_1 production potential. Agronomy Journal 70:986-988.

Effect of Environment on Aflatoxin Development in Preharvest Maize

B.A. Fortnum, Department of Plant Pathology and Physiology, Clemson University, Florence, South Carolina, USA

Abstract

*The effect of environment on preharvest infection, colonization and aflatoxin production in maize (*Zea mays *L.) by* Aspergillus flavus *Link ex Fries is summarized in this paper. Preharvest contamination of maize by aflatoxin is a dynamic process that involves the physiology of the host as well as that of the fungus. This review deals with two major environmental parameters, moisture and temperature, that affect the contamination process. Drought in 1977 and 1980 predisposed maize in the southeastern United States to aflatoxin production. The high temperatures, moisture stress and insect damage during those years provided strong circumstantial evidence that environment had played a significant role in aflatoxin development. Subsequent research results have confirmed the role of weather in preharvest contamination of maize by aflatoxin. Although temperature and moisture have been shown to play a pivotal role in preharvest aflatoxin formation in maize, other weather variables may also influence toxin production.*

Resumen

*En este artículo se resume el efecto del medio ambiente en la infección, colonización y producción de aflatoxinas durante la fase de precosecha en maíz (*Zea mays *L.) con esporas de* Aspergillus flavus *Link ex Fries. La contaminación del maíz con aflatoxinas antes de la cosecha es un proceso dinámico en el que participan la fisiología de la planta huésped y la del hongo. En esta revisión se analizan dos parámetros fundamentales, la humedad y la temperatura, que afectan el proceso de contaminación. Las sequías de 1977 y 1980 predispusieron al maíz del sureste de Estados Unidos a la producción de aflatoxinas. Las altas temperaturas, la falta de agua y el daño producido por los insectos durante esos años brindaron importantes pruebas circunstanciales de que el ambiente había desempeñado un papel importante en el desarrollo de aflatoxinas. Los resultados de investigaciones posteriores confirmaron el papel que desempeña el clima en la contaminación del maíz por aflatoxinas antes de la cosecha. Si bien se ha demostrado que la temperatura y la humedad desempeñan un papel fundamental en la formación de aflatoxinas en el maíz durante la fase de precosecha, es posible que otras variables del clima influyan también en la producción de toxinas.*

Surveys conducted in the southeastern United States demonstrated that *Aspergillus flavus* and aflatoxin are present in preharvest corn (6,12,13). As early as 1920, Taubenhaus observed *A. flavus* growth on developing maize ears (27); however, the potential of the aflatoxin problem was not realized until 100,000 turkey poults died from aflatoxicosis in England in 1960. In a subsequent survey conducted in South Carolina with 297 samples of preharvest maize, 49.5% contained detectable levels of aflatoxin and 62% of these contained more than 20 ppb (12). Severe droughts in 1977 and 1980 predisposed maize in the southern US to aflatoxin formation, and awareness of the possible magnitude of the problem increased. Manwiller and Fortnum (18) estimated that during 1977 90% of the maize crop in South Carolina was contaminated with aflatoxin, rendering some samples unfit for consumption by livestock or humans.

In the midwestern USA, aflatoxin contamination of field maize is considered primarily a storage problem. These regional variations led to early speculation that the environment plays a significant role in preharvest aflatoxin development. Maize grown in the southern USA is exposed to higher temperatures and more moisture stress and insect damage than maize grown in the Midwest. Two decades of research have confirmed that environmental variables affect the colonization of developing maize kernels and the subsequent formation of aflatoxin. This article discusses the two environmental variables, moisture stress and high ambient temperatures, that were most important in the southern USA during the drought years of 1977 and 1980.

Moisture Stress

In evaluating the role of moisture stress in aflatoxin formation in preharvest maize, both the moisture requirements of the fungus and the basic changes in host physiology resulting from drought need to be considered. Widespread drought has occurred every year for the past ten years in South Carolina, and aflatoxin contamination of maize has been a serious problem. Other researchers have noted the association of aflatoxin formation with drought (11), but a clear correlation between rainfall and aflatoxin formation has not been found in all studies. Stoloff and Lillehoj (25) reported that levels of aflatoxin in several southern locations were higher in 1980 than in 1979, in spite of irrigation both years. The variations in research results suggest that the preconditioning of maize to aflatoxin production involves a multitude of factors that may work interactively or independently to alter the level of aflatoxin at harvest.

The viability of *A. flavus* spores is affected by such environmental conditions as moisture and temperature; generally, at a given relative humidity, increasing the temperature lowers the viability of the fungus spores (28). Teitell (28) reports that a narrow band of relative humidity (RH) of around 75% at 29°C had a lethal effect on the conidia of *A. flavus*. As temperatures increased, the lethal RH band shifted from 75% at 29°C to 81% at 45°C. The effect of the lethal RH zone on the epidemiology of *A. flavus* is unknown, and it may play a role in the variability in aflatoxin levels observed between locations.

A popular measure of water availability to microorganisms (water activity or AW) is the ratio of the vapor pressure of the water in the substrate to that of pure water at the same temperature and pressure. Ayerst (2) studied the germination and growth rate of *A. flavus* through a range of temperatures and water activities. The minimum AW for growth was 0.78, with the optimum greater than 0.98. The greatest tolerance of low water activity occurred at the optimum temperature of 33°C. No difference was observed among *A. flavus* strains.

Northolt *et al.* (16,17) evaluated the role of AW on the growth and aflatoxin production of *A. flavus* and *A. parasiticus*. At a high AW, the optimum temperature for aflatoxin B_1 production varied with the strain of *Aspergillus* tested. Optimum temperatures varied from 13° to 16°, 16° to 24° and 24° to 31°C. An AW of 0.95 or less, coupled with moderate or low temperatures, inhibited toxin production more than fungal growth. At high temperatures, substrate and fungal strain have a greater influence on toxin production than AW. Diener and Davis (3) reported an optimum AW of 0.95 to 0.99 for *A. flavus*, depending on the substrate. The differences among fungal strains observed by Northolt *et al.* (16) may explain the different optimum temperatures reported for toxin production. Therefore, substrate, fungal strain and temperature determine the optimum AW. It was also noted that aflatoxin-positive and -negative strains reacted similarly with respect to growth under various conditions.

In stored maize grain, moisture levels must be above 16% (21) to 17.5% (15) to allow colonization by *A. flavus*. However, under field conditions, maize inoculated in the late dough stage of development (when this stage coincides with high temperatures) has the highest aflatoxin levels at harvest (29). Inoculation of maize ears in the field indicated that maximum fungal infection occurred when ears were inoculated in the late milk to early dough stage (20). Aflatoxin production was greatest in ears inoculated 21 to 42 days following midsilk, and harvested from 6 to 9 weeks after silks first appeared (40). Little toxin was formed when ears were inoculated at the time silks first appeared. Widstrom *et al.* (31) also reported that the greatest toxin production was in ears inoculated 20 days following midsilk. The moisture content of maize 21 to 42 days following first silk ranged from 27% to 35% (10).

Several authors have shown a negative correlation between irrigation and preharvest aflatoxin production in maize (5,9). Jones *et al.* (8) found higher aflatoxin levels in maize in the more drought-prone soils of the coastal plain than in the tidewater region of North Carolina, where the soils have a greater water-holding capacity. Kernel infection in irrigated and nonirrigated maize was compared for 1978 and 1979; irrigated plots contained significantly lower levels of kernel infection (9). Subsoiling with a hardpan has been shown to reduce aflatoxin, again implicating moisture availability in the soil as a factor in the tendency of maize to produce aflatoxin (19).

There is strong evidence to support the moisture stress-aflatoxin hypothesis; however, several studies have shown that irrigation does not always reduce aflatoxin levels. Stoloff and Lillehoj (25) reported that aflatoxin levels in preharvest maize remained high at five locations throughout the southeast in spite of irrigation to reduce moisture stress. Fortnum and Manwiller (5) could not demonstrate a reduction in aflatoxin after irrigating maize in which *A. flavus* conidia were applied to intact kernels. Irrigation did reduce the level of aflatoxin in kernels receiving *A. flavus* inoculation plus mechanical injury. These results suggest positive interactions between insect damage and drought stress.

One role of irrigation that is frequently overlooked is its effect on the epidemiology of *A. flavus* and levels of airborne fungal propagules. Jones *et al.* (9) found a significant reduction in airborne propagules of *A. flavus* in irrigated maize. In center-pivot irrigation systems covering large areas, the reduction in airborne *A. flavus* may play a significant role in decreasing preharvest infection of maize and subsequent aflatoxin production.

The role of moisture stress on host-plant physiology and its influence on toxin formation is a new area of investigation. Although a direct cause and effect relationship may be difficult to prove, there is ample evidence to show an association between moisture stress and the increased occurrence of aflatoxin. Changes in host physiology, such as alterations in the composition of carbon or nitrogen compounds or their concentrations in developing kernels, may enhance the establishment of *A. flavus* or the biosynthesis of aflatoxin (11). Increased aflatoxin production in nitrogen-stressed maize underscores the importance of host physiology in the process (8).

Temperature

In the southeastern USA temperature may be the single most important environmental factor influencing the preharvest infection of maize by *A. flavus* and the production of aflatoxin. Aflatoxin outbreaks observed in the south during the hot dry seasons of 1977 and 1980 provide strong evidence for this. Most fungi flourish between 20° and 30°C, but *A. flavus* has a much broader temperature range and may be classified as mesophilic, since its optimum growth temperature is in the range of 35°C. Schindler *et al.* (22) evaluated the effect of temperature on the growth of *A. flavus* and aflatoxin production from 2° to 52°C. They observed the maximum levels of toxin production on Wort media at 24°C; however, the maximum growth of *A. flavus* occurred from 29° to 35°C, and no relationship was observed between the growth rate of *A. flavus* and the levels of toxin produced. Little or no toxin was produced at temperatures below 18°C or above 35°C. A similar report by Ayerst (2) listed the optimum temperature for spore germination and growth of four isolates of *A. flavus* to be between 30° and 40°C. If a different substrate of *A. flavus* isolate is used, considerable variation in results can be obtained. Sorenson *et al.* (24) reported the highest levels of toxin production on rice media at 28° to 32°C. Diener and Davis (4) observed similar levels of aflatoxin production in peanuts stored for 21 days at temperatures ranging from 25° to 35°C.

Preharvest development of aflatoxin in maize occurs in an environment of constantly changing temperatures, which may play a major role in the cumulative levels of aflatoxin at harvest. The effect of temperature cycling on *A. parasiticus* growth has been the subject of several investigations. West *et al.* (30) reported that increasing the incubation temperature from 15° to 21°C after 24 hours and then to 28°C after 48 hours resulted in a fourfold increase in aflatoxin compared to cultures held constant at 28°C. Schroeder and Hein (23) noted that short periods of high temperature (40° to 50°C) in each 24-hour diurnal temperature cycle reduced the growth of *A. parasiticus* and the production of aflatoxin; also, the ratio of aflatoxin B_1 to G_1 increased. A diurnal cycle of 25°C with short exposure to temperatures as low as 10°C had no effect on toxin or growth of *A. parasiticus*, and the ratio of B_1 to G_1 remained the same.

The effect of temperature cycling of aflatoxin B_1 and G_1 was also studied by Lin *et al.* (14). It was observed that cycling the temperature between 33° and 15°C favored aflatoxin B_1 production, whereas a temperature cycle of 25° to 15°C favored the production of G_1. The authors suggested that the enzyme responsible for the conversion of B_1 to G_1 might be more efficient at 25°C than at 33°C. Stutz and Krumperman (26) related total heat input to toxin production under conditions of diurnal and nocturnal time-temperature sequencing. Thermal input (degree hours/day) could be related to fungus sporulation and toxin biosynthesis. Three responses were described: with inputs of less than 208 degree hours/day, no growth was detected; with inputs ranging from 208 to 270 degree hours/day, mycelial growth and pigments were observed; and with an input above 270 degree hours/day, heavy sporulation and aflatoxin biosynthesis occurred.

Temperature has a direct effect on *A. flavus* infection in maize. Jones *et al.* (7) showed warm temperatures (32° to 38°C) favored the infection of kernels more than cool temperatures (21° to 26°C). Thompson *et al.* (29) showed the highest levels of aflatoxin occurred in a 30°/26°C and a 9/15 hour respective day/night temperature regime. Infection of developing kernels occurred through the silks independent of insect vectors. The key factor in the infection of kernels through silks appears to be temperature. At higher temperatures (30°C) the growth rate of *A. flavus* gives it a competitive advantage over other ear-inhabiting fungi. It has also been suggested that *A. flavus* may have an increased parasitic ability at higher temperatures (7).

References

1. Agnihotri, V. 1964. Studies on aspergilli XVI. Effect of pH, temperature, and carbon and nitrogen interaction. Mycopathologia et Mycologia Applicata 24:305-314.

2. Ayerst, G. 1969. The effect of moisture and temperature on growth and spore germination in some fungi. Journal of Stored Products Research 5:127-141.

3. Diener, U.L., and N.D. Davis. 1967. Limiting temperature and relative humidity for growth and production of aflatoxin and free fatty acids by *Aspergillus flavus* in sterile peanuts. Journal of the American Oil Chemists' Society 44:259-263.

4. Diener, U.L., and N.D. Davis. 1970. Limiting temperature and relative humidity for aflatoxin production by *Aspergillus flavus* in stored peanuts. Journal of the American Oil Chemists' Society 47:347-351.

5. Fortnum, B.A., and A. Manwiller. 1985. Effect of irrigation and kernel injury on aflatoxin B_1 production in selected maize hybrids. Plant Disease 69:262-265.

6. Hesseltine, C.W., O.L. Shotwell, W.F. Kwolek, E.B. Lillehoj, W.K. Jackson and R.J. Bothast. 1976. Aflatoxin occurrence in 1973 corn at harvest. II. Mycological studies. Mycologia 68:341-353.

7. Jones, R.K., H.E. Duncan, G.A. Payne and K.J. Leonard. 1980. Factors influencing infection by *Aspergillus flavus* in silk-inoculated corn. Plant Disease 64:859-863.

8. Jones, R.K., and H.E. Duncan. 1981. Effect of nitrogen fertilizer, planting date, and harvest date on aflatoxin production in corn inoculated with *Aspergillus flavus*. Plant Disease 65:741-744.

9. Jones, R.K., H.E. Duncan, and P.B. Hamilton. 1981. Planting date, harvest date, and irrigation effects on infection and aflatoxin production by *Aspergillus flavus* in field corn. Phytopathology 71:810-816.

10. LaPrade J.C., and A. Manwiller. 1976. Aflatoxin production and fungal growth on single cross corn hybrids inoculated with *Aspergillus flavus*. Phytopathology 66:675-677.

11. Lillehoj, E.B. 1983. Effect of environmental and cultural factors on aflatoxin contamination of developing corn kernels. *In* Aflatoxin and *Aspergillus flavus* in Corn, U.L. Diener, R.L. Asquith and J.W. Dickens, eds. Southern Cooperative Series Bulletin 279. Alabama Agricultural Experiment Station, Auburn, Alabama, USA. Pp. 27-34.

12. Lillehoj, E.B., W.F. Kwolek, G.M. Shannon, O.L. Shotwell and C.W. Hesseltine. 1985. Aflatoxin occurrence in 1973 corn at harvest. I. A limited survey in the southeastern U.S. Cereal Chemistry 52:603-611.

13. Lillehoj, E.B., W.F. Kwolek, M.S. Zuber, E.S. Horner, N.W. Widstrom, W.D. Guthrie, M. Turner, D.B. Sauer, W.R. Findley, A. Manwiller and L.M. Josephson. 1980. Aflatoxin contamination caused by natural fungal infection of preharvest corn: Variation between hybrids developed for specific regions but grown at diverse locations. Plant and Soil 54:469-475.

14. Lin, Y.C., J.C. Ayres and P.E. Koehler. 1980. Influence of temperature cycling on the production of aflatoxin B_1 and G_1 by *Aspergillus parasiticus*. Applied and Environmental Microbiology 40:333-336.

15. Lopez, L.C., and C.M. Christensen. 1967. Effect of moisture content and temperature on invasion of stored corn by *Aspergillus flavus*. Phytopathology 57:588-590.

16. Northolt, M.D., H.P. van Egmond and W.E. Paulsch. 1977. Difference between *Aspergillus flavus* strains in growth and aflatoxin B_1 production in relation to water activity and temperature. Journal of Food Protection 40:778-781.

17. Northolt, M.D, C.A.W. Verhulsdonk, P.S.S. Soentoro and W.E. Paulsch. 1976. Effect of water activity and temperature on aflatoxin production by *Aspergillus parasiticus*. Journal of Milk Food Technology 39:170-174.

18. Manwiller, A., and B. Fortnum. 1979. A comparison of aflatoxin levels on preharvest corn from the South Carolina coastal plain for 1977-78. South Carolina Agricultural Experiment Station Agronomy and Soils Research Series 201. Florence, South Carolina, USA.

19. Payne, G.A., D.K. Cassel and C.R. Adkins. 1985. Reduction of aflatoxin levels in maize due to irrigation and tillage. Phytopathology 75:1283.

20. Rambo, G.W., J. Tuite and P. Crane. 1974. Preharvest inoculation and infection of dent corn ears with *Aspergillus flavus* and *A. parasiticus*. Phytopathology 64:797-800.

21. Sauer, D.B., and R. Burroughs. 1980. Fungal growth, aflatoxin production and moisture equilibration in mixtures of wet and dry corn. Phytopathology 70:516-521.

22. Schindler, A.F., J.G. Palmer and W.V. Eisenberg. 1967. Aflatoxin production by *Aspergillus flavus* as related to various temperatures. Applied Microbiology 15:1006-1009.

23. Schroeder, H.W., and H. Hein, Jr. 1968. Effect of diurnal temperature cycles on the production of aflatoxin. Applied Microbiology 16:988-990.

24. Sorenson, W.G., C.W. Hesseltine and O. Shotwell. 1967. Effect of temperature on production of aflatoxin on rice by *Aspergillus flavus*. Mycopathological Applications 33:49-55.

25. Stoloff, L., and E.B. Lillehoj. 1981. Effect of genotype (open-pollinated vs hybrid) and environment on preharvest aflatoxin contamination of maize grown in southeastern United States. Journal of the American Oil Chemists' Society 58:976A-980A.

26. Stutz, H.K., and P.H. Krumperman. 1976. Effect of temperature cycling on the production of aflatoxin by *Aspergillus parasiticus*. Applied and Environmental Microbiology 32:327-332.

27. Taubenhaus, J.J. 1920. A study of the black and the yellow molds of ear corn. Texas Agricultural Station Bulletin 270:3-10. College Station, Texas, USA.

28. Teitell, L. 1958. Effects of relative humidity on viability of conidia of *Aspergillus*. American Journal of Botany 45:748-753.

29. Thompson, D.L., E.B. Lillehoj, K.J. Leonard, W.F. Kwolek and M.S. Zuber. 1980. Aflatoxin concentration in corn as influenced by kernel development stage and postinoculation temperature in controlled environments. Crop Science 20:609-612.

30. West, S., R.D. Wyatt and P.B. Hamilton. 1973. Improved yield of aflatoxin by incremental increases of temperatures. Applied Microbiology 25:1018-1019.

31. Widstrom, N.W., D.M. Wilson and W.W. McMillian. 1981. Aflatoxin contamination of preharvest corn as influenced by timing and method of inoculation. Applied and Environmental Microbiology 42:249-251.

Aflatoxin and Aflatoxin-Producing Fungi in Soil

J.S. Angle, Department of Agronomy, University of Maryland, College Park, Maryland, USA

Abstract

Presence of aflatoxin in the soil raises a number of serious environmental concerns regarding plant uptake, leaching and antagonistic effects on the soil microbial population. Studies were conducted to determine the number of aflatoxin-producing fungi in soil, as well as the fate of aflatoxin in soil. The occurrence of Aspergillus flavus *and* A. parasiticus *in soils under various kinds of cultivation and cropping systems and their ability to produce aflatoxin was determined. Conventional tillage practices increased the soil population of* A. flavus *and* A. parasiticus. *Under conventional tillage practices, a soil cropped in a red clover-wheat-maize rotation was found to have 256* A. flavus *and* A. parasiticus *propagules per gram of soil, whereas no propagules of the two species were detected in a virgin prairie soil.* Aspergillus parasiticus *isolates generally possessed a greater ability to produce aflatoxin than did* A. flavus *isolates. Aflatoxin B_1 added to soil rapidly decomposed to aflatoxin B_2, a much less toxic form. Because the reaction was complete within four days, the conversion was thought to be chemically mediated. Aflatoxin B_2 degraded more slowly, with a detectable presence in soil 77 days after the original aflatoxin B_1 was added. The complete microbial decomposition of ^{14}C-aflatoxin B_1, as measured by CO_2 evolution, accounted for 14% of the added ^{14}C in 112 days of incubation. When aflatoxin B_1 decomposition was monitored in several soil types, decomposition rates were found to vary. Aflatoxin degraded most quickly in a fertile silt loam soil and slowest in a silty clay loam soil. In the silty clay loam soil, it was shown that a conjugate was formed which reduced the rate of decomposition. Therefore, few if any adverse environmental consequences would be expected from the introduction of aflatoxin into soil.*

Resumen

La presencia de aflatoxinas en el suelo plantea diversos problemas ambientales graves respecto a la captación por parte de las plantas, lixiviación y efectos antagónicos en la población microbiana del suelo. Se llevaron a cabo estudios para determinar el número de hongos que producen aflatoxinas en el suelo, así como el destino de las aflatoxinas en el suelo. Se determinó también la presencia de Aspergillus flavus *y* A. parasiticus *en suelos bajo diferentes sistemas de siembra y cultivo, y su capacidad de producir aflatoxina. Las prácticas de cultivo convencionales incrementaron la población de* A. flavus *y* A. parasiticus *en el suelo. Se encontró que, con prácticas de cultivo convencionales, un suelo en el que se rotaban cultivos de trébol rojo, trigo y maíz presentaba 256 propágulos de* A. flavus *y* A. parasiticus *por gramo de suelo, en tanto que no se detectaron propágulos de ninguna de las dos especies en un suelo de pradera virgen. En general, los cultivos puros de* Aspergillus parasiticus *presentaron una capacidad mayor de producir aflatoxinas que los de* A. flavus. *La aflatoxina B_1 agregada al suelo se descompuso rápidamente a aflatoxina B_2, una forma mucho menos tóxica. Dado que la reacción se completó en cuatro días, se pensó que la conversión se llevaba a cabo gracias a una mediación química. La aflatoxina B_2 se degradó con mayor lentitud y fue posible detectar su presencia en el suelo 77 días después de agregar la aflatoxina B_1 original. La descomposición microbiana completa de ^{14}C-aflatoxina B1, medida por la evolución del CO_2, representó el 14% de la ^{14}C en 112 días de incubación. Se descubrió que los índices de descomposición de la aflatoxina B_1 variaban al supervisar dicha descomposición en distintos tipos de suelo.*

La aflatoxina se degradó con mayor rapidez en un suelo margoso limoso fértil y con mayor lentitud en un suelo margoso de arcilla limosa. En el suelo margoso de arcilla limosa, se demostró que se formaba un conjugado que reducía el índice de descomposición. En consecuencia, la introducción de aflatoxinas en el suelo produciría poco o ningún efecto ambiental negativo.

Recent studies have demonstrated that *Aspergillus flavus* Link ex Fries and *Aspergillus parasiticus* Spear often infect field maize prior to harvest (14,18,20). The source of inoculum for field infection is thought to be the soil where these organisms are commonly found. Spores in the soil could be transmitted by wind or insects to the infection site on the standing crop (6).

The soil population of *A. flavus* and *A. parasiticus* is therefore important. Griffin and Garren (7) have quantified *A. flavus* in Virginia field soils where peanuts were grown and found from 0 to 105 *A. flavus* propagules per gram dry soil. Bell and Crawford (4), using a Bortran-amended medium, found 5×10^3 to 2×10^4 *A. flavus* propagules per gram soil in a peanut field in Georgia. In Missouri, the soil population of *A. flavus* has been reported to range from 0 to 2×10^3 propagules per gram soil (22). Swedish researchers have also reported high quantities of *A. flavus* in soil (5).

Previous studies have shown that many soil fungi are significantly influenced by cultural practices (8,9,15). Presumably, the soil population of *A. flavus* and *A. parasiticus* may also be influenced by such practices as fertilizer application, crop rotations and manure application.

The possibility exists that aflatoxin is produced directly in soil. While aflatoxin may be produced *in situ* by the indigenous population, it could also arise from the disposal of aflatoxin-contaminanted maize into soil. When the aflatoxin concentration in a crop exceeds the US Food and Drug Administration limit of 20 $\mu g/kg^{-1}$, the crop is considered unfit for interstate transport. Crops too contaminated to be transported are often disposed of by burial in the soil (3).

The presence of aflatoxin in soil raises a number of serious environmental concerns. Aflatoxin has previously been shown to affect many aspects of the soil microbial population (2). Adverse effects on plant growth, as well as aflatoxin uptake by plants, have also been documented (16,19). Aflatoxin leaching through the soil profile and into the groundwater is another potential concern.

To study some of these problems and questions, several experiments were conducted. Initially, the population of *A. flavus* and *A. parasiticus* in soil under various cultural and cropping systems was determined. These isolates were then screened for their ability to produce aflatoxin both on artificial media and in soil. Aflatoxin was subsequently added to soil to determine the toxin's persistence and subsequent potential as an environmental problem.

Materials and Methods

Soil populations of *A. flavus* and *A. parasiticus*

The population of *A. flavus* and *A. parasiticus* was first determined in eleven soils. Ten samples were collected from Sanborn Field at the University of Missouri, Columbia, Missouri. The soil of this field is a Mexico silt loam (Udollic Ochraqualf; fine, montmorillonitic, mesic). Tillage, rotation, fertilizer and irrigation treatments are shown in Table 1. The soil fungal population was also studied in a virgin, tall-grass prairie (Tucker Prairie) located near Hatton, Missouri.

Table 1. Population and aflatoxin production of *A. flavus* and *A. parasiticus* in soil under various cropping systems

Cropping system	*Aspergillus* isolates per g dry soil	Relative aflatoxin production (% of total isolates)						Total fungi x 10^8 per g dry soil
		A. flavus			*A. parasiticus*			
		None	Medium	High	None	Medium	High	
Conventional (Wheat-red clover-maize rotation)								
Full treatment	271 a[a/]	13	14	9	0	35	29	4.9
No treatment	217 a,b	4	44	9	0	30	13	1.8
Manure	279 a	7	17	0	3	52	21	6.4
Continuous maize								
No treatment	103 c,d	4	42	0	0	47	7	1.0
Manure	108 c,d	7	47	5	1	30	10	1.9
No-till, full treatment (Soybeans-wheat and soybeans-maize)								
Soybeans, irrigated	25 e	0	11	0	0	61	28	5.5
Wheat and soybeans, irrigated	124 b,c,d	20	42	0	0	27	11	5.7
Soybeans	65 d,e	17	39	8	0	30	6	9.0
Wheat and soybeans	147 b,c	14	33	2	5	25	21	5.2
Continuous maize								
Full treatment	177 b,c	6	23	0	0	56	15	9.6
Virgin praire soil	0 e	0	0	0	0	0	0	4.5
Mean		9	31	3	1	40	16	

a/ Values followed by the same letter are not significantly different at the 0.05 level according to Duncan's multiple range test using triplicate sampling of soil composites collected from each plot at monthly intervals from May to October

This soil had never been cultivated or received any soil amendments. The prairie soil is a Putnam silt loam (Mollic Albaqualf; fine montmorillonitic, mesic).

Soil samples were collected once a month from May to October. Five random soil cores from each plot were collected at depths of 0 to 3 cm, placed in sterile plastic bags and immediately transported to the laboratory for microbial analysis. Three separate soil composites were taken from each plot sample.

Representative 10-g units of soil were aseptically transferred to a bottle containing 95 ml of a phosphate buffer solution. The soil suspension was shaken for ten minutes on a reciprocal shaker, serially diluted and the appropriate concentration was spread plated onto a Bortran-amended medium developed by Bell and Crawford (4). The selective media was specifically designed for the isolation of organisms in the *A. flavus* group. Fungal plates were incubated at 30°C for four days.

Aspergillus flavus or *A. parasiticus* colonies were counted, isolated and transferred to a coconut agar medium. This medium, designed by Lin and Dianese (11), was used to distinguish *A. flavus* from *A. parasiticus* isolates and to determine their ability to produce aflatoxin. Isolates were divided into three separate groups according to their ability to produce aflatoxin. The accuracy of this technique was tested by periodically extracting the agar media with acetone.

Aflatoxin decomposition

To significantly affect the soil environment, aflatoxin must persist. Aflatoxin persistance and decomposition were first determined in a Mexico silt loam soil. Aflatoxin B_1, uniformly ring labeled with ^{14}C, was obtained from Moravek Biochemicals (City of Industry, California). Twenty-five grams of dry soil were weighed into flasks for incubation. One set of soil samples was treated with ^{14}C-aflatoxin B_1 (53 mCi $mmol^{-1}$) at a rate of 4.7 μg/kg dry $soil^{-1}$. Another set received the same aflatoxin B_1 plus 50 mg of wheat straw. The straw and glucose additions approximated annual increments of crop residues incorporated into soil. A fourth set remained untreated. Each treatment was replicated three times.

Soils were incubated for 112 days, during which time moist CO_2-free air was passed through the incubation flask at a flow rate of 60 ml/min^{-1}, and the CO_2 was collected in 1.0N KOH. At specified intervals, the CO_2 trapped by the KOH was precipitated with 3.0N $BaCl_2$. Each precipitate was filtered through a sintered glass funnel, washed free of salts with hot distilled water and dried at 105°C. A portion of the precipitate was finely ground and suspended in a scintillation counting solution. Cab-O-Sil gel was used as a suspending agent.

A separate experiment monitored changes in the structure of aflatoxin B_1 added to soils. Aflatoxin was initially added to 100 g of soil at a rate of 10,000 $\mu g/kg^{-1}$ aflatoxin B_1 (Applied Science, State Park, Pennsylvania), using the method described in the previous experiment. The high rate was used to provide for parent and derivative compounds that could be easily extracted and quantified.

Periodically, 10 g of soil was removed from the incubation flask and analyzed for aflatoxins. The acetone extraction and analytical techniques followed those of the Association of Official Analytical Chemistry (1). Extracts were compared against the standard aflatoxin B_1, B_2, G_1 and G_2.

Another study was conducted on the influence of soil type on aflatoxin decomposition. Uniformly ring-labeled aflatoxin was added to several soils to determine the rate of decomposition. The ^{14}C-labeled aflatoxin B_1 was incorporated into the soil at a concentration of 57.2 $\mu g/kg^{-1}$. The aflatoxin was added to the soils using the methods described previously. Moist CO_2-free air was passed over the soil, and the resulting CO_2 produced in the soil was collected in 0.5N KOH. At specific intervals, 5.0 ml of the KOH was added to 10.0 ml of a scintillation cocktail (Beckman Ready-Solv MP^R) for counting.

The ^{14}C-labeled aflatoxin was added to three soil types, as Beltsville silt loam (fine loamy, mixed, mesic typic Fragiudult), a Sassafras sandy loam (mixed, mesic typic Hapludult) and a silty clay loam collected from the B horizon of a Fauquier silt loam (fine, mixed, mesic ultic Hapludalf). In addition, the ^{14}C-labeled aflatoxin was added to the silt loam soil which had been amended to 50,000 $\mu g/kg^{-1}$ aflatoxin B_1. The purpose of this treatment was to determine whether a high concentration of aflatoxin was capable of inhibiting its own rate of decomposition. Pertinent soil characteristics are presented in Table 2.

Preliminary results indicated that the rate of decomposition in the silty clay loam soil was much slower than in the other soils. Therefore, an additional experiment periodically extracted the aflatoxin from the silty clay loam soil to identify the aflatoxin decomposition products. The adsorption rate of aflatoxin in the test soils was assayed in an isotherm study. A mixture of labeled and unlabeled aflatoxin B_1 was used to obtain the desired concentrations. A 10-μl aliquot of ^{14}C-labeled aflatoxin B_1 was added to glass centrifuge tubes. The amounts of unlabeled aflatoxin B_1 were 0.1, 1.0, 5.0 and 10.0 mg/liter^{-1} aflatoxin B_1. Three replications per concentration were used for each soil, plus a control containing no aflatoxin. The methanol was evaporated from the tubes and 10 ml of distilled and demineralized water was added to the aflatoxin residue. A 0.5-g sample of soil was added to each tube and placed on a shaker-incubator at 25°C for two hours, a time period shown to be adequate for completion of the adsorption reaction. The tubes were subsequently centrifuged for one hour at 22,000 g and 25°C. A 1-ml aliquot of the supernatant was removed and placed into 9-ml scintillation cocktail (Beckman Ready Solv HP) for counting. After determination of remaining aflatoxin, the aflatoxin adsorbed to the soil was calculated by subtracting the total amount of aflatoxin added to the centrifuge tubes from the amount in solution. Freundlich's adsorption coefficient (K) and l/n values were determined.

Table 2. Characterization of soils used in aflatoxin decomposition and adsorption studies

Soil	pH	Mechanical analysis (g/k^{-1})			Organic matter (g/k^{-1})	Cation exchange capacity (cmol (+)/kg^{-1})
		Sand	Sand	Clay		
Silt loam	5.8	267	397	336	29	11.72
Sandy loam	6.0	647	232	121	15	5.35
Clay loam	4.2	358	367	275	18	6.60
Silty clay loam	7.3	229	392	378	6	18.04

Results and Discussion

Soil population of *A. flavus* and *A. parasiticus*

The total number of *A. flavus* and *A. parasiticus* propagules isolated from individual plots of Sanborn Field was determined (Table 1). Since no significant population differences were observed through the growing season, population values were combined over time. The combined results of 20 plates for each of three composite soil samples and six sampling dates provided a plot sample which generally totaled at least 35 isolates of *A. flavus* and *A. parasiticus*.

The largest population of *A. flavus* and *A. parasiticus* was observed in soils cultivated in a three-year rotation of wheat-red clover-maize (265 propagules/gram). The conventional plots differed from the others in their cultivation by nominal tillage practices, which incorporated large particles of maize plant residue into the soil. Presumably the residue from the previous year's crop was the growth site of *A. flavus* and *A. parasiticus*. The population was lower with no-till cultivation, because the maize residue was not incorporated, but remained on the soil surface. This hypothesis is supported by a supplemental study conducted by the authors in which no-till and conventionally cultivated plots were compared for *A. flavus* and *A. parasiticus* populations by examining the residue *per se*. Under no-till cultivation, 85% of the total population of *A. flavus* and *A. parasiticus* was associated with plant residue.

The other conventionally tilled plots under study were continuous maize without treatment and a similar plot which received an annual application of manure. The population of *A. flavus* and *A. parasiticus* in these plots was 103 and 108 propagules per gram dry soil, respectively. Presumably, the population in these soils was lower than for the other conventionally cultivated soils because of the extremely low maize yield. The simple addition of manure was inadequate for good yield, and the maize stover had been removed from the two plots each year along with the grain.

The number of isolates of *A. flavus* and *A. parasiticus* was notably lower in the plots where soybeans were planted as a full-season crop than where soybeans were double cropped with wheat. These results indicate that rotation of maize with soybeans rather than wheat provides better control of these fungi. During the entire course of the study, no *A. flavus* or *A. parasiticus* isolates were found in prairie soil. Apparently the residues of prairie grasses and other species are not appropriate substrates for *A. flavus* and *A. parasiticus* development.

The relative proportion of *A. flavus* and *A. parasiticus*, showed that *A. flavus* comprises approximately 43% of the total and *A. parasiticus*, approximately 57%. The distribution was unexpected because preharvest maize grain is infected almost exclusively by *A. flavus* (10). The large population of *A. parasiticus* observed in soils apparently does not result in a significant infection of preharvest maize kernels. A large soil population of *A. parasiticus* has also been observed by Lillehoj *et al.* (10). These authors studied the population of *A. flavus* and *A. parasiticus* in soil and from soil insects. Of the 30 isolates identified, 17 were *A. flavus* and 13 were *A. parasiticus*.

Aflatoxin production relative to the various isolates of the two species is characterized in Table 1. *Aspergillus parasiticus* isolates generally produced higher quantities of aflatoxin than *A. flavus*. *Aspergillus parasiticus* isolates producing a high amount of aflatoxin averaged 16% of the total number of isolates, whereas *A. flavus* isolates producing high amounts of aflatoxin accounted for 3% of the total.

Aspergillus parasiticus isolates that failed to produce aflatoxin accounted for only 1% of the total number of isolates, whereas 9% of *A. flavus* isolates were nonproducers.

Aflatoxin production directly in soil was also investigated. All the test soils containing significant populations of aflatoxin-producing *Aspergillus* were extracted with acetone and analyzed for the presence of aflatoxin. No aflatoxin was detected. The absence of toxin was probably due to the low concentrations and highly localized production, prohibiting detection by conventional procedures.

Aflatoxin decomposition

The evolution of $^{14}CO_2$ from the aflatoxin B_1-amended soil during incubation was determined (Figure 1). During the first week, all treatments exhibited a relatively high rate of $^{14}CO_2$ evolution; this may have been due to the presence of ^{14}C contaminants with simple carbon structures in the aflatoxin B_1 or microbial stimulation associated with rewetting an air-dried soil.

Following the initial flush of $^{14}CO_2$ from all treatments, $^{14}CO_2$ evolution decreased to a slower rate. The soil treated exclusively with aflatoxin B_1 exhibited a constant rate of release of $^{14}CO_2$ from 7 to 112 days. After 112 days, 14% of the ^{14}C-labeled aflatoxin added to the soil was recovered as $^{14}CO_2$. Soil amended with aflatoxin B_1 and wheat straw showed a somewhat reduced but relatively constant rate of $^{14}CO_2$ evolution from 7 to 112 days. At the conclusion of the experiment, 9.7% of the total labeled ^{14}C had been released. The lower rate of $^{14}CO_2$ evolution in the straw-amended soil may have resulted from a binding of

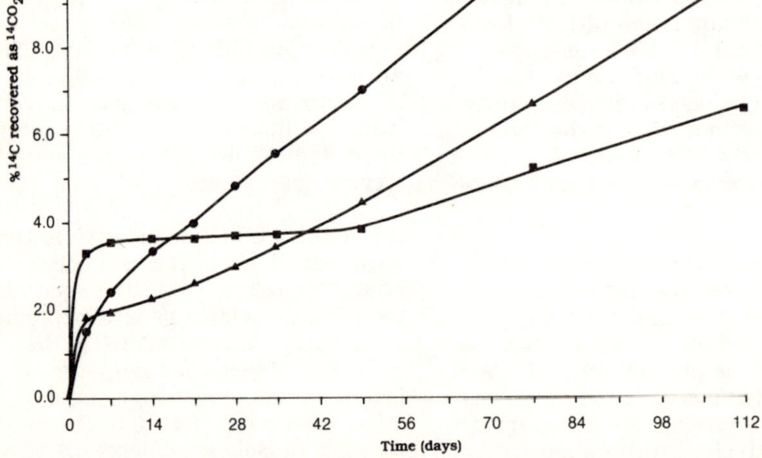

Figure 1. Percent C recovered as $^{14}CO_2$ during incubation of soil treated with 4.7 $\mu g/kg^{-1}$ of ^{14}C-labeled aflatoxin B_1

aflatoxin or its degradation intermediates to compounds produced by microorganisms decomposing the straw.

Soil treated with aflatoxin B_1 plus glucose demonstrated the highest initial release of $^{14}CO_2$. However, the rapid initial activity subsided abruptly, and very little $^{14}CO_2$ was evolved from 7 to 49 days. The rate of $^{14}CO_2$ evolution increased after this time but remained less than that for either of the two treatments, and at 112 days only 6.3% of the total ^{14}C had been released. Glucose is known to greatly stimulate development of the soil microbial population and to be utilized in a short time (17). Presumably, the glucose was exhausted in one week, leaving a large population without the necessary energy for maintenance. The initial population died and decayed in the following several weeks, while the secondary population may have bound aflatoxin or its degradation intermediates, thereby preventing its complete mineralization to CO_2. After the microbial population stabilized, the evolution of $^{14}CO_2$ resumed at a relatively slow rate.

Structural alterations in aflatoxin B_1 occurred when it was added to the soil. Changes in concentration of aflatoxin B_1 and the decomposition compounds that resulted from the chemical and microbial attack of aflatoxin B_1 during an incubation of 11 weeks were elucidated (Figure 2). After being added to the soil, the aflatoxin B_1 concentration rapidly decreased. The rate indicated the involvement of a chemical mechanism. Aflatoxin B_2 was the first observed product after adding aflatoxin B_1 to the soil; this compound results from a reductive attack of the vinyl ether double bond of aflatoxin B_1. Approximately 50% of the extractable aflatoxin B_1 was immediately reduced to aflatoxin B_2. The concentration of aflatoxin B_2 rapidly increased, reaching a maximum at three days and then

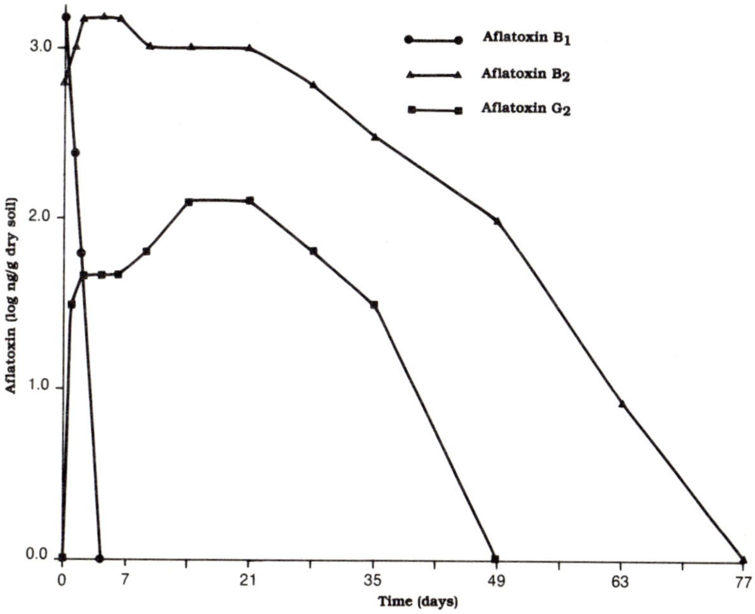

Figure 2. Concentration of extractable aflatoxin species during incubation of soil treated with 10,000 $\mu g/kg^{-1}$ aflatoxin B_1

gradually decreasing. Aflatoxin B_2 was no longer detectable 77 days after adding aflatoxin B_1 to the soil.

Low levels of aflatoxin G_2 were observed one day after the addition of aflatoxin B_1 to the soil. The concentration of this compound slowly increased for 14 days, then gradually decreased until the 35th day; at 49 days no aflatoxin G_2 was detected. Presumably, the aflatoxin G_2 formed in this soil came from microbial transformation of aflatoxin B_1 or B_2. The decomposition of ^{14}C-aflatoxin in four soils is presented in Figure 3. An initial flush of $^{14}CO_2$ was observed from all soils, perhaps because of the decomposition of labeled contaminants or the microbial stimulation associated with the rewetting of an air-dried soil. The initial release of $^{14}CO_2$ lasted for five days and then slowed. Aflatoxin decomposed most quickly in the silt loam soil. After 120 days of incubation, 8.1% of the aflatoxin was liberated as $^{14}CO_2$. While only a small percentage of the aflatoxin was converted to $^{14}CO_2$, this does not suggest that only 8.1% of the aflatoxin had been degraded. Most of the degraded aflatoxin was probably resynthesized into new microbial protoplasm or soil organic matter.

Aflatoxin decomposition in the same soil amended with 10,000 μg/kg^{-1} aflatoxin B_1 indicated that the high concentration of aflatoxin reduced its decomposition rate only during the first 20 days of incubation. It is likely that after this time, much of the aflatoxin had been bound to the exchange sites of the soil, thereby effectively reducing its biotoxicity.

During the first ten days of the study, the decomposition of ^{14}C-aflatoxin proceeded most rapidly in the sandy loam soil, probably because the aflatoxin did not bind to the exchange sites of the soil. Note from Table 2 that the cation exchange capacity (CEC) of the sandy loam soil was low (5.4 cmol/kg^{-1}). This soil contained low quantities of organic matter and clay and thus could not bind aflatoxin. Therefore, much of the added aflatoxin remained in solution where it was readily degraded. After ten days of incubation, however, the rate of aflatoxin decomposition in the sandy loam soil declined, probably as a result of the soil being unable to support a

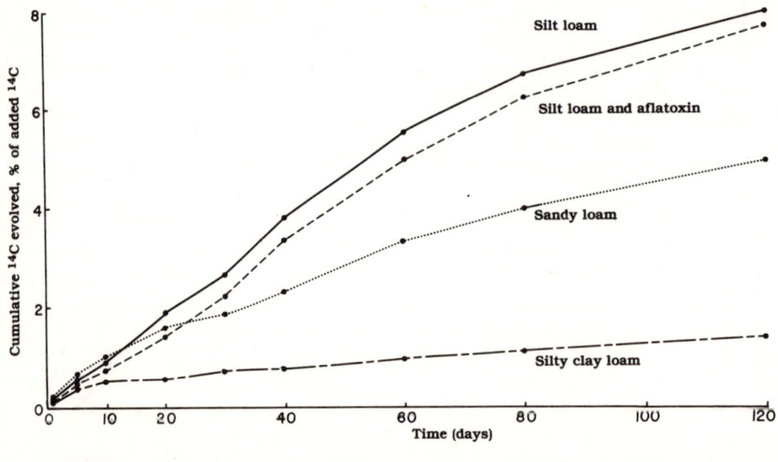

Figure 3. Cumulative ^{14}C evolved over time from a silt loam, silt loam plus 10,000 μg/kg^{-1}, sandy loam and silty clay loam soil

substantial and active microbial population. After 120 days of incubation, approximately two-thirds less aflatoxin had been degraded in the sandy loam soil than in the silt loam soil.

The slowest rate of aflatoxin decomposition was observed in the silty clay loam soil. By the conclusion of the study, only 1.4% of the aflatoxin was released as $^{14}CO_2$, probably as a result of the binding of aflatoxin to the exchange sites of the soil. Aflatoxin bound to the exchange sites of soil appears generally resistant to microbial decomposition.

Since the aflatoxin persisted in silty clay loam soil for a significantly longer period of time, it was investigated in more detail. Aflatoxin extraction from the soil over time revealed that the original aflatoxin B_1 added to the soil rapidly disappeared. A concurrent appearance of aflatoxin B_2 and G_2 was noted. Aflatoxin B_1 was no longer detected in the soil after six days of incubation. The concentrations of aflatoxins B_2 and G_2 rapidly increased from day zero to six days. After this time, the concentration of aflatoxin gradually declined, although even after 20 days of incubation much of the fluorescent activity of aflatoxins B_2 and G_2 remained.

It was also noted that after one day of incubation, a nonmobile (TLC), intensely fluorescing spot was observed. The fluorescent intensity of the spot increased up to four days of incubation and then declined very slowiy; this suggests that a conjugate was formed with a clay or organic component of the soil. The nonmobile form of aflatoxin was treated with 1.ON HCl to attempt to break the conjugate. The aflatoxin became partly soluble after hydrolysis for one hour, thereby supporting the contention of conjugate formation. The chromatographic movement of the hydrolyzed sample, however, was significantly less than any of the aflatoxin standards. These data suggest that either the conjugate was only partially broken or the liberated aflatoxin species was a relatively immobile degradation product.

Adsorption to clay may also be responsible for the reduced rate of aflatoxin decomposition observed in the silty clay loam soil. This soil has a relatively high clay content, which could adsorb the aflatoxin or its degradation products. Masimango *et al.* (13) demonstrated that aflatoxin could bind very tightly to clay.

In further investigations, adsorption isotherms for aflatoxin were established in the soils. Freundlich's isotherm adsorption coefficient K, l/n values, and their 95% confidence limits were determined from regression analysis (Table 3). The K value represents the amount of aflatoxin adsorbed per kilogram of soil when the equilibrium concentration equals 1 mg/kg^{-1}. The l/n values are an indication of the linearity of the adsorption process. According to Freundlich's theory, deviations of the l/n values from one reflect a nonlinear relationship between adsorption and solution equilibrum concentration. The l/n 95% confidence limits of all soils include the value of one.

The silty clay loam soil had a significantly greater K value than the other soils. The lowest K value occurrred in the sandy loam soil. The differences occurring among the K values of the various soils used in this study can be attributed to soil texture, organic matter and the corresponding CEC of the soils. The CEC and K values of the soils used in this experiment were positively correlated; the greater the CEC of the soil, the greater the K value. Soil texture also influenced aflatoxin adsorption. For example, increasing clay content in the soils was reflected in greater K values. Hence, the silty clay loam and sandy loam soils represented the extremes in clay content and K values.

The type of clay in the soil also appeared to have an influence on aflatoxin adsorption. The silty clay loam soil contained predominately 2:1 expanding clays. These clays are very efficient in adsorbing and retaining aflatoxin (13). The 2:1 expanding clays by nature have a larger internal and external surface area that would be exposed for aflatoxin adsorption; these clays also provide for better retention of aflatoxin once it is adsorbed. In contrast, the clay loam soil contained predominately 1:1 clays. This clay has a smaller surface area and, as a result, a lower adsorptive capacity.

Conclusions

The data presented indicate that significant numbers of aflatoxin-producing fungi can be present in soil. Aflatoxin production by soil isolates of *A. flavus* and *A. parasiticus* was readily demonstrated on agar media. Whether or not these fungi produce aflatoxin directly in the soil is still in question.

Aflatoxin in soil, produced by indigenous soil *Aspergillus* or artificially incorporated into the soil via burial of contaminated crops, decomposes relatively quickly. Data presented clearly demonstrate that aflatoxin persists in soil for only a limited period of time. Even in soils where decomposition was inhibited, the formation of conjugates probably reduced the toxicity of aflatoxin. Therefore, few if any adverse environmental consequences can be anticipated with the presence of aflatoxin in soil.

References

1. Anonymous. 1972. Changes in official method of analysis: Natural Poisons. BO1-26. BO3. Journal of the Association of Official Analytical Chemists 55:426-432.

2. Arai, T., T. Ito and Y. Koyama. 1967. Antimicrobial activity of aflatoxins. Journal of Bacteriology 93:59-64.

3. Ayres, J.C., and D.W. Brooks. 1979. Aflatoxin and other mycotoxins: An agricultural perspective. Council for Agricultural and Scientific Technology Report 80. Iowa State University, Ames, Iowa, USA.

4. Bell, D.K., and J.L. Crawford. 1967. A Bortran-amended medium for isolating *Aspergillus flavus* from peanuts and soil. Phytopathology 57:939-941.

5. Clevstrom, L., and H. Ljuggren. 1984. Occurrence of storage fungi, especially *Aspergillus flavus*, in soil, greenstuff, and prepared hay. Journal of Stored Products Research 20:71-82.

Table 3. Freundlich's adsorption and 1/n constants of aflatoxin B_1

Soil	Log K	K (mg/kg^{-1} soil)	Log 1/n
Silty clay	2.38 ± 0.08[a]	238.49 a[b]	1.02 ± 0.06
Silty loam	1.88 ± 0.06	76.19 b	0.94 ± 0.06
Clay loam	1.67 ± 0.30	46.92 c	1.22 ± 0.42
Sandy loam	1.24 ± 0.22	17.40 d	0.93 ± 0.26

a/ Significant at the 5% level

b/ Values followed by the same letter are not significantly different at the 5% level

6. Fennell, D.I., W.F. Kowlek, E.B. Lillehoj, G.L. Adams, R.B. Bothast, M.S. Zuber, O.H. Calvert, W.D. Guthrie, A.J. Bockholt, A. Manwiller and M.D. Jellum. 1977. *Aspergillus flavus* presence in silks and insects from developing and mature corn ears. Cereal Chemistry 54:770-778.

7. Griffin, G.L., and K.H. Garren. 1974. Population levels of *Aspergillus flavus* and the *A. niger* group in Virginia peanut soils. Phytopathology 64:322-325.

8. Herr, L.J. 1957. Soil mycoflora associated with continuous cropping of corn, oats, and wheat. Ohio Journal of Science 57:203-211.

9. Kaufman, D.D., and L.E. Williams. 1964. Effect of mineral fertilization and soil fungi. Phytopathology 54:134-139.

10. Lillehoj, E.B., W.W. McMillian, W.D. Guthrie and D. Barry. 1980. Aflatoxin-producing fungi in preharvest corn: Inoculum source in insects and soils. Journal of Environmental Quality 9:691-699.

11. Lin, M.T., and J.C. Dianese. 1976. A coconut-agar medium for rapid detection of aflatoxin production by *Aspergillus* spp. Phytopathology 66:1466-1469.

12. Martyniuk, S., and G.H. Wagner. 1978. Quantitative and qualitative examination of soil microflora associated with different management systems. Soil Science 125:343-350.

13. Masimango, N., J. Remade and J. Ramaut. 1979. Elimination par des argiles gonflantes de l'aflatoxine B_1 des milieux contamines. Annales de la Nutrition et de l'Alimentation 33:137-147.

14. McMillian, W.W., D.M. Wilson, N.W. Widstrom and R.C. Gueldner. 1980. Incidence and level of aflatoxin in preharvest corn in South Georgia in 1979. Cereal Chemistry 57:83-84.

15. Menon, S.K., and L.E. Williams. 1957. Effect of crop, crop residues, temperature, and moisture on soil fungi. Phytopathology 47:559-564.

16. Misra, R.S., and R.K. Tripathi. 1980. Effect of aflatoxin B_1 on germination, respiration and alpha-amylase in maize. Zeitschrift für Pflanzenkrankheiten und Pflanzenschutz 87:155-160.

17. Wagner, G.H., and K.S. Chahal. 1966. Decomposition of carbon-14 labeled atrazine in soil samples from Sanborn Field. Proceedings of the American Society of Soil Science 30:752-754.

18. Wilson, D.M., W.W. McMillian and N.W. Widstrom. 1979. Field aflatoxin contamination of corn in South Georgia. Journal of the American Oil Chemists' Society 56:798-799.

19. Young, J.W., W.V. Dashek and G.C. Llewellyn. 1978. Aflatoxin B_1 influence on excised soya-bean root growth, ^{14}C-leucine uptake and incorporation. Mycopathology 66:91-97.

20. Zuber, M.S., O.H. Calvert, E.B. Lillehoj and W.F. Kwolek. 1976. Preharvest development of aflatoxin B_1 in corn in the United States. Phytopathology 66:1120-1121.

The Relationship Between Aflatoxin Formation and Kernel Damage in Costa Rican Maize

R. Echandi Z., Grain and Seed Research Center, University of Costa Rica, San Jose, Costa Rica

Abstract

Although results from a preliminary survey of grain sold on the market in Costa Rica suggest that aflatoxin contamination of maize is a major problem in the country, four widely used maize varieties grown in three ecological zones showed low levels of contamination. Bending of maize plants in the field to reduce the penetration of water into the ear, a common practice among Costa Rican farmers, seemed to have no effect on Aspergillus flavus *kernel infection. In laboratory studies with the same varieties, it was found that inoculum must be present and is equally as important as kernel damage for the formation of aflatoxin. The relationship between kernel damage and* A. flavus *infection and aflatoxin production was confirmed by results of laboratory inoculation of kernels on the ear. The lower aflatoxin level of intact, inoculated kernels on the ear compared to that of intact, inoculated shelled kernels suggests that kernels on the ear are less susceptible to aflatoxin contamination.*

Resumen

Si bien los resultados de un estudio preliminar de grano vendido en el mercado en Costa Rica sugieren que la contaminación del maíz con aflatoxinas es un problema grave en ese país, cuatro variedades de maíz de uso extendido que se cultivaron en tres zonas ecológicas mostraron niveles de contaminación bajos. El doblar las plantas en el campo con el fin de reducir la penetración de agua en la mazorca, una práctica común entre los agricultores costarricenses, al parecer no tuvo efecto alguno en la infección con Aspergillus flavus. *En los estudios que se realizaron en el laboratorio con las mismas variedades, se encontró que la presencia de inóculo es un factor tan importante en la formación de aflatoxinas como la integridad de los granos. Los resultados de la inoculación en el laboratorio de granos adheridos a la mazorca confirmaron la relación entre el daño al grano y la infección con* A. flavus *y la producción de aflatoxinas. El nivel bajo de aflatoxinas que se observó en granos íntegros adheridos a la mazorca e inoculados, en comparación con el nivel de granos intactos, desgranados e inoculados sugiere que los granos adheridos a la mazorca son menos susceptibles a la contaminación con aflatoxinas.*

In the tropical zones of the American continent, especially in the Caribbean basin, climatic conditions are highly favorable for the development of microorganisms in grain. These conditions, combined with inadequate technology for grain production and management, are a concern to researchers working on mycotoxins and the problems they cause in food and feed grains.

Worldwide mycotoxin research over 25 years has shown that the presence of mycotoxins in grain can be associated with genetic differences among cultivars, climatic conditions occurring during the production cycle and grain handling procedures during and after harvest.

The presence of mycotoxins in grain produced in tropical America is influenced by a fourth factor, i.e., farmers' agronomic practices. In most cases, the technology applied to grain production, especially that for home consumption or for the local market, is inadequate, resulting in low yields. Local varieties and traditional agricultural practices are the norm, as are farms of only a few hectares. Harvesting, threshing and shelling are done by hand, and the grain is usually dried in the open air and/or in the sun. Grain is often stored in sacks or other containers with inadequate ventilation until it is used for food or taken to market, a period often extending over several months. As a result, the quality of grain found in the market is poor, and since most of the countries do not have (or do not enforce) quality regulations for commercial grain, mycotoxin-contaminated grain may enter the marketing system (3).

This study came about as a result of the desire to establish guidelines for the presence of mycotoxins in grain produced in Costa Rica. The findings are probably relevant to other countries in the region as well, due to the similarity of climates and agronomic systems.

Aflatoxin in Costa Rican Grain

A preliminary study was conducted (4) using 364 samples of maize (*Zea mays*), beans (*Phaseolus vulgaris*) and rice (*Oryza sativa*). Samples were collected monthly over a period of four months during two consecutive years from grain on the market in the three largest cities in Costa Rica. The study showed that aflatoxin contamination is a problem; aflatoxin levels of more than 20 μg/kg were found in 40% of the samples the first year, and 30% of the samples the second year. Significant differences were observed in the frequency of aflatoxin contamination in white- and yellow-grain maize. Beans had lower levels of aflatoxin, and rice samples were free of contamination (Table 1).

Aflatoxin Formation in Preharvest Maize

As the first step in detecting the source of aflatoxin contamination in Costa Rican maize, experimental plots were planted in three different ecological zones. Two had annual average rainfall of more than 2000 mm and average mean temperatures above 23°C. The

Table 1. Aflatoxin content of maize grain in the commercial market, Costa Rica

Grain and year of sampling	Number of samples	Aflatoxin content (μg/kg) 20	21-50	51-100	101-500	500	Maximum aflatoxin
Maize							
First year	52	32	9	4	4	3	1000
Second year	57	40	10	1	4	2	3500
Common bean							
First year	72	69	2	-	1	-	150
Second year	66	64	1	1	-	-	90
Rice[a]							
First year	64	64	-	-	-	-	-
Second year	53	53	-	-	-	-	-
Total	364	322	22	6	9	5	

[a] Shelled and polished rice

third area averaged more than 4000 mm of rainfall and 28°C. Two F_1 hybrids were tested, the white-grain X-105 and the yellow X-304, as well as two open-pollinated varieties, Alajuela Blanco and Alajuela Amarillo. All four are widely used by Costa Rican farmers. Plots of 450 m^2 of each cultivar were grown at each location using intermediate technology.

A common practice among Costa Rican farmers is to bend maize plants 1.25 meters above the ground three weeks after flowering; this is done to reduce the penetration of water into the ear and the resulting growth of microorganisms. In this study 50% of the plants in each plot were bent, and 50% were left in the normal upright position.

One hundred days after planting, 30 ears of each variety were randomly selected from each site, 15 from bent plants and 15 from upright plants. Sampling was continued every two weeks over a four-month period. Ears were dried with forced air at 45°C to a moisture content of approximately 13%. Ears were shelled and ground, and the sample material was passed through a 0.8-mm mesh screen. Aflatoxin content was determined using a modified version of the Velasco method (4,7).

Aflatoxin levels were low for all cultivars in the study (Table 2). No differences in aflatoxin contamination were found between ecological zones or as a result of the bending of the plants. The literature on aflatoxin contamination in the field is abundant, and in almost all cases it emphasizes the variation in varietal behavior in different years (2). The extreme climatic conditions encountered in this study make it of special interest.

Table 2. Aflatoxin formation in preharvest maize, Costa Rica

		Aflatoxin content							
		1 (110 days)		2 (114 days)		3 (128 days)		4 (142 days)	
Variety	**Location**[a]	**Erect**	**Bent**	**Erect**	**Bent**	**Erect**	**Bent**	**Erect**	**Bent**
X-105	A	6	8	7	6	8	7	6	6
	B	9	9	7	6	5	5	6	7
	C	0	0	11	10	5	5	5	6
X-306	A	8	6	8	8	6	2	12	9
	B	9	9	6	11	6	7	8	9
	C	0	0	23	30	7	7	6	7
Alajuela blanco	A	6	6	7	15	9	8	6	7
	B	6	8	8	6	5	8	6	9
	C	0	0	9	27	4	4	6	5
Alajuela amarillo	A	5	5	8	6	4	7	7	8
	B	14	7	10	13	6	6	15	15
	C	0	0	12	10	5	6	5	5

a/ A = Buenos Aires, B = Orotina, C = Alajuela
b/ Samples taken every two weeks beginning 100 days after planting, 50% plants bent and 50% left erect

Aflatoxin Formation in the Laboratory

Kernel inoculation

Laboratory testing was conducted to investigate *Aspergillus flavus* infection and aflatoxin formation in maize. Twelve 4-kg samples each were prepared of the F_1s of the hybrids X-105 and B-666 (white grain) and X-306 (yellow grain), and the open-pollinated varieties Tico V-1 (white) and Tocumen (yellow). After kernel surface sterilization with sodium hypochlorite, kernel moisture was adjusted to 28%. The kernels from six of the samples were mechanically broken into relatively large pieces and sieved to assure uniform size. The kernels of the other six samples were left intact. The samples then received one of four treatments: intact kernels without inoculation; intact kernels with inoculation; damaged kernels without inoculation; and damaged kernels with inoculation. Inoculation was done with conidia of the isolate NRRL 2999 (1×10^6/ml). The samples were kept in well-ventilated containers at 30°C for 96 hours and then autoclaved before evaluation for aflatoxin contamination.

The results shown in Table 3 clearly indicate a significant difference between the inoculated and uninoculated samples. Although mechanical damage of uninoculated kernels resulted in a slight increase in toxin formation, the increase was little greater than that of uninoculated, intact kernels when inoculum was present in the atmosphere. These results paralleled observations of the low incidence of contamination of maize in the field.

Ear inoculation

Subsequent tests included ear inoculation; the kernels of some treatments were left intact and others

Table 2. Cont'd.

(μg/kg) by treatment b/

5 (156 days)		6 (170 days)		7 (184 days)		8 (198 days)	
Erect	**Bent**	**Erect**	**Bent**	**Erect**	**Bent**	**Erect**	**Bent**
7	6	0	0	0	0	0	0
8	18	7	13	0	0	8	30
6	7	5	5	8	9	7	6
8	8	0	0	0	0	0	0
6	8	17	7	0	0	11	12
8	18	6	5	6	6	8	6
5	12	0	0	0	0	0	0
6	7	7	8	0	0	7	7
8	7	5	5	5	6	7	6
20	6	0	0	0	0	0	0
6	10	7	8	0	0	8	8
8	8	5	5	8	6	10	8

were damaged. A plot of the F_1 hybrid X-105 was planted, with special care taken to avoid any type of damage to the kernels, either in the field or during handling; the best-looking ears were selected for the study. The ears had a moisture content in the field of approximately 20%, and were air dried at 45°C to a moisture content of about 13%.

Each treatment consisted of 12 ears divided into two groups of six, which were placed in 4-liter containers. Each container had a false bottom over water, so that a high relative humidity was maintained in the container. Holes in the covers permitted the free exchange of gases in the containers, which were kept at a temperature of 28°C. The treatments consisted of ears with undamaged kernels without inoculation, ears with all kernels damaged (scraped) and without inoculation, inoculated ears with kernel damage in three adjacent rows, and inoculated ears with all kernels damaged. Inoculation was carried out by quickly submerging each treated ear in a suspension of spores of the isolate NRRL 2999 (1 x 10^6/ml).

Ears with undamaged kernels and inoculated ears with kernel damage in three adjacent rows were sampled at eight-day intervals over a 32-day period. Ears with all kernels damaged, both with and without inoculation, were sampled only after 32 days. The containers were autoclaved for one hour and the ears evaluated for aflatoxin by using the modified version of the Velasco method (7).

Aflatoxin levels in response to each of the various treatments were determined (Table 4). No aflatoxin formation was found in the absence of both inoculum and kernel damage. These results support the findings from the field tests. Damaging the kernels resulted in large-scale aflatoxin contamination from what must have been natural inoculum sources. Aflatoxin levels in ears with undamaged kernels were very low, except in two cases where divergence from the mean was attributed to the germination of kernels on the cob. As could be expected, aflatoxin contamination levels were very high in inoculated ears with damaged kernels.

In ears with only three rows of damaged kernels, special assays for aflatoxin content were performed separately on kernels in damaged rows, undamaged rows adjacent to damaged ones and undamaged rows away from the damaged ones. Massive aflatoxin contamination developed in the damaged kernels under the favorable conditions of temperature and humidity in which the ears were

Table 3. Aflatoxin formation (μg/kg) in maize kernels in the laboratory, Costa Rica

Treatment	B-666 (white)	X-105 (white)	X-306 (yellow)	Tocumen (yellow)	Tico V-1 (white)
Intact kernels					
Inoculated[a]	51	317	580	67	23
Uninoculated	16	22	24	19	25
Damaged kernels					
Inoculated[a]	58,000	104,000	121,000	29,000	24,000
Uninoculated	21	42	42	26	27

a/ NRRL 2999 (1 x 10^6/100g)

kept (Figures 1 and 2). Aflatoxin accumulation was less in adjacent, undamaged rows, and even less in more distant rows.

Results

Preharvest aflatoxin contamination

The results presented here show low levels of preharvest aflatoxin contamination in maize under Costa Rican climatic conditions, even when the grain remains in the field for several months, exposed to the high temperature and humidity of a tropical climate. These findings differ from those reported in the USA (2), where aflatoxin contamination in preharvest maize was highest in maize grown in the hot, humid climate of the southern part of the country (1,2).

Currently, the behavior of the four maize varieties tested in relation to aflatoxin accumulation cannot be explained. The high amount of rainfall might be one factor, since this affects existing levels of inoculum. Jones *et al.* (5) reported an inverse relationship between precipitation and leaf moisture and the quantity of live spores present.

Bending of maize plants in the field, a common practice among Costa Rican farmers, seemed to have no effect on *A. flavus* kernel infection, even though the ears of erect plants showed moisture-induced damage.

Kernel inoculation studies

The relationship between kernel damage and contamination by saprophytic microorganisms, such as the *Aspergillus* spp., is shown by the contrast between aflatoxin contamination found in intact and damaged kernels in the presence of equal amounts of inoculum-producing toxins. Inoculum must be present and is equally as important as kernel damage for the formation of aflatoxin; aflatoxin was also found in intact kernels in the presence of inoculum. The difference in aflatoxin levels found in the four maize varieties in the study suggests that they differ in their susceptibility to the toxin.

Ear inoculation studies

The relationship between kernel damage and *Aspergillus* spp. infection and production of aflatoxin was reaffirmed by the results of laboratory inoculation of kernels on the ear. The

Table 4. Aflatoxin formation (μg/kg) in X-105 maize ears in the laboratory, Costa Rica

Treatment	Days after inoculation			
	8	116	24	32
Uninoculated				
Healthy	0 - 0	0 - 0	0 - 0	0 - 0
Damaged	NS[b/]	NS	NS	16,250 - 28,750
Inoculated[a/]				
Healthy	0 - 0	7 - 15	19	10
Damaged	NS	NS	NS	54,200 - 72,600

a/ Inoculated by submersion in a suspension of NRRL 2999 spores (1 x 10^6/ml)

b/ NS = not sampled

Note: Ears kept at a temperature of 28°C, RH 95%

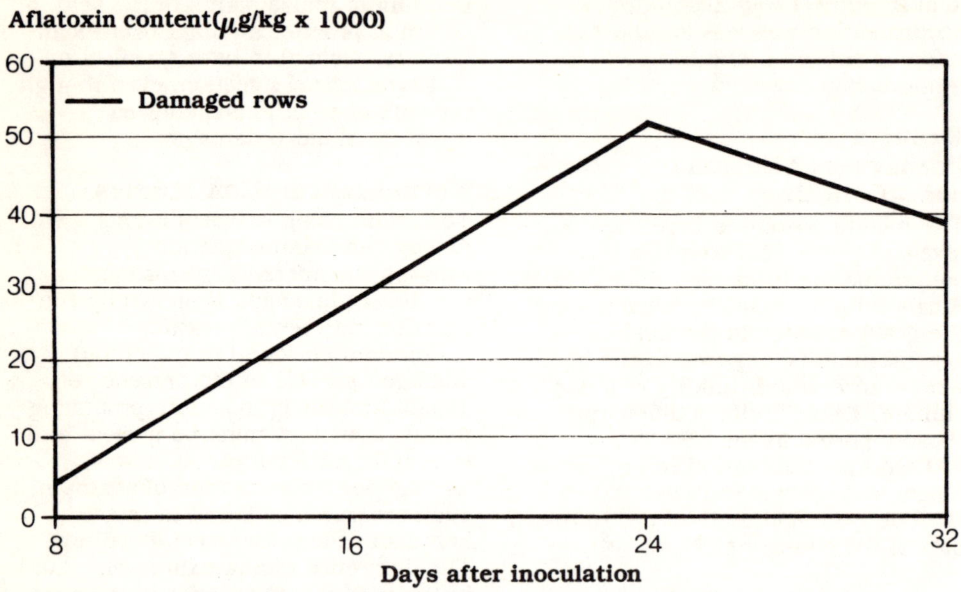

Figure 1. Aflatoxin formation in kernels on ears of maize variety X-105 inoculated in the laboratory, Costa Rica

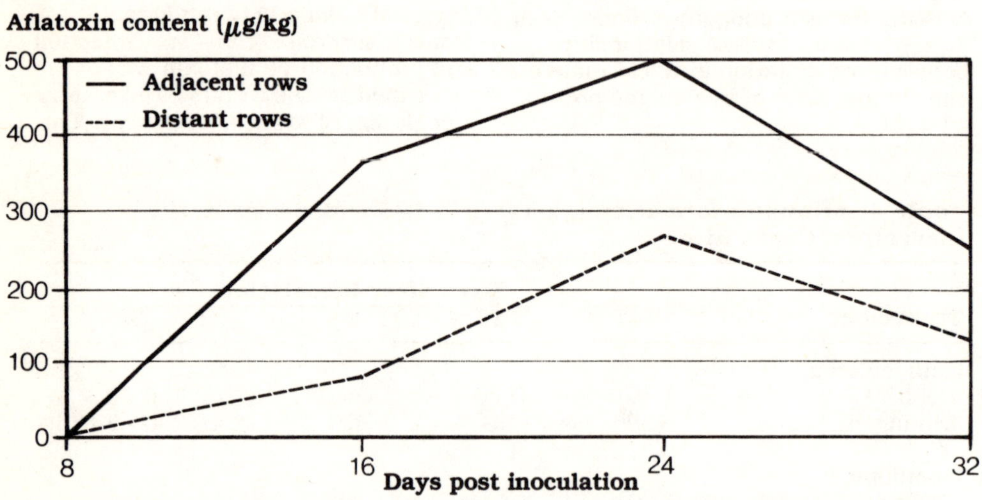

Figure 2. Aflatoxin formation in kernels on ears of maize variety X-105 inoculated in the laboratory, Costa Rica

lower aflatoxin level of intact, inoculated kernels on the ear compared to that of intact, inoculated shelled kernels suggests that kernels on the ear are less susceptible to aflatoxin contamination, even when very close to contaminated kernels. Lee *et al.* (6) found that *A. flavus* does not seem to spread easily from contaminated to healthy kernels on the same ear. This fact suggests that it might be advisable for Costa Rican farmers to store maize on the cob instead of as shelled grain. The storage of maize in ear form is not new; it is already a practice among some farmers. Therefore, it is important that the economics of handling maize on the ear be studied, at least until it is possible to reduce the moisture content of the grain to levels unfavorable for the growth of toxin-producing molds.

Conclusions

Results from the preliminary survey of grain being sold on the market in Costa Rica suggest that aflatoxin contamination in maize is a major problem in the country. A large percentage was found to have aflatoxin levels higher than are acceptable. Although a definitive solution to this problem is unlikely, considering prevailing production and handling methods, a reduction in aflatoxin contamination in maize grain could be achieved by developing appropriate cultural practices and encouraging their adoption by the small-scale farmers who produce a large part of the maize grown in Costa Rica.

References

1. Anderson, H.W., E.W. Nehring and W.R. Wichser. 1975. Aflatoxin contamination of corn in the field. Journal of Agricultural and Food Chemists 23:775-782.

2. Davis, N.D., C.G. Currier and U.L. Diener. 1985. Response of corn hybrids to aflatoxin formation by *Aspergillus flavus*. Alabama Agricultural Experiment Station, Auburn, Alabama, USA.

3. Echandi, R., and F. Camacho. 1986. Assessment of the problem of mycotoxins as grain contaminants in the Caribbean Basin through a monitoring study carried out in Costa Rica. Paper presented at the S-175 group meeting, Mayaguez, Puerto Rico, January 1986.

4. Echandi, R., and F. Camacho. 1986. Sampling and detection methods for aflatoxin used in Costa Rica. Paper presented at the S-175 group meeting, Mayaguez, Puerto Rico, January 1986.

5. Jones, R.K., H.E. Duncan, G.A. Payne and K.J. Leonard. 1980. Factors influencing infection by *Aspergillus flavus* in silk-inoculated corn. Plant Disease 64:859-863.

6. Lee, L.S., E.B. Lillehoj and W.E. Kwolek. 1980. Aflatoxin distribution in individual kernels from intact ears. Cereal Chemistry 57:340-343.

7. Velasco, J. 1975. Fluorometric measurement of aflatoxin absorbed in florisil in minicolumns. Journal of the Association of Official Analytical Chemists 58:757-763.

The Effect of Climatic Conditions on the Incidence and Severity of Aflatoxin in the USA

P.F. Sisson, Department of Agricultural Economics, The Quaker Oats Company, Chicago, Illinois, USA

Abstract
Climatic conditions in 1980 and 1981 were correlated with the incidence and severity of aflatoxin in maize at nine locations in the USA. Regression equations were developed that explained 80 to 89% of the variability of aflatoxin incidence and severity. The incidence of aflatoxin was primarily related to average high humidity in July and average minimum temperature in August. The relevant variables in a second equation were the number of weeks exceeding 32°C for weekly maximum temperature, and the number of weeks in which average weekly high humidity exceeded 95%. The severity of aflatoxin was related to July average maximum temperature, August average minimum temperature and July average high humidity.

Resumen
Las condiciones climáticas que imperaron en 1980 y 1981 se correlacionaron con la incidencia y severidad de la contaminación de maíz con aflatoxina en nueve localidades de Estados Unidos. Se elaboraron ecuaciones de regresión que explicaron del 80 al 89% de la variabilidad de la incidencia y severidad de las alfatoxinas. La incidencia de las aflatoxinas se relacionó principalmente con la humedad alta media de julio y con la temperatura mínima media de agosto. Las variables más importantes de una segunda ecuación fueron el número de semanas que presentaron una temperatura semanal máxima de más de 32°C y el número de semanas en las que la humedad alta media semanal excedió el 95%. La severidad de las aflatoxinas se relacionó con la temperatura máxima media de julio, la temperatura mínima media de agosto y la humedad alta media de julio.

In a comparison of eight open-pollinated maize varieties and four hybrids for preharvest aflatoxin contamination in the USA, Zuber *et al.* (3) found that plant stress, especially during the grain-filling stages, affected the level of aflatoxin. High temperatures were more important than lack of moisture, although both enhanced aflatoxin levels.

Lillehoj *et al.* (1) documented the incidence and severity of aflatoxin in kernel samples from hybrids grown in the crop years 1980 and 1981 at nine locations in the USA. They found high, medium and low incidence of aflatoxin among locations. High incidence (90 to 100%) was found in Florida, Georgia, North Carolina, South Carolina and Mississippi, medium incidence (30 to 45%) in Kentucky and Tennessee, and low incidence (2 to 25%) in Ohio and Indiana.

The current study was conducted to determine if the incidence and severity of aflatoxin in maize kernels, as reported by Lillehoj *et al.* (1), could be correlated with the climatic conditions that occurrred in 1980 and 1981 at the respective locations.

Weekly averages of climatic conditions were developed for each location within states for the two years, based on data from the Weekly Weather and Crop Bulletin (2). In most instances, climatic data were obtained from a

National Weather Service (NWS) station in the approximate vicinity of the test site (Table 1).

Regression analysis was used to assess the impact of climatic conditions on aflatoxin development. A number of variables were generated (Table 2) and used to develop regression models. Incidence of aflatoxin (A_p) and aflatoxin levels (A_b) were considered dependent variables and the other parameters, independent variables.

Results

Climatic conditions recorded in 1980 and 1981 correlate well with the incidence of aflatoxin at the nine locations. Two regression equations were developed that explained 80 to 89% of the variability in aflatoxin incidence. A third regression was developed for aflatoxin severity.

In Regression Equation 1 (Table 3), the R^2 value was .8896; all of the variables used were statistically significant. The two most significant variables were average minimum temperature for August and average high humidity for July.

In Equation 2 (Table 4) the number of weeks of selected temperatures and humidity events were determined to correlate with aflatoxin incidence, R^2 = .802. High temperature and high humidity were the major determinants of aflatoxin incidence; again, weekly lows of 21°C and below were negatively correlated. The greater the number of weeks with temperatures of 21°C and below, the less the quantity of aflatoxin.

In Equation 3 (Table 5), July average maximum temperature, August average minimum temperature, July average high humidity and July average low humidity were the primary variables that correlated with aflatoxin levels. Comparisons of the fitted regression to actual observations

Table 1. Maize plot locations and weather stations, Quaker Oats aflatoxin tests, 1980 and 1981

Test plot location	Nearest weather station	Number of plots	
		1980	1981
Gainesville, Florida	Orlando, Florida	1	8
Tifton, Georgia	Macon, Georgia	3	10
Raleigh, North Carolina	Raleigh, North Carolina	7	16
Florence, South Carolina	Columbia, South Carolina	4	11
Mississippi State, Mississippi	Meridian, Mississippi	—	15
Lafayette, Indiana	Indianapolis, Indiana	5	12
Wooster, Ohio	Akron/Canton, Ohio	2	9
Knoxville, Tennessee	Knoxville, Tennessee	6	13
Manhattan, Kansas	Topeka, Kansas	—	14

Table 2. Climatic variables used in regressions, Quaker Oats aflatoxin tests, 1980 and 1981

A_p	= Incidence of aflatoxin positives (percent)
A_b	= Severity of aflatoxin $B_1 + B_2$(ng/g^{-1})
A_{lt}	= August average minimum temperature
J_{hh}	= July average high humidity
W_{32t}	= Number of weeks during growing season in which weekly average maximum temperature exceeded 32°C
J_{ht}	= July average maximum temperature
J_{lh}	= July average low humidity
W_{90h}	= Number of weeks during growing season in which weekly average high humidity exceeded 90%
D_1	= Dummy variable indicating locations with 0 to 5 weeks in which weekly maximum temperature exceeded 32°C
D_2	= Dummy variable indicating location with 11 to 15 weeks in which weekly maximum temperature exceeded 32°C
C	= Constant
W_{95h}	= Number of weeks during growing season in which weekly average high humidity exceeded 90%
W_{38t}	= Number of weeks during growing season in which weekly average maximum temperature exceeded 38°C
W_{21t}	= Number of weeks during growing season in which weekly average minimum temperature exceeded 21°C
W_{55h}	= Number of weeks during growing season in which weekly average low humidity exceeded 55%

Table 3. Ordinary least squares regression of the percent of aflatoxin incidence, Quaker Oats aflatoxin tests, 1980 and 1981

Independent variable	Estimated coefficient	Standard error	T-statistic
Equation 1 [$A_p = f(A_{lt}, J_{hh}, D_1, D_2, W_{32t})$ $R^2 = .8896$]			
C	-272.312	154.2	-1.77
A_{lt}	-21.495	5.7	-3.79
J_{hh}	6.828	2.1	3.26
D_1	37.232	22.9	1.63
D_2	-93.496	25.1	-3.72
W_{32t}	20.502	4.4	4.66

$R^2 = .8896$
Adjusted $R^2 = .8344$
F-statistic (5,10) = 16.1138
Durbin-Watson statistic (adjusted for 0 gaps) = 1.5794
Number of observations = 16
Sum of squared residuals = 2570.61
Standard error of the regression = 16.0331

Table 4. Ordinary least squares regression of the percent of aflatoxin incidence, Quaker Oats aflatoxin tests, 1980 and 1981

Independent variable	Estimated coefficient	Standard error	T-statistic
Equation 2 [$A_p = f(W_{32t}, W_{95h}, W_{38t}, W_{21t}, W_{55h})$ R^2 = .8021]			
C	40.293	22.1	1.82
W_{32t}	4.613	2.5	1.82
W_{95h}	5.669	2.3	2.94
W_{38t}	24.513	10.7	2.29
W_{21t}	-6.489	3.5	-1.86
W_{55h}	-2.494	2.1	-1.17

R^2 = .8021
Adjusted R^2 = .7032
F-statistic (5,10) = 8.10854
Durbin-Watson statistic (adjusted for 0 gaps) = 1.8629
Number of observations = 16
Sum of squared residuals = 4606.35
Standard error of the regression = 21.4624

Table 5. Ordinary least squares regression of the level of aflatoxin severity, Quaker Oats aflatoxin tests, 1980 and 1981

Independent variable	Estimated coefficient	Standard error	T-statistic
Equation 3 [$A_b = f(J_{ht}, A_{lt}, J_{hh}, J_{lh}, D_1, W_{90h})$ R^2 = .8198]			
C	54.659	1077.7	.51
J_{ht}	17.409	16.9	1.03
A_{lt}	-61.635	16.7	-3.69
J_{hh}	13.864	8.3	1.55
J_{lh}	-14.597	6.1	-2.39
D_1	-70.318	66.1	-1.06
W_{90h}	14.439	7.2	1.99

R^2 = .8198
Adjusted R^2 = .6997
F-statistic (6,9) = 6.82377
Durbin-Watson statistic (adjusted for 0 gaps) / 2.1819
Number of observations = 16
Sum of squared residuals = 46364.3
Standard error of the regression = 71.7745

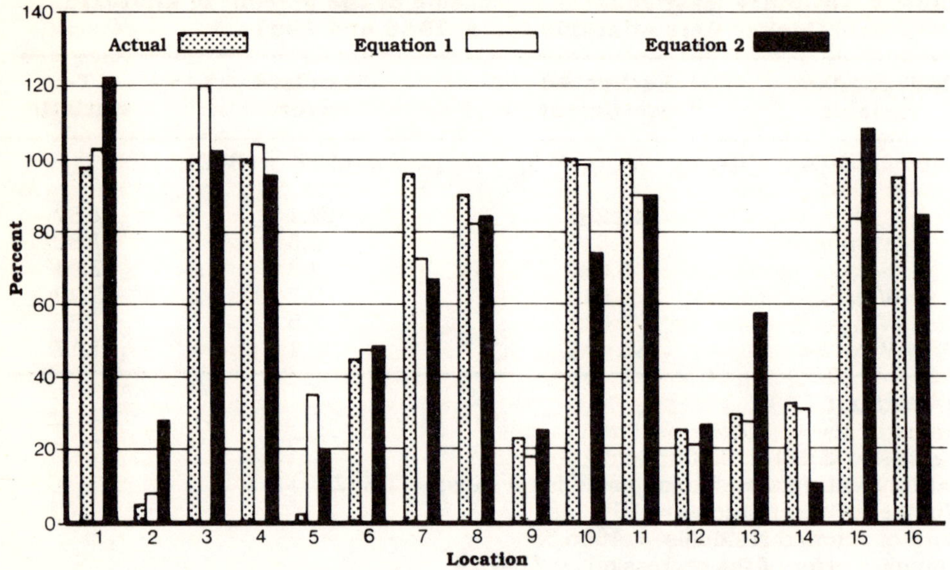

Figure 1. A comparison of the percentage of aflatoxin occurrence under varied climatic conditions at diverse locations in the USA

Note: Location numbers refer to sites listed in Table 1

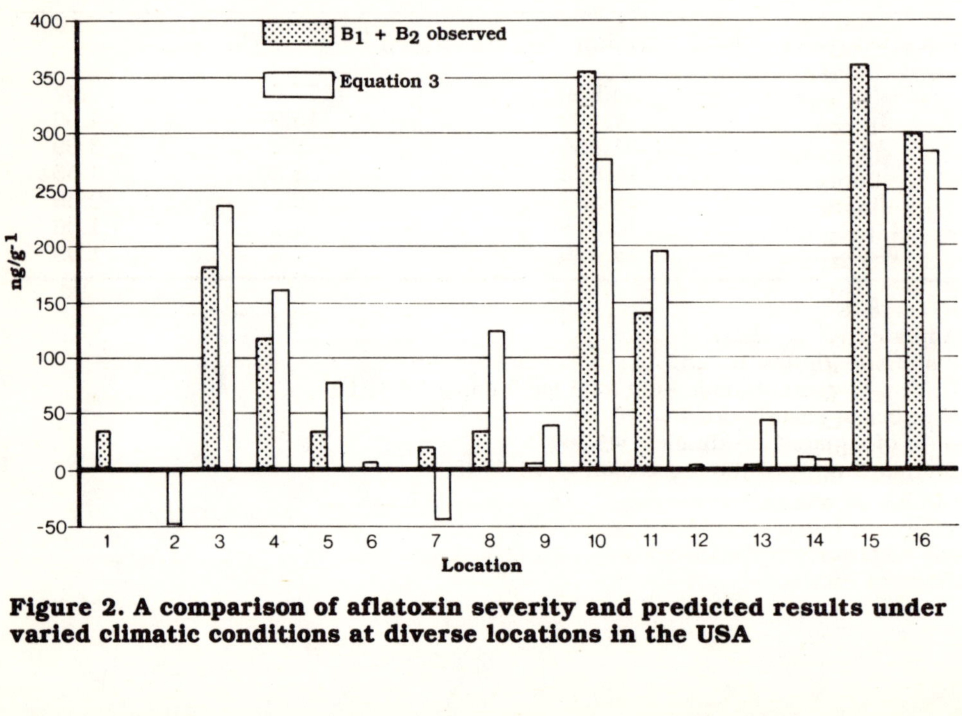

Figure 2. A comparison of aflatoxin severity and predicted results under varied climatic conditions at diverse locations in the USA

indicate that the variables explain much of the variability in incidence and severity of aflatoxin documented at the locations studied (Figures 1 and 2).

Discussion

Quaker Oats produces a number of high-quality maize products. In a number of recent years aflatoxin has been a major problem that has resulted in significant rejections of maize by the Quality Assurance Department. Testing procedures for aflatoxin are costly and take time. The company is interested in developing means of forecasting aflatoxin severity to determine if testing for aflatoxin is required before a new crop of maize is purchased. These studies represent initial attempts to develop relevant production equations.

High average July temperature and high average July humidity levels would be expected to cause stress in maize plants and should also be conducive to aflatoxin development. Surprisingly, both Equation 1 (Table 3) and Equation 3 (Table 5) indicate an association between higher monthly average minimum temperature in August, and reduced incidence and severity of aflatoxin. The observation seems to contradict current theory that higher temperatures increase stress on maize plants and enhance aflatoxin contamination processes. Since maize in a number of the test locations reached maturity in August, the higher temperature could accelerate maize maturation. Also, elevated minimum temperature might result in a lower incidence of dew formation, which could be a factor in fungal development and aflatoxin development. Equation 2 (Table 4) shows a similar situation in which the number of weeks that average minimum temperature exceeded 21°C was negatively correlated with aflatoxin incidence.

More studies of climatic conditions using a broader range of observations of aflatoxin incidence and severity are necessary to confirm initial conclusions. A time-series analysis for specific locations over a longer time period would be valuable. Dew formation as a potential explanation for the significance of August average minimum temperature requires further study.

References

1. Lillehoj, E.B., M.S. Zuber, L.L. Darrah, W.F. Kwolek, W.R. Findley, E.S. Horner, G.E. Scott, A. Manwiller, D.B. Sauer, D. Thompson, H. Warren, D.R. West and N.W. Widstrom. 1983. Aflatoxin ocurrence and levels in preharvest corn kernels with varied endosperm characteristics grown at diverse locations. Crop Science 23:1181-1184.

2. National Oceanic and Atmospheric Administration. 1980, 1981. Weekly weather and crop bulletin. USDA Agricultural Weather Facility, Washington, D.C., USA.

3. Zuber, M.S., L.L. Darrah, E.B. Lillehoj, A. Manwiller, G.E. Scott, R.T. Gudauska, E.S. Horner, N.W. Widstrom, D.L. Thompson, A.J. Bockholt and J.L. Brewbaker. 1983. Comparison of open-pollinated maize varieties and hybrids for preharvest aflatoxin contamination in the southern United States. Plant Disease 67:185-187.

Maize Plant Stress Inheritance

S.D. Jensen, Pioneer Hi-Bred International, York, Nebraska, and A.J. Cavalieri, Pioneer Hi-Bred International, Johnston, Iowa, USA

Abstract

Regression analysis of 339 commercial and experimental hybrids showed that 34% were significantly different from the mean for yield stability, the b value. In a group tested with over 500 replicates, 69% were significantly different from the mean, a surprising degree of variability. There were more values of <1.0 than >1.0, indicating stronger selection for stress tolerance than for high yield only. The effect on b values of testing in different regions of the country and in different years was investigated. Some crosses were stable, and others were not. The b value x year interactions were lower than those for b value x region, but both interactions appeared important. Paired stability regressions, paired comparisons at low, medium and high yield levels, and paired comparisons on controlled dryland versus irrigated conditions are discussed as methods of comparing the relative stress tolerance of hybrids. A study comparing 20 inbred lines crossed to four unrelated testers under dryland and irrigated conditions showed 13 out of 20 significantly different from the mean for combining ability and seven out of 20 significantly different from the mean for b values. A breeding scheme for drought-tolerant, single-cross hybrids is described. Procedures include utilizing elite inbred lines as source material; early testing and selection for nonadditive effects; side-by-side dryland and irrigated testing; and selection for b values as well as combining ability. It is suggested that estimation of the b values is important for the program's success.

Resumen

El análisis de regresión de 339 híbridos comerciales y experimentales indicó que el 34% difería bastante de la media en cuanto a la estabilidad del rendimiento, el valor b. En un grupo sometido a prueba en más de 500 repeticiones, se encontró que el 69% difería de manera significativa de la media, lo cual constituye un grado de variabilidad sorprendente. Se observaron más valores de <1.0 que de >1.0, lo cual indica que hubo una mayor selección para obtener tolerancia a la sequía que para obtener tan sólo alto rendimiento. Se investigó el efecto sobre los valores b de las pruebas en diferentes regiones del país. Algunas cruzas fueron estables y algunas no lo fueron. Las interacciones del valor b x año fueron inferiores a las del valor b x región, aunque ambas resultaron importantes. Se analizaron las regresiones apareadas de estabilidad, las comparaciones apareadas a niveles bajos, medios y altos de rendimiento, y las comparaciones apareadas sobre tierras áridas controladas frente a las condiciones de irrigación como métodos para comparar la tolerancia relativa a la sequía que presentan los híbridos. Un estudio en el que se comparaban 20 líneas endogámicas que se cruzaron con cuatro probadores no relacionados bajo condiciones de sequía y con irrigación indicó que 13 de las 20 variaban de manera significativa de la media en cuanto a la capacidad de combinación y que 7 de las 20 variaban significativamente de la media en cuanto a los valores b. Se describe un esquema de mejoramiento para híbridos simples tolerantes a la sequía. Los procedimientos incluyen el empleo de líneas endogámicas elite como material original; prueba y selección precoz para efectos no aditivos; pruebas colaterales en tierras áridas e irrigadas, y selección para obtener los valores b y la capacidad dc combinación. Se propone que la estimación de los valores b es importante para el éxito del programa.

This paper will deal with some experiences, covering more than 25 years, with selection and testing for drought tolerance in US Corn Belt maize. Genetic variability for drought tolerance in modern maize lines and hybrids will be considered, as well as the problems of measuring and capitalizing on that variability. A breeding scheme designed for the selection of drought-tolerant hybrids will be proposed, with the hope that this will generate ideas for incorporating selection for aflatoxin tolerance as well.

Soil water deficits, especially when accompanied by excessively high temperatures, are probably the most common yield-limiting factors in maize production. This is true in most maize-growing areas around the world, whether in the tropics or in temperate zones, and especially at the southern and western fringes of the US Corn Belt. Drought is a factor in maize production most years, even in the central part of the Corn Belt (8,9).

Despite the common occurrence of drought and the advances in breeding technology over the last half century, very little breeding for drought tolerance *per se* has taken place. Breeders have discovered that it is difficult to select for a trait when conditions necessary for its expression cannot be controlled. In the case of drought, yields can be too high or too low to be useful in selection. Error variances in drought-stressed environments are generally higher than normal, and differing rainfall and temperature patterns often lead to genotype x environment interactions that are higher than expected.

Breeders have good evidence that, despite the limited direct effort, modern maize hybrids are genetically superior to hybrids of past decades in their ability to withstand drought stress. Drought tolerance can be defined as the relative ability of a genotype to withstand drought stress compared to other materials being tested. Russell (10) and Duvick (1) both show that there has been considerable genetic gain for drought tolerance over years. Apparently drought stress has been an effective factor in most maize breeding and testing programs. Regression analysis is a useful tool for measuring such relative differences (2,3,4,5).

Phenotypic stability (b value) for a hybrid is the slope of the linear regression of its yield at a given location against the mean yield of all hybrids grown at that location. The mean yield of a hybrid is expressed as a percent of mean yield of the location to characterize its relative yield level. The average b value for the group of hybrids is 1.0. We will define a hybrid with $b > 1.0$ as one with lower than average stress tolerance, and one with $b < 1.0$ as higher than average. If low yield is the result of drought, then low b values coupled with above-average mean yields are indications of drought tolerance. Values of r^2 for the regression may also be indications of stability.

Genetic Variability for B Values among Maize Hybrids

Regression analysis was used to measure relative yield stability of genotypes over a range of environmental conditions. In experiments grown at locations scattered over most of the maize-growing areas of the US and Canada, 339 hybrids were compared over a three-year period (6). This group included commercial and experimental Pioneer hybrids, commercial hybrids from other private companies and crosses of important public lines.

Despite the select nature of these hybrids, considerable variation for yield stability existed among them. Of the 339 hybrids, 34% had b values significantly different from the mean. In a group that was tested in over 500 replicates and with a high degree of precision, 69% had b values that were significantly different from 1.0. The data also suggest that the distribution for b was skewed toward the low side, with 22% having b values < 1.0 and 12% > 1.0, a ratio of nearly two to one in favor of stress-tolerant types. The results suggest that standard testing programs have been effective in selecting stable, stress-tolerant hybrids more often than high-yield types only.

Twelve hybrids were selected to obtain a better understanding of the effects of geographical region and year on b values; these hybrids had been tested in over 600 replications at many sites over the USA for three years. The hybrids represented a range of b values and were grouped as b < 0.95, 0.96 < b < 1.04 and b > 1.05; four hybrids were in each group. The r^2 values ranged from .70 to .80 and indicated a good fit of the regression line in all cases.

The overall, regional and year effects for b value are shown in Table 1. Some hybrids were reasonably stable regardless of location, whereas others interacted with regions. Drought appeared to be a major yield-limiting factor in all regions, but other yield-eroding effects, such as temperature, relative humidity, diseases and insects, would be confounding elements in the response of hybrids to moisture stress. In these data, b x year interactions appeared to be somewhat less than b x region interactions. However, both types of environmental interactions are felt to be important.

Table 1. Phenotypic stabilities (b values) of 12 widely tested maize hybrids in different regions of the USA, 1979-1981

		Region			Year		
Maize hybrid	**Overall**	**East**	**Central**	**West**	**1979**	**1980**	**1981**
Hybrid with b > 1.05							
3323	1.21**	1.21**	1.19**	1.28**	1.14**	1.20**	1.28**
B73/Mo17	1.13**	.98	1.08*	1.26**	—	1.12**	1.12**
3541	1.07**	1.08	1.05*	1.06*	1.03	1.14	.95
3183	1.06**	1.05	1.05	.94	1.09**	1.03	1.10**
Hybrid with 0.96 < b < 1.04							
3389	1.04	1.13	1.08*	1.04	—	.99	.99
3780	.98	.97	.72**	1.03	.96	1.01	.96
3509	.97	1.03	.98	.97	1.03	1.02	.92
3377	.97	1.17	1.09*	.79**	—	1.04	.93*
Hybrid with b < 0.95							
3358	.93**	1.08	.90**	.82**	—	.94*	.97
3901	.93**	.70**	.71**	.94**	.94	.91**	.93**
3720	.90**	1.04	.77**	.90**	.96	.90**	.92**
3382	.93**	.96	.92**	.98	.89**	.98	1.01

* P > 0.95 that b is different than 1.0
** P > 0.99 that b is different than 1.0

The most critical comparison between two hybrids is obtained when both are grown together at the same locations. Such paired comparisons allow direct evaluation of two hybrids over a range of yield levels. Pioneer brand hybrids 3377 and 3378 are widely sold commercial hybrids. Figure 1 illustrates three techniques for documenting their individual responses to different yield levels. Figure 1a shows the paired regression values for the two hybrids. Comparisons of 1061 replications demonstrated an overall yield advantage of 0.32 t/ha for 3377. However, with a b value of 1.10 for 3377 and 0.96 for 3378, this difference is expected to increase at higher yield levels and be reduced or reversed at lower yield levels. Another technique for showing response to yield levels is presented in Figure 1b. The same two hybrids, in side-by-side tests, are compared at low-, moderate- and high-yielding sites. Conclusions are the same as observations obtained from the comparison of paired regression values, i.e., 3377 and 3378 yield nearly the same at low yield levels, but 3377 is superior at high yield levels.

A technique that has been used for several years to measure hybrid response to moisture stress under more controlled conditions is available in irrigated areas. A sketch of the testing procedure is shown in Figure 2. Water is withheld from the dryland test site until the desired level of drought is obtained. A potential problem with the system is that scheduling to obtain the desired stress level cannot be controlled. Consistent, improper scheduling could bias the selection to fit the watering pattern. However, multiple locations and years of testing with various rainfall patterns tend to prevent or at least reduce such a bias. Figure 1c shows an advantage for 3377 over 3378 under irrigation, but 3378 excells by a small margin under dryland conditions.

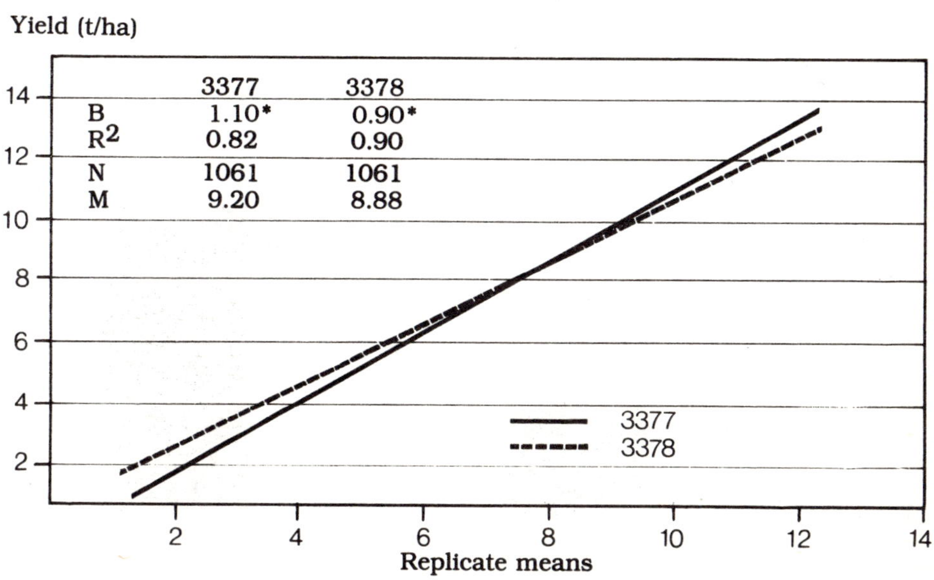

Figure 1a. Paired stability regressions of two hybrids over locations and years

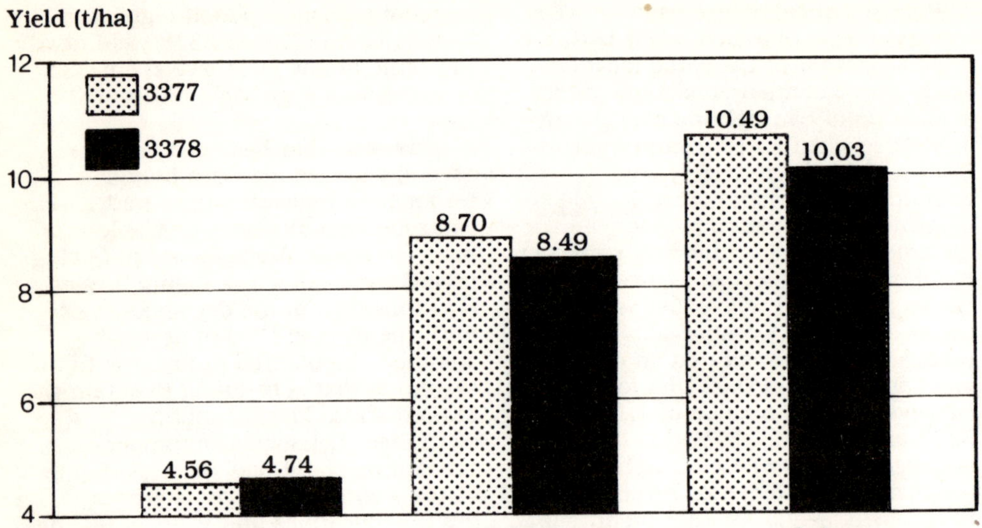

Rep mean	<6.5	6.8<9.4	>9.4
Reps	80	189	112
Diff	.18	.21	.44
Prob/sig/diff	.173	.029	.001

Figure 1b. Paired comparisons of two hybrids at three yield levels over locations and years

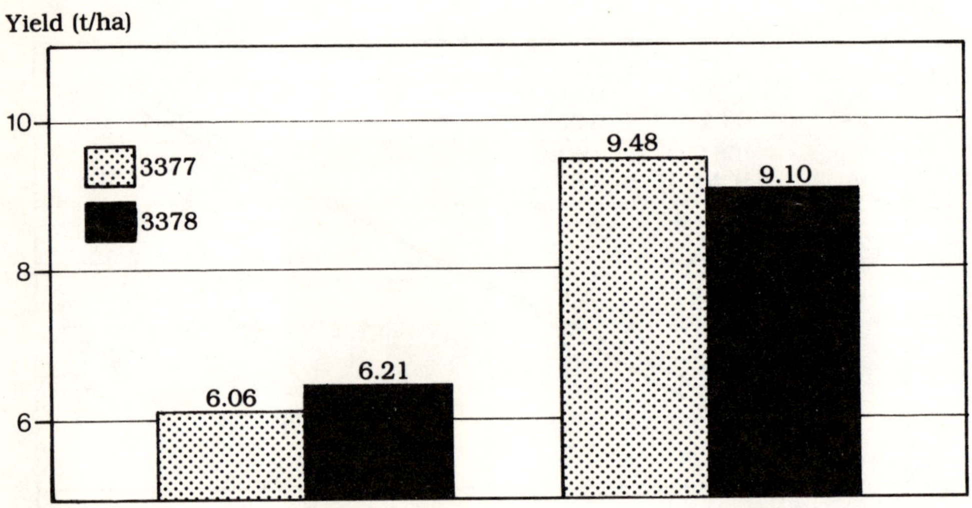

Area	Dryland	Irrigated
Reps	127	280
Diff	.13	.38
Probs/sig/diff	.233	.000

Figure 1c. Paired comparisons of two hybrids grown under dryland and irrigated conditions over locations and years

Genetic Variability for B Value Among Inbred Lines

Information is needed on the variability among inbred lines of maize for b, and whether inbred lines can be classified as above or below average in their ability to transmit stability in crosses. For that reason, two experiments were conducted during the years 1982, 1983 and 1984 in south central Nebraska. Ten elite lines of stiff-stalked origin, each crossed to four nonstiff-stalked testers, were grown at two locations under dryland and irrigated conditions. In a second experiment, 10 elite nonstiff-stalked lines crossed to four stiff-stalked testers were tested in the same manner. The average yield, b and r^2 values are shown in Table 2 for the two groups of lines. Yield values for the stiff-stalked lines are listed as percents of the test mean; they range from 90 to 104. Eight of the 10 lines are significantly different from the mean. For b value, much larger differences are required for significance, and only two of the 10 are significantly different from 1.0. The r^2 values are quite high, ranging from .80 to .90. Although the range in yield differences is about the same for the nonstiff-stalked lines, only five of the 10 are significantly different from the mean for yield, and five of the 10 are statistically different from the mean for b. The r^2 values again are high. In the two experiments, 13 out of 20 lines were significantly different from the mean for yield, and seven out of 20 were significantly different from the

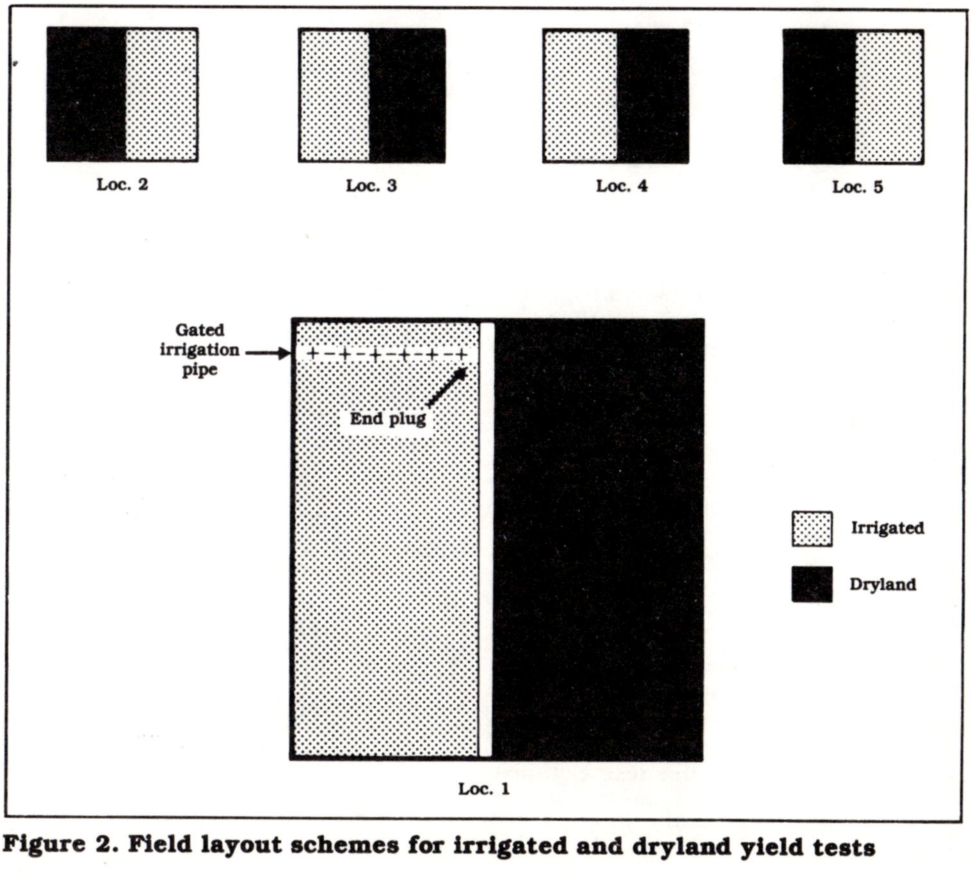

Figure 2. Field layout schemes for irrigated and dryland yield tests

mean for b. The data suggest that an estimate of b as well as combining ability is important in evaluating inbred line performance. Knowledge of b is also important in evaluating a line as a source of stress tolerance in a breeding program.

A Breeding Scheme for Drought-Tolerant, Single-Cross Maize Hybrids

A system for identifying drought-tolerant, single-cross maize hybrids has been developed and used with reasonable success for a number of years (Figure 3). Some of the important concepts of this program are:

- Elite inbred lines are utilized as genetic material;
- A relatively high percentage of the total genetic variance surrounding such elite germplasm is nonadditive in nature (7); and
- Exploitable variability for b as well as for combining ability is available.

Selection is based on a sequential scheme of five years.

Year 1

One hundred S_1s, 50 from stiff-stalked synthetics and 50 from nonstiff-stalked origins, are grown under controlled drought conditions and advanced from S_1 to S_2. The most elite lines available are involved in these S_1 populations. An understanding of b for each line would be helpful, but such information is generally not available. Approximately 240 plants of each S_1 are grown for selfing and advancing to S_2. A total of 1000 S_2 ears is the goal at harvest.

Year 2

1000 S_2s are testcrossed to an appropriate elite unrelated tester in an isolated crossing block. A second replication of these S_2s is grown in a dryland yield trial. In this test, S_2 lines can be observed for agronomic traits and harvested or observed for yield. Twenty percent (200) are selected for topcross testing in the following season. The S_2s are not yet advanced to S_3s.

Year 3

This year features yield testing of 200 S_2 testcrosses. There are two locations of controlled dryland and two locations of irrigated testing, with two replications per location. The S_2 lines are also advanced to S_3. Forty S_3 lines are selected, based on performance under both dryland and irrigated conditions, for advancement in the winter.

Table 2. Combining ability and b values of elite inbred maize lines

Line	Yield (% $\bar{x}$)	B value	r^2
Group I, stiff-stalked lines			
SS1	99	.85**	.80
SS2	101	1.03	.87
SS3	90**	.93	.86
SS4	102*	.99	.86
SS5	102*	.97	.86
SS6	104**	1.04	.88
SS7	98*	1.09	.87
SS8	98*	1.03	.90
B73	103**	1.12**	.90
B84	103**	.95	.86
	$\bar{x}$ = 7.58 t/ha		
Group II, nonstiff-stalked lines			
NSS1	99	.88*	.82
NSS2	98*	1.21**	.86
NSS3	100	.86*	.79
NSS4	101	.96	.85
MO17	94**	1.34**	.90
NSS5	99	.86*	.80
NSS6	100	1.10	.88
NSS7	102*	.90	.80
NSS8	104**	.91	.85
NSS9	104**	.99	.85
	$\bar{x}$ = 7.57 t/ha		

* P>0.95
** P>0.99

Year 3 (winter)
Forty S_3 families, each with two subfamilies, are advanced to S_4 and crossed to two unrelated elite testers.

Year 4
The S_4 lines from the winter nursery are advanced to S_5, and S_3 testcrosses are tested under dryland and irrigated conditions. Lines with good combining ability and b values are selected.

Year 4 (winter)
Depending on performance relative to elite checks, up to eight S_5 lines are selected, advanced to S_6 and crossed to four or more testers.

Year 5
The selected lines are advanced to S_7, and single crosses are tested under dryland and irrigated conditions. Combining ability and b values are calculated. Selected lines are recombined among themselves and with other elite lines that have been contributed from other programs.

One of the negative aspects of this selection program is the long cycle time. Due to high genotype x location and genotype x year interactions, the utilization of three years of testing for combining ability and b value is suggested before recombining. In some environments where drought stress is more predictable, this might be reduced to two years. It should be pointed out, however, that new breeding populations can be formed each year with the most current elite set of lines available.

It is felt that future success of this procedure will be dependent not only on successful selection within the program, but also on the careful integration of materials from other programs. Genetic materials from broad-based synthetic varieties, or other pedigree selection programs in which drought stress is not emphasized, would provide good sources of test materials.

Figure 3. Selection scheme for the development of drought-resistant, single-cross maize hybrids

Year	Procedure	No.	Testing	Reps Dry	Reps Irrig.	No. of plots
1	S1 → S2	100 F2s				
2	TX S2s in Iso	1000	S2 *per se*	1		1000
3	S2 → S3	200	S2 x tester[a]	6		1200
3 (winter)	S3 → S4 S3 x 2 testers	40				
4	S4 → S5	80 (40x2)	S3 x 2 testers	4	4	1280
4 (winter)	S5 → S6 S5 x 4 testers Code	8				
5	S6 → S7	8	S5 x 4 testers	4	4	256

[a] Testers are always elite inbred lines

References

1. Duvick, D.N. 1977. Genetic rates of gain in hybrid maize yields during the past 40 years. Maydica 22:187-196.

2. Eberhart, S.A., and W.A. Russell. 1966. Stability parameters for comparing varieties. Crop Science 6:36-40.

3. Finlay, K.W. 1964. Adaptation: Its measurements and significance in barley breeding. *In* Barley Genetics I, Proceedings of the First International Barley Genetics Symposium, Wageningen, Netherlands, 1963. Pp. 351-359.

4. Finlay, K.W. 1971. Breeding for yield in barley. *In* Barley Genetics II, Proceedings of the Second International Barley Genetics Symposium, 1969. Washington State University Press, Pullman, Washington, USA. Pp. 338-345.

5. Finlay, K.W., and G.N. Wilkinson. 1963. The analysis of adaptation in a plant breeding programme. Australian Journal of Agricultural Research 14:742-754.

6. Jensen, S.D., and A.J. Cavalieri. 1983. Drought tolerance in U.S. Maize. *In* Plant Production and Management under Drought Conditions, J.F. Stone and W.O. Willis, eds. Elsevier Science Publishers, Amsterdam, Netherlands.

7. Jensen, S.D., W.E. Kuhn and R.L. McConnell. 1983. Combining ability studies in elite U.S. maize germplasm. *In* Proceedings of the Annual Corn Sorghum Research Conference 38:87-96.

8. National Climatic Center. 1982. Palmer Index Environmental Data Service. NOAA, Ashville, North Carolina, USA.

9. Newman, J.E. 1978. Drought impacts on American agricultural productivity. *In* North American Droughts, N.J. Rosenberg, ed. AAAS Selected Symposia Series 15. Westview Press, Boulder, Colorado, USA. Pp.43-62.

10. Russell, W.A. 1974. Comparative performance for maize hybrids representing different eras of maize breeding. *In* Proceedings of the Annual Corn and Sorghum Research Conference 29:81-101.

Systemic Infection of Maize Plants by *Aspergillus flavus*

S.M. Kelly, Department of Plant Pathology, and J.R. Wallin, Agricultural Research Service, US Department of Agriculture, and Department of Plant Pathology, both of the University of Missouri, Columbia, Missouri, USA

Abstract

Systemic growth of Aspergillus flavus *Link ex Fries in plant tissue has been demonstrated by recovering the fungus from material sampled from three different parts of the maize (*Zea mays *L.) plant and at three different stages of growth. Inoculation with* A. flavus *was done by decapping kernels, surface disinfecting, plating with an inoculum density of 3 x* 10^3 *conidia/ml of the fungus and incubating 13 days at 26°C. Uninoculated checks were included. Following incubation, the kernels were again surface disinfected and scrubbed free of all visible mycelia and conidia, planted in steam-sterilized soil and grown in sterile growth chambers. Germination rate and seedling survival rate were recorded. Tissues were sampled, surface disinfected and plated on* Aspergillus *Differential Media (ADM). Characteristic orange-pigmented colonies were observed growing on the ADM, indicating the presence of the fungus in the various plant tissue segments sampled. Significant differences in the distribution of* A. flavus *colonies occurred among the leaf, stem and root sections sampled at the three-leaf stage; differences in the distribution of the fungal colonies also occurred among the tissue samples obtained at different stages of growth as represented by the leaf, stem and root sections.*

Resumen

Se ha demostrado la existencia de crecimiento sistémico de las esporas de Aspergillus flavus *Link ex Fries en tejidos vegetales mediante la recuperación del hongo en material obtenido de tres partes diferentes de la planta de maíz (*Zea mays *L.) en tres etapas diferentes de crecimiento. La inoculación de* A. flavus *se efectuó cortando la parte superior de los granos, desinfectando la superficie, recubriéndola con una densidad de inóculo de 3 x* 10^3 *conidios/ml del hongo e incubando durante 13 días a 26°C. Se incluyeron testigos no inoculados. Después de la incubación, se volvió a desinfectar la superficie de los granos y se les cepilló hasta eliminar todos los micelios y conidios visibles; se les sembró en tierra esterilizada con vapor y se les cultivó en cámaras de crecimiento estériles. Se registró el índice de germinación y el índice de supervivencia de plántulas. Se muestrearon los tejidos, se les desinfectó la superficie y se les recubrió con medio diferencial de cultivo de* Aspergillus. *Se observó que las colonias características, pigmentadas de color naranja, crecían en este medio, lo cual indica la presencia del hongo en los segmentos de los diversos tejidos que se muestrearon. Se observaron diferencias significativas en la distribución de las colonias de* A. flavus *entre las secciones de hoja, tallo y raíz muestreadas en la etapa de tres hojas; también se observaron diferencias en la distribución de las colonias de hongos entre las muestras de tejidos obtenidas en diferentes etapas de crecimiento y representadas por las secciones de hoja, tallo y raíz.*

This study was conducted to investigate the nature of colonization, as well as the coexistence of maize (*Zea mays* L.) and *Aspergillus flavus* Link ex Fries, previously classified as a saprophytic fungus. Field stress of maize plants enhances *A. flavus* invasion in developing ears. Colonization of kernels damaged by feeding birds and insects, or senescing silks (4,5,6,7,9), appear to be avenues of entrance of the fungus.

Misra and Tripathi (8) demonstrated that maize seeds infected with aflatoxin-producing *A. flavus*, naturally contaminated with aflatoxins or treated with purified toxin resulted in reduced germination. Brodnik *et al.* (2) compared the toxic effects of *A. flavus* metabolites with those of aflatoxin B_1 on maize embryo growth and found the toxic metabolites had a more pronounced effect than those of B_1.

Schoental and White (10) and Slowatizky *et al.* (11) observed virescent leaf patterns on maize seedlings grown from seeds infected with *A. flavus*. Gardner and Wallin (3) subjected four inbred lines to *A. flavus* and B_1 and noted their response. Kernels were subjected to a nonaflatoxin-producing strain (NRRL 5565), a toxin producer (NRRL 3357), NRRL 5565 plus toxin, toxin *per se* and a nontreated control. Seedling emergence and plant height were found to be the best variables for measuring treatment effects and inbred line differences.

Given the results of the studies discussed in the preceding paragraphs, the investigators attempted to determine whether *A. flavus* infects the maize seedling systemically from a contaminated kernel, and how the distribution of the live fungus occurs in leaf, stem and/or root tissues. This was observed with scanning electron microscopy (SEM), light microscopy and fungal isolations from infected maize tissue.

Materials and Methods

In 1982, a preliminary experiment was conducted to develop a technique, i.e., a method of treatment and sampling, and to estimate the plant tissue sample size. Treated and nontreated plants were grown to maturity, and plant tissue and kernels were assayed for *A. flavus*.

Two methods of treatment were compared to determine the best inoculation technique. Kernels of *Zea diploperennis* (Iltis, Doebley and Guzman) were inoculated in petri dishes with 2 x 10^3 conidial/ml aqueous suspension of a toxin-producing strain of *A. flavus*, NRRL 3357. The samples were incubated at 26°C for eight days. Twelve percent of the treated seeds germinated and survived. The second method used kernels of the inbred line Mo17 and the cross Mo17 x *Z. diploperennis* that were planted in soil and then saturated with *A. flavus* conidial suspension at 2 x 10^4 conidial/ml. Sixty-one percent of the Mo17 kernels and 67% of the Mo17 x *Z. diploperennis* kernels germinated.

The treated and untreated control plants were transplanted into 15-cm clay plots with steam-sterilized soil and placed in the greenhouse. At physiological maturity, the tissue from the stem and leaves of the lower, middle, and top area of each plant, and the ear, including the cob, were harvested, surface sterilized with a 1% sodium hypochlorite (NaOCl) solution for 90 seconds and rinsed twice with sterile, distilled water. (Preliminary experiments showed that soaking maize tissue for 90 seconds in 1% sodium hypochlorite, followed by two rinses in sterile, distilled water, removed essentially all surface contaminants, but did not kill *A. flavus* within the plant tissue.) Leaf tissue is difficult to surface disinfect because the waxy cuticle is not readily wetted. Tissues were rubbed to reduce surface tension. The disinfected samples were plated on *Aspergillus* Differential Media (ADM), a media developed for rapid identification and enumeration of *A. flavus* and related fungi (1). Some of the plants failed to produce kernels, but those kernels produced were surface disinfected, split longitudinally with an alcohol-flamed razor blade and placed cut side down on ADM. After three days at 26°C, the *A. flavus* colonies were counted.

The experimental techniques described in the previous paragraphs were repeated in 1983. Kernels from three ears of an open-pollinated white maize originating from a bulk of commercial white-maize hybrids were used. One hundred surface-disinfected, decapped kernels were inoculated in sterile petri dishes with 3 x 10^3 conidial/ml and incubated at 26°C for 13 days. Kernels were decapped to enhance infection. An equivalent number of controls were incubated in sterile, distilled water.

Kernels that germinated in the petri dishes were counted and individual seedlings were hand-scrubbed and surface disinfected with a 1% NaOCl solution to remove visible mycelia and conidia. The treated and untreated control seedlings were planted in individual 7.5-cm peat pots containing steam-sterilized soil and placed in growth chambers (disinfected with approximately 2.5% NaOCl) under incandescent and fluorescent lights. The lighting was timed for photoperiods of 14 hours of light and ten hours of darkness with resulting temperatures of 29° and 24°C, respectively. An Andersen six-stage viable particle sampler was operated in the chambers periodically to monitor the extent of airborne conidia. The peat pots were watered from below to reduce contamination from one pot to another by splashing.

Survivors were counted during the first week after planting. Eleven treated and 33 untreated plants were selected at the three-leaf stage and sectioned into eight pieces. Pieces one to three consisted of the leaves, pieces four to six consisted of the stem, and seven and eight, the roots and hypocotyl. Each piece was surface disinfected with a 1% solution of NaOCl for 90 seconds, rinsed twice in sterile, distilled water and dissected into smaller pieces with an alcohol-flamed scalpel before plating on ADM. After three days at 26°C, the plates were removed from the incubator and tissue samples were examined for the presence of viable *A. flavus*.

Another group of 11 treated and 33 untreated control plants was selected in the four-leaf stage, surface disinfected and plated on ADM. The last group to be assayed consisted of ten treated and 32 untreated plants in the five-leaf stage.

Positive samples were prepared for the scanning electron microscope (SEM), using ethanol cryofracture techniques. Samples were fixed in 1% aqueous osmium tetroxide and prepared for SEM by dehydration in a graded absolute ethanol series with three changes for 30 minutes each. Specimens were plunged into liquid nitrogen, fractured with a chilled razor blade, replaced in absolute ethanol and critical-point dried in carbon dioxide. Fragments were mounted on specimen holders with a piece of electrically conductive tape and placed vertically with the adhesive side up. Specimens were spatter coated with two angstroms of gold paladium and viewed in a JEOL JSM 35U scanning electron microscope.

Samples were viewed at 600X and 3000X magnification to observe propagule distribution. Tissue was set aside that contained *A. flavus* and was cleared of chlorophyll by boiling in a 1:1 glacial acetic acid and ethanol. The transparent tissue was then stained with lactophenol cotton blue and viewed under 200X magnification with a light microscope.

Results

Plants of inbred Mo17, Mo17 x *Zea diploperennis*, and *Z. diploperennis* that survived *A. flavus* inoculation were grown to maturity and the leaves sampled for the presence of live systemic *A. flavus*. *Aspergillus flavus* was detected particularly in leaves from the middle areas and top portions of mature plants, but was not isolated from the control leaves. Only a few ears were assayed, but *A. flavus* was recovered from 89% of the kernels plated from those ears. Minimal contamination of other fungi was observed on the plates. The preliminary study showed that *A. flavus* grew systemically and that decapped kernel inoculation in petri dishes was an efficacious technique for inoculating kernels.

In a second experiment, 76% of the *A. flavus*-inoculated kernels germinated in petri dishes compared to 98% for the control. Although *A. flavus* is commonly thought to be a saprophyte, it appears to be a weak pathogen, since it caused seedling death for 44% of the 76% germinated seedlings soon after planting in peat pots.

Data were grouped in two classes according to presence or absence of *A. flavus* growth obtained from tissue samples. A row-by-column Chi-square test of contingency was computed. Originally, the plant was divided into eight sections. The X^2 test revealed that some of the expected values (row frequency x column frequency/sample size) were less than 1. Since the expected values of the individual pieces (1 to 8) were less than 1, the data were combined: 1 to 3 represented leaf area, 4 to 6 represented the stem and 7 and 8 represented the roots and hypocotyl. Another row-by-column Chi-square was computed on the combined pieces.

Table 1 is a summary of six row-by-column X^2 tests. The X^2 values found in the last row of the table represent three of the six row-by-column X^2s and test the presence or absence of *A. flavus* in the plant part for each leaf stage. The other three row-by-column X^2 values found in the last column of the table represent presence or absence

Table 1. Distribution of *Aspergillus flavus* recovered in plant pieces sampled at the three-, four- and five-leaf stages of growth

Plant piece	Three-leaf stage (%)	Total no. samples	Four-leaf stage (%)	Total no. samples	Five-leaf stage (%)	Total no. samples	X^2 a/
Leaves	92.3	13	85.7	21	20.6	34	31.50 b/
Stems	46.7	15	85.7	21	33.3	33	14.34 b/
Roots	80.0	10	85.7	14	39.1	23	9.79 b/
X^2 c/	7.58 b/		0 d/		2.53 d/		

a/ Row X^2 lists testing differences between percentages due to stage of development
b/ Significant at P = < 0.05
c/ Column X^2 lists testing differences between percentages due to plant pieces
d/ Not significant

of *A. flavus* at different leaf stages for each plant part. Values are the percentages of *A. flavus* present and total number or tissue pieces sampled. The analyses revealed that significant differences for presence or absence of *A. flavus* existed between the leaf, stem and root pieces sampled during the three-leaf stage of development. *Aspergillus flavus* was located 80% of the time in the root area, but 92.3% of the pieces sampled indicated the fungus was located in the leaf tissues. In the four-leaf stage, the distribution of the fungus was equal among leaf, stem and root areas within the plant, but in the five-leaf stage the distribution of the fungus appeared to shift to the root and lower areas of the plant. When results of the different sampling stages were pooled, the overall distribution of *A. flavus* in leaves (54.4%), stems (50.3%) and roots (61.7%), after pooling the different stages of sampling, did not show significant differences.

Comparisons between the developmental stages sampled were also made using a row-by-column Chi-square test of contingency. The row Chi-square test showed significant differences in all areas of the plants due to stage of development. The largest amount (92.3%) of *A. flavus* occurred in the leaves at the three-leaf stage of development and only a slight decrease was observed in the four-leaf stage; a significant decrease occurred in the leaves at the five-leaf stage.

The stem area analyses indicated a shift in the distribution of the fungus. The stems of the three-leaf stage showed 46.7%, increasing significantly to 85.7% at the four-leaf stage; the percentage fell to 33.3% for the five-leaf stage.

The root area in the three-leaf stage revealed 80.0% fungal recovery and the four-leaf stage, 85.7%. This dropped to 39.1% in the five-leaf stage. In all instances, a decrease in percent distribution of the fungus in all areas of the plant in the five-leaf stage may be explained by the physiological maturing of the plant, resulting in a stronger resistance response.

The overall distribution of *A. flavus* in the different leaf stages sampled after pooling the leaf, stem and root pieces indicated significant differences among the different stages of maturity of the plant. Data from the three-leaf stage indicated 71.1% fungal recovery, with an increase to 85.7% observed for the four-leaf stage. Again the five-leaf stage was much lower (30.0%).

The control plants were used for comparing plant height, germination rate and appearance. These plants were healthy and vigorous, with very uniform characteristics, i.e., plant height, vigor and deep green color.

Intercellular hyphae were observed within the leaf tissue under the SEM as shown in Figure 1. Since hyphae were seen in the treated sample and none in the untreated samples, the probability was high that the hyphae shown in the

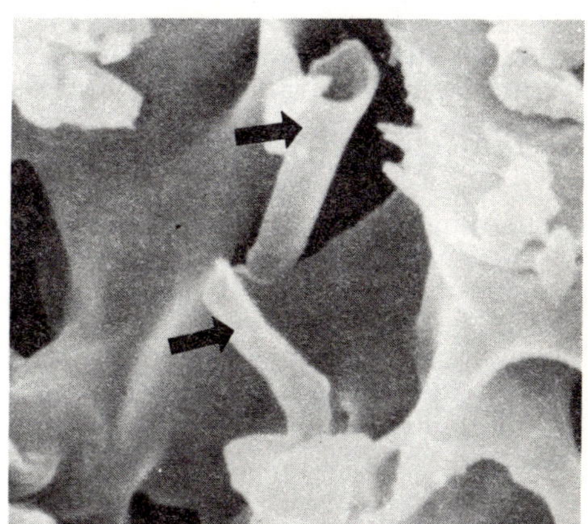

Figure 1. Scanning electron micrograph of hyphae in intercellular space in corn leaf tissue (3000 x magnification)

photo were *A. flavus*. Recommended staining techniques were employed, but no fungal hyphae were observed in sections from tissue samples prepared for light microscopy. Presumably the fungus is present only in trace amounts, barely subsisting in intercellular spaces. But the results of these experiments indicate that *A. flavus* can be present in a developing maize plant originating from infected seed.

Conclusions

Aspergillus flavus was shown to become systemic in young maize seedlings grown from contaminated seed. The distribution of the organism was highly variable. Germination rates were considerably lower when *A. flavus* entered and contaminated the seed. The isolate appeared to cause seedling death. Distribution of the fungus within the plant suggested that initially the organism may follow the meristem of the plant. As the plant matured, the incidence and distribution of the fungus was suppressed, suggesting possible host tolerance as the plant matures.

Future studies will deal with (1) the inclusion of different genotypes to determine if differences in resistance or susceptibility occur among them; (2) the monitoring of plant and fungal development throughout the life cycle of the plant to determine if this might be a mode of infection for preharvest contamination of maize kernels; and (3) the investigation of this assay as a feasible screen for resistance, eliminating the necessity for the inoculation techniques presently used in field studies and the uncertainty of simulating natural preharvest infection.

References

1. Bothast, R.J., and D.I. Fennell. 1974. A medium for rapid identification and enumeration of *Aspergillus flavus* and related organisms. Mycologia 66:365-369.

2. Brodnik, T., N. Klemenc, P. Vospernick and J. Zust. 1978. Influence of toxins from maize infected by *Aspergillus flavus*, *Penicillium rubrum*, and *Fusarium graminearum* and of aflatoxin B, rubratoxin A and toxin F-2 on maize embryo growth. Seed Science Technology 6:965.

3. Gardner, C.A.C., and J.R. Wallin. 1981. Relationship of fungus in the *Zea mays-Aspergillus flavus* host-pathogen system. Paper presented at the American Phytopathology Society Annual Meeting. Abstract 171.

4. Garman, H., and H.H. Jewett. 1914. The life history and habits of the corn earworm (*Chloridea obsoleta*). Kentucky Agricultural Experiment Station Bulletin 187:1118-1123. Lexington, Kentucky, USA.

5. Jones, R.K., H.E. Duncan and P.B. Hamilton. 1981. Planting date, harvest date, and irrigation effects on infection and aflatoxin production by *Aspergillus flavus* in field corn. Phytopathology 71:810-816.

6. Jones, R.K., H.E. Duncan, G.A. Payne and K.J. Leonard. 1980. Factors influencing infection by *Aspergillus flavus* in silk-inoculated corn. Plant Disease 64:859-863.

7. McMillian, W.W. 1983. Role of arthropods in field contamination. *In* Aflatoxin and *Aspergillus flavus* in Corn, U.L. Diener, R.L. Asquith and J.W. Dickens, eds. Southern Cooperative Series Bulletin 279. Alabama Agricultural Experiment Station, Auburn, Alabama, USA.

8. Misra, R.S., and R.K. Tripathi. 1980. Effect of aflatoxin B in germination, respiration and alpha-amylase in maize. Plant Disease Protection 87:155-160.

9. Rambo, G.W., J. Tuite and P. Crane. 1974. Preharvest inoculation and infection of dent corn ears with *Aspergillus flavus* and *A. parasiticus*. Phytopathology 64:797-800.

10. Schoental, R., and A.F. White. 1965. Aflatoxins and albinism in plants. Nature 205:57. London, England.

11. Slowatizky, I., A.M. Mayer and A. Polijakoff-Mayber. 1969. The effect of aflatoxin on greening of etiolated leaves. Israel Journal of Botany 18:31-36.

Relation of Insects to Aflatoxin Contamination in Maize Grown in the Southeastern USA

W.W. McMillian, Insect Biology and Population Management Research Laboratory, Agricultural Research Service, US Department of Agriculture, Tifton, Georgia, USA

Abstract

Insect damage and aflatoxin contamination are chronic problems in maize (Zea mays *L.) grown in the southeastern USA. Research has demonstrated that insects significantly increase aflatoxin levels in preharvest ears. Insects transport* Aspergillus flavus *Link spores, as well as damaging kernels, which results in increased fungal infection. Infection by* A. flavus *appears to be associated more with foliage-feeding insects, whereas* A. parasiticus *appears to be associated more with soil-inhabiting insects. Maize weevils* (Sitophilis zeamais *Motschulsky) appear to be more important in increasing kernel infection and aflatoxin contamination than the fall armyworm* [Spodoptera frugiperda *(J. E. Smith)], the European corn borer* [Ostrinia nubilalis *(Hübner)], or the corn earworm* [Heliothis zea *(Boddie)]. The application of insecticides to silks of developing ears generally reduces insect damage and aflatoxin contamination. Heavy morning dew, which is common in the southeastern states, appears to favor aflatoxin development. Insect damage and aflatoxin contamination are positively correlated with mean temperature in the field during plant development. Aflatoxin contamination and net evaporation are also positively correlated.*

Resumen

El daño producido por insectos y la contaminación por aflatoxinas representan problemas crónicos en el maíz (Zea mays *L.) que se cultiva en el sureste de Estados Unidos. La investigación efectuada ha demostrado que los insectos hacen que aumenten en forma significativa los niveles de aflatoxina en las mazorcas antes de la cosecha. Los insectos transportan esporas de* Aspergillus flavus *Link ex Fries y también dañan los granos, lo cual da por resultado un incremento en la infección por hongos. Al parecer, la infección con* A. flavus *se asocia más con insectos que se alimentan del follaje, en tanto que la infección con* A. parasiticus *parece relacionarse más con insectos que viven en el suelo. Los gorgojos del maíz* (Sitophilis zeamais *Motschulsky) parecen ser más importantes para aumentar la infección de los granos y la contaminación por aflatoxinas que el gusano cogollero* (Spodoptera frugiperda *[J.E. Smith])), barrenador europeo del maíz* (Ostrinia nubilalis *[Hübner]), o que el gusano elotero* (Heliothis zea *[Boddie]). La aplicación de insecticidas a los estigmas de las mazorcas en desarrollo suele reducir el daño ocasionado por los insectos y la contaminación por aflatoxinas. Al parecer, el abundante rocío matinal, común en los estados del sureste, favorece el desarrollo de aflatoxinas. El daño por insectos y la contaminación por aflatoxinas se relacionan en forma positiva con una temperatura media en el campo durante el desarrollo de la planta. Asimismo, la contaminación por aflatoxinas y la evaporación neta se relacionan en forma positiva.*

Recently, Dr. T. Kinney, Jr., Administrator for the Agricultural Research Service of the US Department of Agriculture, stated that one of the prime challenges facing agricultural research today is to find ways to improve crop quality and to reduce the yield losses in maize as a result of grain invasions by insects and fungi (4). Perhaps the best justification for Dr. Kinney's remarks was the devastating insect damage, *Aspergillus flavus* Link infection and aflatoxin contamination inflicted on the 1977 maize crop grown in the southeastern USA (8).

In 1977, the Maize Host Plant Resistance and Aflatoxin Contamination Project was initiated at the Agricultural Research Service Laboratory at Tifton, Georgia, with three overall objectives:

- Investigate the interactions of insects, plants, fungi and aflatoxins;
- Identify and develop maize germplasm with resistance to insect damage, *A. flavus* infection and aflatoxin formation; and
- Develop management practices that reduce crop losses, production costs and environmental pollution.

This presentation will center mainly on results of research conducted at the Tifton laboratory during the past eight years.

Over the last several years, preharvest maize in Georgia has experienced an increase in insect damage. During the 1930s and 1940s, estimated yield losses in maize due to ear damage by a complex of insects, especially the corn earworm [*Heliothis zea* (Boddie)], averaged 10%. During the 1950s and 1960s, these losses declined to about 2% a year. In the 1970s, however, yield losses began to increase, and by the early 1980s, losses had risen to 6% annually (11,21).

Increased losses in maize from insect damage are probably due to a combination of factors. New, early maturing hybrids with loose, open husks and quick dry-down attributes have recently been introduced in the southeast and have probably contributed substantially to the increase in insect damage. Also, infestations by some insect species have occurred earlier in the season than in past years. The increased acreage of irrigated maize and earlier planting dates may contribute to early-season increases in insect populations that migrate to and severely damage late-planted, nonirrigated maize.

As early as 1920, Taubenhaus (15) published the first paper that specifically associated *A. flavus* infections with insect injury in maize. In 1960, Walsh and Riley (16) reported that a greenish-brown fungus (probably *A. flavus*) grew on the developing ears of maize that had previously been damaged by the earworm. They stated that "the dampness from the exudated sap of injured kernels favored the growth of fungi." When aflatoxins were identified in the 1960s, insect damage was immediately associated with contamination (1,3,14,19). The role that insects play in contributing to aflatoxin contamination in preharvest maize was reviewed by Widstrom in 1979 (17) and updated by McMillian in 1983 (8).

Laboratory studies have demonstrated that *A. flavus* infection and aflatoxin development in maize kernels can occur without insects and the damage they inflict (13). However, a variety of field experiments have demonstrated that insects significantly enhance the aflatoxin problem by transporting fungal spores as well as by damaging kernels.

Research has shown a broad, interregional occurrence of *A. flavus*-contaminated insects on maize plants (8). Insect contamination can be both internal and external. *Aspergillus flavus* appears to be associated more with foliage-feeding insect species, whereas *A. parasiticus* appears to be associated with soil-inhabiting insects (6). An eight-year evaluation of earworm moths captured in Tifton, Georgia, maize fields during the growing season revealed that contamination with *A. flavus* ranged from 30% of the moth population in May, when plants were in the seedling stage, to 70% of the population in August, when plants were mature. Incidence of the fungus fluctuated from 15% in 1982 to 84% in 1981. In general, the level of aflatoxin contamination in preharvest maize appears to follow the degree of earworm moth contamination with *A. flavus*.

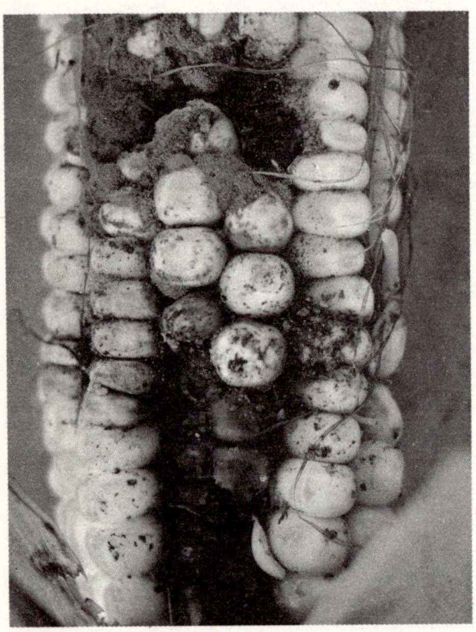

Aspergillus flavus infection of maize ear after kernel damage by the corn borer.

In field tests, maize weevils (*Sitophilus zeamais* Motchulsky) were exposed to spores of toxin-producing *A. flavus* NRRL 3357 and *A. parasiticus* ATCC 24690, and applied to silks of developing ears; fungal spores were also applied to the silks. The study demonstrated that certain combinations of maize weevils and fungal isolates were significantly more effective in increasing levels of infection and aflatoxin contamination (W.W. McMillian, unpublished data). Specifically, the combination of *A. flavus* spores and weevils resulted in 81% kernel infection, compared to 24% kernel infection for the *A. parasiticus* spores and weevils combination. *Aspergillus flavus* spores applied to silks with no weevils resulted in 61% kernel infection, compared to 16% kernel infection for the treatment of *A. parasiticus* spores with no weevils. Aflatoxin levels were considerably higher in the *A. flavus* spores plus weevils treatment (1287 ng/g^{-1}) than in the *A. parasiticus* plus weevils treatment (412 ng/g^{-1}). Aflatoxin levels for *A. flavus* without weevils were 299 ng/g^{-1} and for *A. parasiticus*, 220 ng/g^{-1}.

Surveys of preharvest maize growing in randomly selected fields in Georgia during the last eight years (1977 to 1984) revealed that the percent of ears with visible insect damage ranged from 55% in 1979 to 100% in 1977 (W.W. McMillian, unpublished data). A corresponding grain yield loss of 20% in Georgia in 1977 compared with a 2% grain loss in 1979. During the same period, average aflatoxin contamination ranged from 37 ng/g^{-1} in 1984 to 622 ng/g^{-1} in 1977. It was noted each year that regardless of the severity of aflatoxin contamination, a few fields had no aflatoxin or very low levels. For example, in 1977 some fields produced grain samples with up to 4708 ng/g^{-1} of aflatoxin, whereas samples from neighboring fields contained only 40 ng/g^{-1} of aflatoxin. Double checking ruled out sampling error; therefore, it appeared that certain field conditions and/or crop management practices could have been responsible for minimizing insect damage as well as aflatoxin contamination.

Some of the first studies at Tifton identified maize insect species in Georgia that contributed to *A. flavus* infection and aflatoxin development under field conditions. The primary species that traditionally inflict economic damage to preharvest ears in the area are the corn earworm, the fall armyworm [*Spodoptera frugiperda* (J.E. Smith)], the European corn borer [*Ostrinia nubilalis* (Hübner)] and the maize weevil. A three-year field study in which developing maize ears were infested with these insects (previously exposed to *A. flavus* spores) demonstrated that the maize weevil increased the aflatoxin level (299 ng/g^{-1}) in maize more than other insect species tested (W.W. McMillian, unpublished data). Grain from plots infested with corn earworm (55 ng/g^{-1}), fall armyworm (77 ng/g^{-1}) and European corn borer (65 ng/g^{-1}) was about equally contaminated with aflatoxin but was significantly more contaminated than grain from the untreated check (31 ng/g^{-1}). The untreated check sustained slight damage from natural insect infestations, which may account for the background contamination in check plots. Other studies (2) have suggested that mites are not significantly involved in increasing aflatoxin contamination in preharvest maize in Georgia.

Correlations (* = statistically significant at the 5% probability level, ** = statistically significant at the 1% probability level) among data means from a six-year study of maize fields in Georgia revealed a positive association between the percent of ears with visible *A. flavus* and the percent of ears with visible insect damage (0.17**) and between the percent of ears with visible *A. flavus* and the amount of insect damage to ears measured in centimeters of feeding penetration down the ear (0.50**) (12). There was also a positive correlation between the level of aflatoxin accumulation and the percent of ears with visible insect damage (0.15**), as well as between the level of aflatoxin accumulation and the amount of insect damage measured in centimeters of feeding penetration (0.32**).

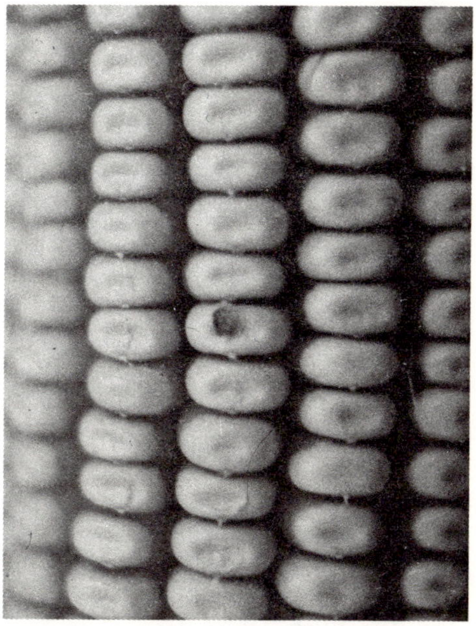

Maize kernel damage by the maize weevil; fungal growth will begin here.

Relationships were predictably closer for years when contamination was most severe, but trends still emerged for years in which levels of contamination were low.

Experiments to reduce aflatoxin contamination by controlling field insects have produced mixed results (5,18). Insecticides applied to silks of developing ears generally reduced, but did not completely eliminate, either insect damage or aflatoxin contamination. Research conducted in Georgia, Florida and South Carolina demonstrated that maize treated with numerous insecticide applications (three times a week for six weeks) yielded grain with an average of 70% less aflatoxin than the untreated check. Normally, however, the use of insecticides to protect field maize is not economically practical. Limited field studies at Tifton involving the application of fungicide to developing silks have failed to show a significant degree of control of either *A. flavus* or aflatoxin formation (7).

Other studies have demonstrated that *A. flavus* and aflatoxin can be detrimental to insect biology. When 250 ng/g^{-1} of aflatoxin B_1 were incorporated into a standard laboratory diet and fed to second instars of the corn earworm, fall armyworm and European corn borer, larval mortality was 96%, 44% and 70%, respectively (9,20). Aflatoxin concentrations of 25 or 2.5 ng/g^{-1} showed no measurable effect on larval biology. In another study, maize weevil mortality was significantly higher (22%) when weevils were reared on grain contaminated with *A. parasiticus* than when reared on grain contaminated with *A. flavus* (12%) (9).

Aflatoxin development in preharvest ears is also increased by simulated, early morning dew (10). When silks were sprayed with a mist of water on three mornings a week for four weeks to simulate heavy, early-morning dew, aflatoxin accumulation averaged 299 ng/g^{-1}, compared to 74 ng/g^{-1} in plots receiving no spray. A trend also appeared for increased insect damage in plots sprayed with water (8 cm per ear) compared to unsprayed plots (7 cm per ear).

An evaluation of six years of weather data collected at Tifton demonstrated a positive correlation (0.86*) between mean temperature and aflatoxin accumulation in the grain. Also, a positive correlation (0.83*) was obtained between the net evaporation and aflatoxin accumulation (12). Macro- and micro-climate appear to be intimately involved in aflatoxin contamination of preharvest maize.

In summary, research results demonstrate that multiple factors influence the ultimate degree of insect damage and aflatoxin contamination in preharvest maize. Solutions to minimize losses in maize should involve a multidisciplinary approach toward total management, from seed selection to product consumption.

References

1. Anderson, H.W., E.W. Nehring and W.R. Wichser, 1975. Aflatoxin contamination of corn in the field. Journal of Agricultural and Food Chemistry 23:775-782.

2. Barry, D., M.S. Zuber, E.B. Lillehoj, W.W. MicMillian, N.J. Adams, W.F. Kwolek and N.W. Widstrom. 1985. Evaluation of two arthropod vectors as inoculators of developing maize ears with *Aspergillus flavus*. Environmental Entomology 14:634-636.

3. Fennell, D.I., E.B. Lillehoj and W.F. Kwolek. 1975. *Aspergillus flavus* and other fungi associated with insect-damaged field corn. Cereal Chemistry 52:314-321.

4. Kinney, T.B., Jr. 1985. World food and agricultural prospects. Remarks at the Brookings Institution Executive Leadership Seminar on Critical Public Policy Issues, Washington, D.C., June 13, 1985.

5. Lillehoj, E.B., W.F. Kwolek, A. Manwiller, J.A. Durant, J.C. LaPrade, E.S. Horner, J. Reid and M.S. Zuber. 1976. Aflatoxin production in several corn hybrids grown in South Carolina and Florida. Crop Science 16:483-485.

6. Lillehoj, E.B., W.W. McMillian, W.D. Guthrie and D. Barry. 1980. Aflatoxin producing fungi in preharvest corn: Inoculum source in insects and soil. Journal of Environmental Quality 9:691-694.

7. Lillehoj, E.B., W.W. McMillian, W.D. Guthrie, J.L. Jarvis, D. Barry and W.F. Kwolek. 1984. Aflatoxin contamination of maize kernels before harvest: Interaction of *Aspergillus flavus* spores, corn earworm larvae, and fungicide application. Mycopathologia 86:77-81.

8. McMillian, W.W. 1983. Role of arthropods in field contamination. *In* Aflatoxin and *Aspergillus flavus* in Corn, U.L. Diener, R.L. Asquith and J.W. Dickens, (eds.) Southern Cooperative Series Bulletin 279. Alabama Agricultural Experiment Station, Auburn, Alabama, USA. Pp. 20-22.

9. McMillian, W.W., N.W. Widstrom and D.M. Wilson. 1981. Rearing the maize weevil on maize genotypes when aflatoxin-producing *Aspergillus flavus* and *A. parasiticus* isolates were present. Environmental Entomology 10:760-762.

10. McMillian, W.W., N.W. Widstrom and D.M. Wilson. 1985. Insect damage and aflatoxin contamination in preharvest corn: Influence of genotype and ear wetting. Journal of Entomological Science 20:66-68.

11. McMillian, W.W., N.W. Widstrom and B.R. Wiseman. 1976. Yield losses in south Georgia field corn resulting from damage by several insects. Journal of the Georgia Entomological Society 11:208-211.

12. McMillian, W.W., D.M. Wilson and N.W. Widstrom. 1985. Aflatoxin contamination of preharvest corn in Georgia: A six-year study of insect damage and visible *Aspergillus flavus*. Journal of Environmental Quality 14:200-202.

13. Payne, G.A. 1983. Nature of field infection of corn by *Aspergillus flavus*. *In* Aflatoxin and *Aspergillus flavus* in Corn, U.L. Diener, R.L. Asquith and J.W. Dickens, eds. Southern Cooperative Series Bulletin 279. Alabama Agricultural Experiment Station, Auburn, Alabama, USA. Pp. 16-19.

14. Shotwell, O.L., M.L. Goulden, E.B. Lillehoj, W.F. Kwolek and C.W. Hesseltine. 1977. Aflatoxin occurrence in 1973 corn at harvest. III. Aflatoxin distribution in contaminated, insect-damaged corn. Cereal Chemistry 54:620-626.

15. Taubenhaus, J.J. 1920. A study of the black and yellow molds of ear corn. Texas Agricultural Experiment Station Bulletin 270. College Station, Texas, USA.

16. Walsh, B.D. and C.V. Riley. 1869. The bollworm or cornworm. Transactions of the American Entomological Society 2:42-44.

17. Widstrom, N.W. 1979. The role of insects and other plant pests in aflatoxin contamination of corn, cotton, and peanuts: A review. Journal of Environmental Quality 8:5-11.

18. Widstrom, N.W., E.B. Lillehoj, A.N. Sparks and W.F. Kwolek. 1976. Corn earworm damage and aflatoxin B_1 on corn ears protected with insecticides. Journal of Economic Entomology 69:677-679.

19. Widstrom, N.W., A.N. Sparks, E.B. Lillehoj and W.F. Kwolek. 1975. Aflatoxin production and lepidopteran injury on corn in Georgia. Journal of Economic Entomology 68:855-856.

20. Wilson, D.M., W.W. McMillian and N.W. Widstrom. 1984. Differential effect of *Aspergillus flavus* and *A. parasiticus* on survival of *Heliothis zea* (Boddie) and *Spodoptera frugiperda* (J. E. Smith) (Lepidoptera: Noctuidae) reared on inoculated diet. Environmental Entomology 13:100-104.

21. Wiseman, B.R., N.W. Widstrom and W.W. McMillian. 1984. Increased seasonal losses in field corn to corn earworm. Journal of the Georgia Entomological Society 19:41-43.

Insects of Maize and Their Association with Aflatoxin Contamination

D. Barry, Agricultural Research Service, U.S. Department of Agriculture, University of Missouri, Columbia, Missouri, USA

Abstract

Many insect species are associated with maize and several of these may be directly or indirectly important in aflatoxin contamination. Pertinent insects can be broadly divided into above- and below-ground feeders, but the divisions are not mutually exclusive. Aspergillus flavus *isolates can be obtained from most soil samples or insects associated with maize. Geographic location significantly influences aflatoxin contamination of preharvest maize kernels. Genotypes of maize with resistance or susceptibility to the European corn borer (ECB) and corn earworm (CEW) do not appear to affect the amount of aflatoxin produced. However, insects such as ECB, CEW and the maize weevil may be responsible for increased levels of* A. flavus *infection and aflatoxin in preharvest maize kernels, depending on location and climatic conditions.*

Resumen

Un sinnúmero de especies de insectos se asocian con el maíz y muchas de ellas pueden ser directa o indirectamente importantes en la contaminación por aflatoxinas. Los insectos más importantes se pueden clasificar en forma general en insectos que se alimenten sobre el suelo e insectos que se alimenten dentro del suelo, aunque estas clasificaciones no se excluyen mutuamente. Se pueden obtener aislamientos de Aspergillus flavus *de casi todas las muestras de suelo o de insectos asociados con el maíz. La situación geográfica influyen de manera significativa en la contaminación por aflatoxinas de los granos de maíz en la fase de precosecha. Al parecer, los genotipos de maíz resistentes o susceptibles al barrenador europeo del maíz o al gusano elotero no afectan la cantidad de aflatoxinas producidas. Sin embargo, insectos tales como el barrenador europeo del maíz, el gusano elotero y el gorgojo del maíz suelen ser los responsables de que aumenten los niveles de infección con* A. flavus *y de aflatoxinas en los granos de maíz antes de la cosecha, dependiendo de la situación geográfica y de las condiciones climatológicas.*

There is little question that insects play a significant role in aflatoxin contamination in preharvest maize. Insects can transport spores to the developing kernels, and along with inflicting damage to the grain, they facilitate colonization by *A. flavus*. The latter appears to contribute most to aflatoxin contamination. Of course, *A. flavus* inoculum must be present when environmental conditions are favorable for the development of the fungus.

One of the first considerations in approaching the insect and fungi relationships contributing to the aflatoxin problem in preharvest maize is the diversity of insect species associated with maize production. Broadly, these insects can be characterized as above- or below-ground feeders, but these categories are not mutually exclusive.

Although this paper deals primarily with insects found in the US Corn Belt, some other insect species will be considered in the context of *A. flavus* infection of developing maize kernels and aflatoxin contamination.

Above-Ground Maize Insects

Above-ground maize insects are probably of most importance in relation to *A. flavus* infection of developing maize kernels.

Armyworm, *Pseudaletia unipuncta* (Haworth)

Description-This insect is called the true armyworm and can be recognized by its grayish-brown forewings which have a white spot in the center. The eggs are greenish white, globular and laid in rows. Freshly hatched pale green larvae become yellowish- or brownish- green with age and have greenish-brown heads mottled with dark brown. The larval body reaches 3 to 4 cm in length, and has two longitudinal orange stripes along each side and a pale orange, white-bordered stripe both above and below the spiracles.

Biology-The true armyworm has a wide range of hosts, primarily grasses. It is found throughout the USA east of the Rocky Mountains. The larvae are nocturnal feeders, capable of devouring tender young plants or skeletonizing older ones. The armyworm overwinters as a partly grown larva, with the first-generation moths emerging about May. A female moth is capable of laying 2000 eggs. There may be two to three generations per year in the northern USA and five or more generations in the south.

Fall armyworm, *Spodoptera frugiperda* (J.E. Smith)

Description-The forewings of the fall armyworm are dark gray, mottled with lighter and darker splotches and with a whitish spot near the tip. The wing span is about 2.5 to 4.2 cm. The eggs are light gray and laid in clusters. Fully grown larvae vary in color from light tan or green to nearly black and are 3 to 4 cm long.

Biology-This insect is very common in the US states around the Gulf of Mexico and in the tropics of North, Central and South America. It migrates northward or is blown northward from the Gulf area in the spring, often resulting in large populations in the Corn Belt by midsummer to late August. Females normally lay their eggs at night in clusters of fifty to several hundred on maize leaves or other vegetation. Depending on the temperature, the eggs hatch within two to ten days. The larvae feed in the maize whorl and developing tassel or ear. They are voracious, feeding any time during the day or night. After two to three weeks continuously on plants, they pupate in the soil. Fall armyworms have one generation per year in the northern USA, but several in the south.

Corn earworm, *Heliothis zea* (Boddie); tobacco budworm, *H. virescens* (Fabricius); Old World corn earworm *H. armigera* (Hübner)

Description-The forewing color of these noctuids varies between and within sexes of each species. The corn earworm forewings may be light yellowish olive, yellowish brown or pinkish brown with a dark spot near the center. The egg is white, dome-shaped and appears sculpted; it develops a reddish-brown band before hatching. Full-grown larvae are moderately hairy and range in color from brownish to greenish and are sometimes purplish. They have yellowish lateral stripes and alternating light and dark longitudinal stripes on the dorsal surface of the body.

Biology-Maize is one of the preferred hosts, but these insects have over 200 recorded hosts throughout the world. The first generation larvae feed in the whorl of maize, giving the plant a ragged appearance. Second generation larvae initially feed on the ear silks, but go through the silk channel to the developing kernels, either feeding on the kernels at the tip of the ear or feeding down a line of kernels on the cob. Larvae are cannibalistic so only one insect generally develops in an ear. In the southern USA, corn earworms overwinter in the soil as pupae, emerging as moths in the spring and depositing eggs on seedling maize if available. The eggs are laid singly. Larvae feed two to four weeks, going through five to six instars before dropping to the ground and pupating. The second generation of moths generally emerge about the time maize is silking, and they usually deposit their eggs on the silks.

Corn leaf aphid, *Rhopalosiphum maidis* (Fitch)

Description-The oval, soft-bodied, wingless adult is pale bluish green with black antennae, legs and cornicles, with a dark area around the base of the cornicles. The head has two longitudinal dark bands, and the abdomen has a row of black spots on each side.

Biology-This aphid is found throughout the temperate and tropical areas of the world. Heavy colonies cause mottling and discoloration of maize leaves, which may wilt and die; maize tassles and silks may be covered with sticky honydew. By feeding on the stem below the tassel, this aphid can reduce or prevent pollination. It is assumed that the corn leaf aphid migrates into the Corn Belt from the southern USA. Generally, this aphid is parthenogenic, but on rare occasions males are found. In the northern USA, this insect may only have a few generations per year, but in southern Texas there may be as many as 50.

Flea beetle, *Chaetocnema pulicaria* (Melshiemer)

Description-This beetle is small, oval and black, tinged with bronze or bluish green. The hind legs are distinctly enlarged. The larva is a white, slender, cylindrical grub with a brown head and tiny legs.

Biology-Three species of flea beetles that may be found on maize are distributed throughout the states of the north central USA. The corn flea beetle is a general feeder, but it prefers grasses and horticultural crops. Maize plants are attacked as soon as they emerge and as long as the leaves remain tender. The feeding damage itself is usually not an economic factor, but the beetle transmits Stewart's wilt which can be an important disease of maize. Stewart's wilt is usually more of a problem after a mild winter that favors overwintering of the beetle.

Corn earworm on a maize ear

Grasshoppers: redlegged grasshopper, *Melanoplus femurruben* (DeGeer); differential grasshopper, *M. differentialis* (Thomas); two-striped grasshopper, *M. bivittatus* (Say); migratory grasshopper, *M. sanquinipes* (Fabricius)

Description-Grasshoppers are generally brown-green to gray with various markings; they are recognized as jumping insects. They are usually 19 to 38 mm long, with large heads and large compound eyes. The forewings are narrow, leathery and thick, whereas the hind wings are membranous, broadly triangular and often brightly colored.

Biology-Grasshoppers are distributed throughout the USA. They are general feeders, and can be found feeding on most agricultural crops, including maize. They consume the foliage and silks, which reduces pollination. Grasshoppers overwinter as eggs in pods (25 to 150 eggs) which are deposited in the soil in vegetative areas, such as fence rows or waterways. The young are nymphs that resemble the adult except for having incomplete wings. They molt five to six times during the 35 to 50 days before becoming adults.

European corn borer, *Ostrinia nubilalis* (Hübner)

Description-The female moth is pale yellow to light brown and has a robust body; the male is darker and smaller. The outer one-third of the wing area is usually crossed by two dark zigzag lines. The eggs are pearly white when first deposited in masses, shingle fashion. The egg turns darker as it develops and appears as a black head (the head of the larva) just previous to hatching. The larvae become about 25 mm long when full grown and pupate into carmel to dark brown pupae after five instars.

Biology-The European corn borer was introduced into the USA in 1917 and has spread to nearly all areas where maize is grown. Its distribution in Europe and south into Egypt is similar. Maize is the main host, though over 200 other hosts have been identified. The maize plant is usually thought to have two developmental stages that are susceptible to attack, the whorl stage and anthesis. During the whorl stage, the first generation larvae feed on the unfurled leaves and may eventually enter the maize stem. Second generation larvae usually begin to feed on pollen and tender tissue at the leaf axil before proceeding to feed behind the leaf sheath and collar. They also enter the stem during the fourth to fifth instar. Full-grown larvae can overwinter in maize stems in a state of diapause. In the Corn Belt, adult moths emerge in late May and early June, and a single female may deposit 500 to 600 eggs in masses of 20 to 60 on the underside of a maize leaf. Eggs hatch within four to eight days, depending on temperature; the larval stage is usually 16 to 24 days, and the pupal stage, 9 to 14 days.

Spotted stalk borer or pink stalk borer, *Chilo partellus* (Swinhoe)

This is a tropical stem borer resembling the European corn borer, both in appearance and habits. It is distributed throughout India, much of Africa and East Asia.

Maize stalk borer, *Busseola fusca* (Fuller)

This tropical stem borer, found in Africa, is much larger and a more voracious feeder than *C. partellus*. It may feed on maize leaves, but is more often found in stems. The eggs are deposited in batches behind the leaf sheath.

Pink stalk borer, *Sesamia calamistis* (Hamps.); purple stem borer, *S. inferens* (Wlk.); *S. cretica* (Led.)
These stem borers make up a complex and are frequently referred to by a specific name that may or may not be correct. However, the behavior and damage of these insects are similar to those of the maize stem borer.

Southwestern corn borer, *Diatraea grandiosella* (Dyar); southern corn stalk borer, *D. crambidoides* (Grote); sugarcane borer , *D. saccharalis* (Fabricius); sugarcane stalk borer, *Eldana saccharina* (Wlk.)
Biology-Maize is one of the hosts of pyralid stem borers inhabiting the tropics and the more temperate areas of the world. These insects have some similar characteristics and behavior.

Biology-The southwestern corn borer was introduced into the USA from Mexico in 1913. It has since spread west to Arizona, north to Kansas and Missouri and east to Georgia. It is a voracious feeder: the first generation attacks maize in the whorl stage, and the second generation feeds on and around the ear and in the stem. The larva of the summer generation(s) of this insect is spotted, but the diapausing or winter generation is immaculate and creamy white. Generally, only one full-grown larva of the second generation of this insect is found in a plant. This borer prepares an overwintering site in the taproot of the maize plant, frequently girdling the stem about 15 to 25 cm above the ground and causing the top portion of the plant to break.

Leafhoppers: Blackfaced leafhopper, *Graminella nigrifrons* (Forbes); *Dalbulus maidis* (Delong and Wolcott)
The primary problem caused by leafhoppers on maize is diseases, especially viruses.

Sap beetles, *Nitidulidae*
These beetles may be found on maize silks, feeding on sap and some types of fungi, generally as secondary invaders.

Below-Ground Maize Insects
Seed corn beetle, *Stenolophus lecontei* (Chaudoir)
Description-The seed corn beetle is small and dark brown with a tan border on the wings.

Biology-This insect is generally distributed in the USA and Canada. It may feed on the contents of seed, especially seed with low viability or seed left ungerminated in the ground for an extended period of time. Normally, this beetle is a scavenger, feeding on insects and decaying matter. It is a member of the Carabidae family and may have one or more generations per year.

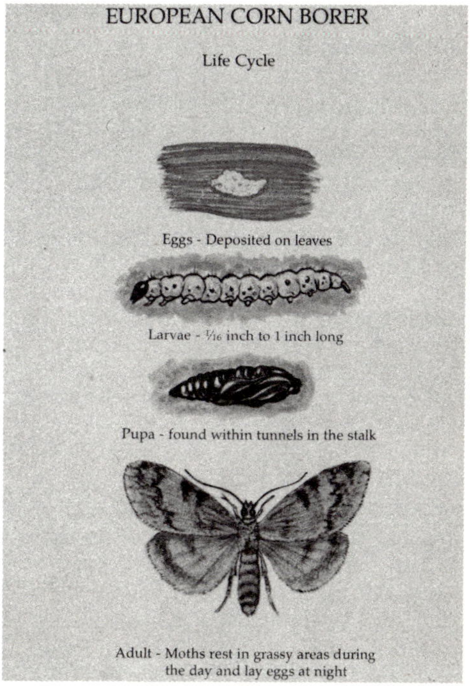

Life cycle of the European corn borer

Seed corn maggot,
***Hylema platura* (Meigen)**

Description-This insect as an adult is a gray fly, about 5 mm long and with scattered bristles on its black body and legs. The larva is a legless, 12-segmented maggot with a pointed head and rounded tail, white to yellow and 5 to 7 mm long.

Biology-This maggot, first discovered in New York in 1856, is widely distributed in the USA, southern Canada and Mexico. The maggot has a variety of hosts, including maize. This insect overwinters as a larva inside a puparium in which the pupa develops. The adult lays eggs singly or in small clusters on decaying vegetation. There may be several generations per year with each generation requiring four to five weeks, depending on temperature. Since the larvae feed on seeds, they reduce germination and weaken seedlings.

Thief ant,
***Solenopsis molesta* (Say)**

Description-The thief ant is very small, about 1.5 mm long, and light brownish yellow. The genus *Solenopsis* differs from other ants by having a 10-segmented antenna with a two-segmented club and a two-segmented pedicel or waist.

Biology-The ant is distributed throughout the USA and feeds on grain by hollowing out the ungerminated seed; its presence is revealed by small starch grains scattered in the soil.

Northern corn rootworm,
***Diabrotica longicornis* (Say)**

Description-The adult beetle is about 6 mm long, with a pale green or green body without any distinguishing marks. Larvae are white and slender, about 12 mm long when full grown, and have a brown head and a dark plate on the dorsal side of the terminal segment. The pupa is white, about 6 mm long, and has the basic form of the beetle. Eggs are white, oval and about 0.1 mm long.

Biology-This species is widely distributed in North America, from Colorado to the eastern Corn Belt and south to Oklahoma. Maize is the preferred host. The larvae are the most serious problem because they damage maize roots, which results in goose-neck, a form of root lodging; adults feed on leaves and silks. The insect overwinters as an egg and has one generation per year.

Southern corn rootworm,
***Diabrotica undecimpunctata howardi* (Barber)**

Description-This beetle is about the same size or slightly larger than the northern corn rootworm. It has a bright greenish-yellow body with twelve black spots on the wing covers and the head; legs and antennae are black. The full grown larva is 15 mm long with a creamy white wrinkled body. The upper side of the terminal segment has two spine-like tubercles. Pupae are white to yellow and about 6 mm long.

Biology-The southern corn rootworm is found throughout the USA, but is not considered a serious maize pest. The beetles prefer broadleaved plants, and larvae may damage maize in the same manner as the northern corn rootworms. The insect overwinters as an adult beetle and becomes active in the spring when temperatures reach 21° C. It requires six to nine weeks to complete the life cycle.

Western corn rootworm,
***Diabrotica virgifera* (Le conte)**

Description-The adult beetle is yellow, about 6 mm long, with a black stripe on the outer side of each wing cover. The stripe is more pronounced in females than in males, as the entire posterior half of the male's wing may be black.

Biology-The western corn rootworm, once distributed in the western Corn Belt, has now established itself as far east as Ohio and Michigan. Maize is the preferred host of the larva and adult. The life cycle and habits of this insect are similar to those of the northern corn rootworm.

Black cutworm, *Agrotis ipsilon* (Hufnagel)

Description-The forewings of the moth are long, narrow, and usually dark, becoming gradually paler near the tips. There are three black lines on each forewing. The eggs are round and white. The larvae vary from very small when freshly hatched to about 3 to 4 cm long when full grown. Above the spiracles, the larvae are light gray to black. They differ from other cutworms by having convex, rounded, distinctly isolated coarse grandules with smaller grandules interspersed between the larger ones. The larvae curl up if disturbed. The pupae are brown, about 20 mm long, with a tapering posterior and a blunt head with evident mouth parts and antennae.

Biology-The black cutworm occurs in most areas of the western hemisphere. Young larvae feed on leaves of young plants by chewing holes. As the larvae mature, they move to the soil where they cut off seedlings at ground level or below. Frequently, these larvae migrate into the field from other vegetation. Some stay at the site of the plant they have cut off and feed on the roots; they may also attempt to drag the upper portion of the plant into a crack or hole in the soil. One larva may cut off four to six plants. There is considerable discussion as to how the black cutworm overwinters, but in the Corn Belt it is probably as larvae or pupae. In the Corn Belt, they begin to lay eggs in March. In Canada, two generations per year are usually produced, whereas further south in Tennessee four generations can be produced.

Other types of soil cutworms are:

Bristly cutworm, *Lacinipolia renigera* (Stephens)
Larva: The body is yellowish gray with diamond-shaped markings on the back. It has very coarse hairs on the body.

Bronzed cutworm, *Nephelodes minians* (Guenee)
Larva: It has three stripes which run from the head to the tail end. The body is dark brown.

Claybacked cutworm, *Agrotis gladiaria* (Morrison)
Larva: The body is greenish to dark brown, with a broad pale dorsal stripe.

Dark-sided cutworm, *Euxoa messoria* (Harris)
Larva: The body is dull gray with several stripes; it has dark gray stripes just above the spiracles.

Dingy cutworm, *Feltia ducens* (Walker)
Larva: The body is brownish tan with a faint dark V-shaped marking on the back of each segment.

Glassy cutworm, *Crymodes devastator* (Brace)
Larva: The body is a translucent, greasy white color; the head and the shield of the head are reddish brown. There are no stripes on the body.

Sandhill cutworm, *Euxoa detersa* (Walker)
Larva: The worm is whitish to pale gray with seven faint chalky white stripes running the length of the body.

White grubs, *Phyllophaga* spp.

Description-White grubs are scarab beetles that are robust, oblong (12 to 20 mm long), hard shelled and tan to reddish brown to black. The eggs are white and spherical and may be found encased in soil particles. The larvae are white, C-shaped grubs with a brown to grayish head capsule. The pupa varies from creamy to dark brown.

Biology-These grubs are found throughout North America and feed on the roots of many grasses and cultivated crops. This insect has a life cycle of two to four years (usually three).

Annual white grub, *Cylocephala* spp.

Biology-This insect overwinters as a larva in the soil, pupates in May, and emerges as an adult, laying eggs in the soil in a few weeks.

Japanese beetle, *Popillia japonica* (Newman)

Description-This scarab is shiny, metallic green and about 13 mm long.

Biology-This insect was first reported in New Jersey in 1916 and now occurs in 20 states along the Atlantic coast and as far west as Missouri. Its primary hosts are turf grasses, but it has over 250 known host plants, including maize. It primarily damages maize by feeding on the silks, thus preventing pollination. The Japanese beetle has a one-year life cycle, overwintering as larvae.

Wireworms, *Melanotus* spp.

Description-The adult wireworm is hard, smooth-bodied, reddish brown to black, elongate and tapers on both ends (more so towards the rear); it is about 6-19 mm long. Larvae are yellow to reddish brown, with a brown flat head and a wire-like appearance when full grown. Pupae are white, soft bodied and have no protecting cover. A dozen or more species of these click beetles are known to attack maize.

Biology-These worms have a six-year life cycle and are distributed throughout the USA. They are abundant in the Midwest and the southern states. *Melanotus* spp. feed on the roots of many grasses, including maize, though they are not limited to grasses. They are frequently a problem in maize planted in no-till or in conventionally cultivated sod fields. The larvae feed on seed, destroying the germ, or on the roots of emerging plants, leaving spotty stands or weak plants. The larger larvae also damage maize by boring into the underground stem and taproot area, causing the plant to wilt and/or die. Generally, larvae spend five years in the ground feeding on plant roots, before emerging as adults. Adults feed on pollen before hibernating in protected areas; in May or June of the next year they start laying eggs.

Sod webworm, *Crambus teterrellus* (Zincken)

Description-This is a small white to brownish moth with a pointed snout (labial palps). At rest the wings are folded around the body. The larva is tannish gray with paired black spots on the top and sides of the abdominal segments.

Biology-Many species of webworm, all native to America, periodically infest maize in the USA and Canada. Damage is prevalent in the eastern USA and sandhills of Nebraska. Partially grown larvae emerge from hibernation in the spring and feed on the roots and into the stem below the ground, damage resembling that of the cutworm. The insect overwinters as a larva in a silk-lined tube in the soil. Sod webworms attacking maize have one generation per year.

Grape Colaspis, *Colaspis brunnea* (Fabricus)

Description-This is an oval, yellowish-brown beetle, 4 to 5 mm long; the wings appear striped. The larvae are white or tan with a dark brown head and prothoracid shield. They have fat grub-like bodies with three pairs of legs.

Biology-The grape colaspis is most commonly distributed in the eastern USA, but is found as far west as New Mexico and Arizona. The adults are usually foliage feeders, and the larvae feed on the roots of the same plants. They have a wide range of hosts, including fruits, legumes, grasses and weeds. A symptom of larval damage is a purpling of plants, similar to phosphate deficiency. The larvae overwinter in the soil and emerge as adults in July. The insect has one generation per year.

Corn root aphid, *Anuraphis maidiradicis* (Forbes)

Description-This is a wingless, blue-green, soft-bodied aphid with a black head and black or reddish-brown eyes. The female, during the egg-laying period, has a gray body with a pink abdomen and a white powdery coating. The eggs are elliptical and yellowish green, turning to black. Nymphs resemble adults, but they are smaller, pale green and have red eyes.

Biology-The corn root aphid can be found in the maize- and cotton-growing areas east of the Rocky Mountains; it may also be found on smartweeds. The damage to maize plants is caused by sucking sap from roots, resulting in stunting when the plants are 15 to 45 cm tall. The life cycle of this insect is attuned to the ant, which transports eggs to the ant nest for overwintering and the young nymphs to the maize plants to establish the aphid colony. The ants feed on the honeydew of the aphids. The eggs all hatch as females, which are parthenogenic and continue to produce 40 to 50 nymphs each until cold weather, when they also produce true wingless males. These aphids mate and produce eggs.

Stored Grain Pests

Angoumois grain moth, *Sitotroga cerealella* (Oliver); maize weevil, *Sitophilus zeamais* (Motschulsky)

Although these two pest are thought of as stored-grain pests, they are also capable of infesting grain in the field. The maize weevil is frequently found in the field in the southern parts of the USA.

Effects of Insects, Maize Genotype and Location on Preharvest Alfatoxin Contamination

Insect samples were collected from maize plants and soil along with soil samples at four locations (Iowa, Illinois, Missouri and Georgia) for identification of *A. flavus* and *A. parasiticus* isolates (6). Visual identifiable isolates were found at all locations, and about 50% of the isolates produced aflatoxin. All *A. parasiticus* isolates were either from the soil or from insects collected from the soil, whereas *A. flavus* was also obtained from the plant-collected insects. Toxin production was noted at all locations, but only Georgia had a wide occurrence of aflatoxin in mature maize.

Regional research (3,4) in Iowa, Missouri and Georgia with European corn borer (ECB), *A. flavus*, maize hybrids resistant and susceptible to ECB and planting dates again indicated the significance of location with regard to aflatoxin production. There was a broad distribution of toxin in Georgia and Missouri, but little in Iowa. Increased toxin levels were found in samples from plants hand-infested with ECB, but planting date, fungal inoculation or different hybrids did not affect toxin levels. It appeared that inoculation with nontoxin-producing isolates did not block toxin propagules from entering kernels of developing ears. The ECB-resistant hybrid had a higher level of aflatoxin in Georgia than the susceptible hybrid; in Missouri the reverse was true.

More fungi were found in larvae from the ECB-susceptible hybrid than the resistant hybrid, but no similar association was found in the grain samples. These results demonstrated the importance of interactions among the various treatment components.

Lillehoj *et al.* (5) studied the interactions among three species of fungi, *A. flavus, Penicillium oxaicium* and *Fusarium moniliforme*, on two maize hybrids with different degrees of resistance to ECB at three locations, Iowa, Georgia and Missouri. A higher incidence of *A. flavus* isolates was obtained from ECB larvae collected from *A. flavus*-inoculated ears than from uninoculated ears. The resistant hybrid had less ECB damage than the susceptible hybrid, but there did not appear to be any hybrid effect on the association of *A. flavus* and ECB. Locations were significant for differences in aflatoxin contamination. Conditions in Missouri were very conducive (419 ppb) to aflatoxin development.

Studies (7) with corn earworm (CEW), *A. flavus* and a resistant and susceptible hybrid (to CEW) indicated a distinct regional variation. Incidence was relatively high in Georgia, intermediate in Missouri and low in Iowa. No treatment effects were noted in Georgia, but *A. flavus* inoculation and CEW infestation increased toxin accumulation in Missouri kernels. The hybrids did not appear to affect the amount of aflatoxin produced, and the application of fungicide (Benomyl) did not significantly reduce aflatoxin levels.

Barry *et al.* (1) determined that the maize weevil, inoculated with *A. flavus* spores, can effectively increase the incidence of aflatoxin in kernel samples in Missouri. The wheat curl mite was not an effective vector. In another study (2), a tight husk effectively reduced the amount of aflatoxin in kernel samples.

References

1. Barry, D., E.B. Lillehoj, N.W. Widstrom, W.W. McMillian, M.S. Zuber, W.F. Kwolek and W.D. Guthrie. 1986. Effect of husk tightness and insect infestation on aflatoxin contamination in preharvest maize. Environmental Entomology. (In press.)

2. Barry, D., M.S. Zuber, E.B. Lillehoj, W.W. McMillian, N.J. Adams, W.F. Kwolek and N.W. Widstrom. 1985. Evaluation of two arthropod vectors as inoculators of developing maize ears with *A. flavus*. Environmental Entomology 14(5):634-636.

3. Guthrie, W.D., E.B. Lillehoj, D. Barry, W.W. McMillian, W.F. Kwolek, A.O. Franz, E.A. Catalano, W.A. Russell and N.W. Widstrom. 1982. Aflatoxin contamination of preharvest corn: Interaction of European corn borer and *Aspergillus flavus*-group isolates. Journal of Economic Entomology 75(2):265-269.

4. Guthrie, W.D., E.B. Lillehoj, W.W. McMillian, D. Barry, W.F. Kwolek, A.O. Franz, E.A. Catalano, W.A. Russell and N.W. Widstrom. 1981. Effect of hybrids with different levels of susceptibility to 2nd-generation European corn borers on aflatoxin contamination in corn. Journal of Agricultural and Food Chemistry 29(6):1170-1172.

5. Lillehoj, E.B., W.F. Kwolek, W.D. Guthrie, D. Barry, W.W. McMillian and N.W. Widstrom. 1982. Aflatoxin accumulation in preharvest maize kernels: Interaction of three fungal species, European corn borer and two hybrids. Plant and Soil 65:95-102.

6. Lillehoj, E.B., W.W. McMillian, W.D. Guthrie and D. Barry. 1980. Aflatoxin-producing fungi in preharvest corn: Inoculum source in insects and soils. Environmental Quality and Safety 9(4):691-694.

7. Lillehoj, E.B., W.W. McMillian, N.W. Widstrom, W.D. Guthrie, J.L. Jarvis, D. Barry and W.F. Kwolek. 1984. Aflatoxin contamination of maize kernels before harvest. Mycopathologia 86:77-81.

Breeding Strategies to Control Aflatoxin Contamination of Maize Through Host Plant Resistance

N.W. Widstrom, Insect Biology and Population Management Research Laboratory, Agricultural Research Service, US Department of Agriculture, Tifton, Georgia, USA

Abstract

*Maize (*Zea mays *L.) with resistance to infection by* Aspergillus flavus *Link ex Fries and subsequent aflatoxin formation became urgently needed when it was determined that preharvest aflatoxin contamination of the kernels is chronic in several maize-growing areas and intermittent in many others. Initial observations, mostly of naturally occurring contamination, suggested that little or no resistance existed in available germplasm. Some refinements of technique, including artificial inoculation, have resulted in the demonstration of genetic differences in plant resistance to infection and contamination. Recent findings also suggest that sufficient genetic variability exists to warrant selection for plants with improved resistance to aflatoxin contamination. The genetic variability is believed to be primarily additive. Since different factors, such as resistance to insects and plant stresses, are probably responsible for resistance to infection and contamination, a strategy of selection based on an index appears desirable. Any plan to control aflatoxin contamination must include genetic and nongenetic management components, since no single component seems to have a major, consistent effect.*

Resumen

*El maíz (*Zea mays *L.) resistente a la infección con esporas de* Aspergillus flavus *Link ex Fries y a la subsecuente formación de aflatoxinas se necesitó con urgencia cuando se determinó que la contaminación por aflatoxinas de los granos antes de la cosecha es crónica en muchas zonas en las que se cultiva el maíz e intermitente en muchas otras. Las observaciones iniciales, gran parte de las cuales se referían a la contaminación que se producía en forma natural, indicaron que el germoplasma existente poseía poca o ninguna resistencia. Ciertos refinamientos de la técnica, incluyendo la inoculación artificial, han dado como resultado la demostración de la existencia de diferencias genéticas en la resistencia de las plantas a la infección y contaminación. Hallazgos recientes también indican que existe suficiente variabilidad genética que justifica la selección para obtener plantas que presenten una mejor resistencia a la contaminación por aflatoxinas. Se cree que la variabilidad genética es fundamentalmente aditiva. Puesto que es probable que diferentes factores, tales como resistencia a los insectos y a las condiciones adversas, sean responsables de la resistencia a la infección y a la contaminación, parece conveniente emplear una estrategia de selección que se base en un índice. Cualquier plan destinado a combatir la contaminación por aflatoxinas debe incluir elementos de manejo genético y no genético, ya que, al parecer, ningún elemento por sí solo ejerce un efecto importante de manera constante.*

Plant resistance to molds by maize (*Zea mays* L.) kernels was suggested more than 65 years ago by Taubenhaus (18). At least two of the molds observed in that early study were *Aspergillus niger* Van Tieghem and *Aspergillus flavus* Link ex Fries, the latter being a producer of aflatoxin. Breeding strategies to control aflatoxin contamination in maize were not seriously considered until the 1970s, when contamination in Georgia was clearly demonstrated to be a preharvest condition (1). At about the same time, minor amounts of preharvest contamination were reported in Indiana and Kentucky (17), and significant amounts were present in maize from Missouri and South Carolina (8,9) and from Georgia and Texas (29).

Naturally occurring infection and contamination are often sporadic, and the variation prohibits delineating resistance among test genotypes (25), even when many test locations are used (2,30). However, when the incidence and amounts of aflatoxin are high, as they were in 1977 and 1980 (16), genotype differentiation is possible (12,13).

A review by Zuber (27) in 1977 indicated that information regarding plant resistance to infection and aflatoxin production was inconclusive. By 1983 (26) and 1984 (21), several studies reported a more favorable prognosis. Most of these studies employed some type of inoculation technique to enhance infection and contamination during years when natural contamination was too low to identify genotype differences (6,11). The percentage of kernel infection was also found to be an effective measurement of resistance if artificial inoculation was used (4).

Inherent differences in the ability to resist infection or to produce aflatoxin have been clearly demonstrated for commercial hybrids (4,7,22), experimental single crosses (5,23,28) and inbreds (20). The differences also now extend to open-pollinated varieties (19,30) and specialty maize, such as sweetcorn (23) and popcorn (15). There is a distinct advantage in waiting until physiological maturity of the kernels before sampling grain for aflatoxin production. Two commercial hybrids differing in aflatoxin production capabilities have been shown to have unique linear aflatoxin accumulation rates during ear maturation, when wound-inoculated at 20 days after silking (24). This suggests that differences in aflatoxin concentration in the grain of hybrids can be most easily detected when sampling occurs at physiological maturity or later. Both field (3,12) and laboratory (14,22) tests proved that genotypes differing in endosperm carbohydrates also supported varying amounts of aflatoxin production. In general, sugary endosperm supports large amounts of aflatoxin production, whereas lesser amounts accumulate in a substrate of high amylose or more complex endosperm starch.

Breeding for Maize Improvement

Certain prerequisites are critical in successfully implementing a breeding strategy to improve any trait. The first is to select a trait for which genetic variability exists. This may seem to be of small importance, but in the case of resistance to infection by *A. flavus* and subsequent aflatoxin production, great difficulties are encountered in demonstrating significant differences among genotypes. The relative amount of total variability attributable to genetic sources is that which determines the heritability of a trait and consequently whether improved plant types can be obtained from the available germplasm. It also has a direct bearing on the ease of screening large numbers of plants and locating those with the desired resistance characteristics.

At least four major genetically controlled traits influence or condition resistance in the maize plant to aflatoxin contamination. The first two factors, which are measured directly, are 1) plant resistance to the infection process itself, and 2) the process of toxin production after infection has occurred. Two additional factors are indirect measures of resistance, since they involve processes widely believed to reduce aflatoxin contamination: 3) plant resistance to insect damage, and 4) resistance to environmental stresses.

Resistance to infection is difficult to measure in the field; it is usually necessary to test individual intact kernels. Failure of the fungus to become established on the ear is often attributed to physical factors, such as a pericarp that is not easily penetrated. However, resistance may also be due to chemical factors in the silk, husk or in the kernel itself; they can prevent spore germination or fungal penetration. Accurate measurement of resistance to infection usually involves precautions against physical damage to the kernels; under field conditions, kernel damage alone dramatically increases infection percentages and the incidence of aflatoxin-positive samples (10,11).

A distinct advantage of using infection percentages to assess resistance is that toxin formation is not necessary. Therefore, the expense of aflatoxin analysis can be avoided. When aflatoxin production is used as the measurement of plant resistance, the need for an aflatoxin-producing isolate of the fungus leads to the expense of chemical analysis. Toxin production may be measured on damaged or intact kernels in the field or laboratory, allowing more flexibility in a testing program. Finally, in view of the heterogeneity in infection rates and aflatoxin accumulation under most naturally occurring circumstances, an obvious need exists for appropriate inoculation and rapid, inexpensive resistance measurement techniques.

Choosing a Breeding Strategy

A breeding strategy can be developed after selecting a trait or traits to measure plant resistance. A trait is

***Aspergillus flavus* infection on maize ears**

deemed appropriate if it characterizes the nature and mechanisms of resistance, and if sufficient variability exists for the trait to demonstrate differences among available germplasm sources. Further testing may be required to determine optimal conditions for eliciting differences so that traits will be useful tools for screening genotypes, but the proportion of variation attributable to various genetic sources is the most important criterion for deciding on a plant improvement procedure.

Assessing sources of variation, or the inheritance of a trait, involves using at least one of several experimental designs so that the variability can be conveniently partitioned. One of the simplest designs used to evaluate a specific group of sources (usually inbred lines) is a mating design, the diallel. This design is useful when germplasm screening identifies a rather limited number of inbred line parents (usually six to twelve). Table 1 shows the results from two diallel experiments, each conducted over a three-year period. The mean squares indicate that tremendous variation existed for contamination from year to year, and that inbreds contributed significantly different amounts of resistance than did single crosses as measured by aflatoxin production in the grain. The general performance of the dent lines was fairly consistent from year to year, but certain sweetcorn crosses performed differently over years.

Aflatoxin contamination means of inbred lines obtained from their F_1 crosses is given in Table 2. Reduced average amounts of contamination for the dent diallel still reflect both differences from year to year and a significant average level of general combining ability for resistance among dent inbreds. Heritability estimates were 27% for the dent and 26% for the sweetcorn crosses, suggesting that stringent selection will be needed to successfully select for improved resistance to aflatoxin contamination. The fact that no sources thus far identified have more than moderate amounts of resistance will add to the difficulty of developing germplasm with improved plant resistance. The high reproductive capacity of maize, and its compatibility and versatility for genetic manipulation, will help make progress in selection possible.

Table 1. Analyses of variance for aflatoxin contamination of single crosses in diallel experiments involving nine maize dent and eight sweetcorn inbreds, Tifton, Georgia, 1978-1981

	Mean squares	
Source of variation[a/]	**Dent**	**Sweetcorn**
Years	847.2**	1299.1**
GCA	7.6*	7.0**
SCA	1.8	2.5
GCA x years	2.5	1.2
SCA x years	2.0	2.0*

a/ CGA = general combining ability, SCA = specific combining ability

*, ** Significant at the 0.05 and 0.01 probability levels, respectively

Source: Widstrom, *et al.* (23)

The breeder has several options when choosing a selection procedure, but the correct choice should be based on results of tests such as the diallel experiments. For example, the results in Table 1 indicate that great differences in contamination occur from one year or one environment to the next. Therefore mass selection, which normally involves selection based on the performance of one genotype in a single environment, would be inappropriate as a selection scheme. The diallel tests clearly indicate some type of selection based on general performance that can be tested at more than one location. In addition, the performance of specific crosses should be monitored for the sweetcorn germplasm. When both general and specific combining abilities are of importance, reciprocal recurrent selection is useful to assure response to selection.

It may be important that the breeder determine more than one trait, such as infection percentage and toxin production, laboratory and field test traits measuring different mechanisms of resistance, and separate measurements for resistance to aflatoxin production, plant stress and insects. Selection for more than one trait has three basic forms:

- Tandem selection (separate successive selection for each trait), which is most often used but is least efficient;
- Independent culling, in which selection intensity is severely reduced with each additional trait; and
- Index selection, the most efficient but also the most expensive procedure.

Because of the need for low selection intensities, and the low heritability associated with traits presently used to measure resistance, index selection would probably be best for aflatoxin studies. The procedure would also maximize information obtained from expensive aflatoxin analyses.

Whether selection is for one trait or more, the progress or genetic gain from selection will be a function of the

Table 2. Amounts of aflatoxin contamination for maize dent and sweetcorn inbreds evaluated as single crosses of diallels, Tifton, Georgia, 1978-1981

Dent Inbred	Aflatoxin (ng/g^{-1})[a]	Sweetcorn inbred	Aflatoxin (ng/g^{-1})[a]
Mp464	10.0 a[b]	259	82 a[b]
Mp313E	10.4 a	Od3	95 a,b
GE317	13.2 a,b	Tex703-338	122 b,c
Sa4(w)	15.2 b	I45	138 c
GE335	15.5 b	230	147 c
Ab18	17.5 b	M119	147 c
9-54C	17.8 b	415	150 c
BJ30	18.0 b	339	156 c
SC413	18.4 b		

a/ Geometic means of aflatoxin concentrations

b/ Numbers followed by different letters are significantly different at the 0.05 probability level.

Source: Widstrom, *et al.* (23)

variability of the trait and the intensity of selection. The gain for each cycle may be mathematically expressed as

$$k\sigma_p H = \text{genetic gain},$$

where k is the selection intensity in standard units, σ_p is the square root of σ_p^2 (the phenotypic or total variation for the trait), $H = \sigma_g^2 / \sigma_p^2$ is the heritability of the trait, and σ_g^2 is the additive variation attributable to genetic sources. Selection intensity (k) is adjusted by the breeder's control of each generation's proportion saved for the next cycle of selection. The breeder also has some control of variability through choice of a mating design.

The selection intensity, selection scheme and mating design chosen may change the response per cycle of selection, but the mating design also influences the length of time required to complete each cycle of selection (Table 3). Ultimately, each choice or adjustment of a procedure depends on accumulated information about the trait in question. In a population generated by the dent inbreds of Table 2, selection would be best accomplished by testing S_1 or S_2 progenies and recombining the selected one from remnant seed, because these systems use a large proportion of the additive genetic variance in the genetic gain formula, the type of variation critical to general combining ability.

A program of recurrent selection for reduced amounts of aflatoxin contamination from such a population among S_1 progenies was begun at Tifton, Georgia, in 1983, and the second cycle of selfed progeny evaluations is now being completed. The estimate of heritability using components of variance from the first cycle of progeny evaluation was 48%. This estimate was higher than estimates from the diallel tests, probably because the σ_g^2 estimate from the S_1 progeny tests contains a genotype x environment interaction that was removed from the diallel estimates. Evaluation of selection progress will be conducted as soon as the selected progenies are recombined to form the second cycle population.

Sweetcorn inbreds (Table 2) used to form a breeding population for reduced aflatoxin present a selection situation somewhat different from that of the dent population. Specific combining ability, in addition to general combining ability, is also important among the sweetcorn inbreds; it

Table 3. Time requirements and coefficients for genetic variance used in genetic gain formulae to estimate progress of various selection schemes[a/]

Selection scheme	Coefficient for σ_g^2	Generations per cycle
Mass[b/]	1/2	1
Half-sib[c/]	1/4	2
Full-sib	1/2	2
Half-sib progeny[d/]	1/2	3
S_1 progeny	1	3

a/ Each selection procedure results in a slightly different method for estimating σ_p^2 and definition of σ_g^2 changes slightly as inbreeding occurs

b/ Selection for one sex only

c/ Remnant half-sib seed is used for recombination

d/ Selfed seed is used for recombination

suggests the use of half-sib progenies, for example, to allow testing of crossbred performance, and places some emphasis on both general and specific combining abilities.

Very little is known about the basic chemical nature of plant resistance to infection and contamination. As this information becomes available, the potential of genetic engineering techniques to provide the resistance needed for transfer to hybrids is increased. Tissue and cell culture techniques will become more useful as the problems associated with plant regeneration in maize are resolved. Since good sources of resistance are not now available for maize, and assuming that sources of resistance can be identified in other species, interspecific gene transfer would seem to be the most promising approach.

Utilization of Available Germplasm

The final stage of a breeding strategy employs the developed germplasm as part of a working program of aflatoxin control. Such a program controls contamination by utilizing as many factors as possible that influence the contamination process. Many of these will be genetically controlled and closely related to traits selected during development of the germplasm. The factors would include the use of adapted hybrids, timely irrigation, good fertilization practices, chemical control of insects and disease when necessary, as well as limited environmental control through the adjustment of planting dates. Each of these factors, some of which are difficult to adjust, becomes less important as the amount of direct genetic control imposed by the available germplasm is increased.

High levels of control have not been achieved with available hybrids, but that seems likely to change as resistant germplasm is developed. However, any present program of control will rely heavily on improved management and must exploit related factors in every way possible. This approach, along with the use of existing hybrids with modest levels of resistance, could substantially reduce contamination of the maize grain crop in many areas. As more resistant germplasm becomes available and we learn more about the nature of resistance, a continual reduction in preharvest contamination can be expected.

References

1. Anderson, H.W., E.W. Nehring and W.R. Wichser. 1975. Aflatoxin contamination of corn in the field. Journal of Agricultural and Food Chemistry 23:775-782.

2. Davis, N D., C.G. Currier and U.L. Diener. 1985. Response of corn hybrids to aflatoxin formation by *Aspergillus flavus*. Alabama Agricultural Experiment Station Bulletin 575. Auburn, Alabama, USA.

3. Gardner, C.A.C., and J.R. Wallin. 1980. Aflatoxin production in maize endosperm mutants. Transactions of the Missouri Academy of Science 14:174.

4. King, S.B., and G.E. Scott. 1982. Field inoculation techniques to evaluate maize for reaction to kernel infection by *Aspergillus flavus*. Phytopathology 72:782-785.

5. King, S.B., and G.E. Scott. 1982. Screening maize single crosses for resistance to preharvest infection of kernels by *Aspergillus flavus*. Phytopathology 72:942.

6. LaPrade, J.C., and A. Manwiller. 1976. Aflatoxin production and fungal growth on single-cross corn hybrids inoculated with *Aspergillus flavus*. Phytopathology 66:675-677.

7. LaPrade, J.C., and A. Manwiller. 1977. Relation of insect damage, vector, and hybrid reaction to aflatoxin B_1 recovery from field corn. Phytopathology 67:544-547.

8. Lillehoj, E.B., W.F. Kwolek, D.I. Fennell and M.S. Milburn. 1975. Aflatoxin incidence and association with bright greenish-yellow fluorescence and insect damage in a limited survey of freshly harvested high moisture corn. Cereal Chemistry 52:403-412.

9. Lillehoj, E.B., W.F. Kwolek, G.M. Shannon, O.L. Shotwell and C.W. Hesseltine. 1975. Aflatoxin occurrence in 1973 corn at harvest. I. A limited survey in the southeastern U.S. Cereal Chemistry 52:603-611.

10. Lillehoj, E.B., W.F. Kwolek, E.E. Vandergraft, M.S. Zuber, O.H. Calvert, N. Widstrom, M.C. Futrell and A. J. Bockholt. 1975. Aflatoxin production in *Aspergillus flavus* inoculated ears of corn grown at diverse locations. Crop Science 15:267-270.

11. Lillehoj, E.B., W.F. Kwolek, M.S. Zuber, A.J. Bockholt, O.H. Calvert, W.R. Findley, W.D. Guthrie, E.S. Horner, L.M. Josephson, S. King, A. Manwiller, D.B. Sauer, D. Thompson, M. Turner and N.W. Widstrom. 1980. Aflatoxin in corn before harvest: Interaction of hybrids and locations. Crop Science 20:731-734.

12. Lillehoj, E.B., M.S. Zuber, L.L. Darrah, W.F. Kwolek, W. R. Finley, E.S. Horner, G.E. Scott, A. Manwiller, D.B. Sauer, D. Thompson, H. Warren, D.R. West and N.W. Widstrom. 1983. Aflatoxin occurrence and levels in preharvest corn kernels with varied endosperm characteristics grown at diverse locations. Crop Science 23:1181-1184.

13. Manwiller, A., and B. Fortnum. 1979. A comparison of aflatoxin levels in preharvest corn from the South Carolina coastal plain for 1977-78. South Carolina Agricultural Experiment Station, Agronomy and Soils Research Series 101. Clemson, South Carolina, USA.

14. McMillian, W.W., N.W. Widstrom and D.M. Wilson. 1981. Rearing the maize weevil on maize genotypes when aflatoxin-producing *Aspergillus flavus* and *A. parasiticus* isolates were present. Environmental Entomology 10:760-762.

15. McMillian, W.W., N.W. Widstrom and D.M. Wilson. 1982. Aflatoxin production on various popcorn genotypes. Agronomy Journal 74:156-157.

16. McMillian, W.W., D.M. Wilson and N.W. Widstrom. 1985. Aflatoxin contamination of preharvest corn in Georgia: A six-year study of insect damage and visible *Aspergillus flavus*. Journal of Environmental Quality 14. (In press.)

17. Rambo, G.W., J. Tuite and R.W. Caldwell. 1974. *Aspergillus flavus* and aflatoxin in preharvest corn from Indiana in 1971 and 1972. Cereal Chemistry 51:595-604.

18. Taubenhaus, J.J. 1920. A study of the black and the yellow molds of ear corn. Texas Agricultural Experiment Station Bulletin 270. College Station, Texas, USA.

19. Tulpule, P.G., R.V. Bhat, V. Nagarajan and E. Priyadarshini. 1977. Variations in aflatoxin production due to fungal isolates and crop genotypes and their scope in prevention of aflatoxin production. Archives of the Pasteur Institute (Tunis) 54:487-493.

20. Wallin, J.R., M.S. Zuber, D.V. Loonan and C.A.C. Gardner. 1980. The response of inbred lines of maize challenged with *Aspergillus flavus*. Transactions of the Missouri Academy of Science 14:174.

21. Widstrom, N.W., W.W. McMillian and D.M. Wilson. 1984. Contamination of preharvest corn by aflatoxin. *In* Proceedings of the 39th Annual Corn and Sorghum Research Conference. Pp. 68-83.

22. Widstrom, N.W., W.W. McMillian, D.M. Wilson, D. L. Garwood and D.V. Glover. 1984. Growth characteristics of *Aspergillus flavus* on agar infused with maize kernel homogenates and aflatoxin contamination of whole kernel samples. Phythopathology 74:887-890.

23. Widstrom, N.W., D.M. Wilson and W.W. McMillian. 1984. Ear resistance of maize inbreds to field aflatoxin contamination. Crop Science 24:1155-1157.

24. Widstrom, N.W., D.M. Wilson and W.W. McMillian. 1986. Differentiation of maize genotypes for aflatoxin concentration in developing kernels. Crop Science. (In press.)

25. Widstrom, N.W., B.R. Wiseman, W.W. McMillian, W.F. Kwolek, E.B. Lillehoj, M.D. Jellum and J.H. Massey. 1978. Evaluation of commercial and experimental three-way corn hybrids for aflatoxin B_1 production potential. Agronomy Journal 70:986-988.

26. Widstrom, N.W., and M.S. Zuber. 1983. Prevention and control of aflatoxin in corn: Sources and mechanisms of genetic control in the plant. *In* Aflatoxin and *Aspergillus flavus* in Corn, U.L. Diener, R.L. Asquith and J.W. Dickens, eds. Southern Cooperative Series Bulletin 279. Alabama Agricultural Experiment Station, Auburn, Alabama, USA. Pp. 72-76.

27. Zuber, M.S. 1977. Influence of plant genetics on toxin production in corn. *In* Mycotoxins in Human and Animal Health, J.V. Rodricks, C.W. Hesseltine and M.A. Mehlman, eds. Pathotox, Park Forest South, Illinois, USA. Pp. 173-179.

28. Zuber, M.S., O.H. Calvert, W.F. Kwolek, E.B. Lillehoj and M.S. Kang. 1978. Aflatoxin B_1 production in an eight-line diallel of *Zea mays* infected with *Aspergillus flavus*. Phytopathology 68:1346-1349.

29. Zuber, M.S., O.H. Calvert, E.B. Lillehoj and W.F. Kwolek. 1976. Preharvest development of aflatoxin B_1 in corn in the United States. Phytopathology 66:1120-1121.

30. Zuber, M.S., L.L. Darrah, E.B. Lillehoj, L.M. Josephson, A. Manwiller, G.E. Scott, R.T. Gudauskas, E.S. Horner, N.W. Widstrom, D.L. Thompson, A.J. Bockholt and J.L. Brewbaker. 1983. Comparison of open-pollinated maize varieties and hybrids for preharvest aflatoxin contamination in the southern United States. Plant Disease 67:185-187.

Host-Plant Resistance: Screening Techniques

G.E. Scott and N. Zummo, Crop Science Research Laboratory, Agricultural Research Service, US Department of Agriculture, Mississippi State University, Mississippi State, Mississippi, USA

Abstract

*Published studies to evaluate response of different maize (*Zea mays *L.) genotypes to infection with* Aspergillus flavus *Link ex Fries and/or aflatoxin production by the fungus in preharvest grain are reviewed. A number of inoculation techniques have been used in an attempt to obtain a relatively high and uniform level of infection with* A. flavus*; some of these techniques damage kernels and others are designed to avoid kernel damage. Kernel infection and/or aflatoxin concentration have been evaluated using undamaged kernels only, kernels from the whole ear or damaged kernels only. Kernel infection with* A. flavus *and aflatoxin concentration have been much higher when only damaged kernels were included in the sample. Significant differences among maize hybrids for kernel infection with* A. flavus *or aflatoxin concentration have been detected in at least some environments with many of the inoculation techniques, but the variability of the results has been greater than desired and is unacceptable for some inoculation techniques. An inoculation technique that produces a consistently high and uniform level of kernel infection and aflatoxin concentration without damaging the kernel has not yet been developed.*

Resumen

*Se revisan estudios publicados cuyo fin fue evaluar la respuestas de diferentes genotipos de maíz (*Zea mays *L.) a la infección con esporas de* Aspergillus flavus *Link ex Fries y/o a la producción de aflatoxinas por parte del hongo en el grano antes de la cosecha. Se han empleado diversas técnicas de inoculación con el objeto de obtener un nivel relativamente alto y uniforme de infección con* A. flavus*; algunas de estas técnicas dañan los granos y otras están diseñadas para evitar el daño a los granos. Se evalúa la infección de los granos y/o la concentración de aflatoxinas utilizando únicamente granos no dañados, granos extraídos de toda la mazorca o únicamente granos dañados. La infección de los granos con* A. flavus *y la concentración de aflatoxinas fue mucho mayor cuando sólo se incluyeron granos dañados en la muestra. Se detectaron diferencias significativas entre los híbridos de maíz respecto a la infección de los granos con* A. flavus *o a la concentración de aflatoxinas en algunos ambientes con muchas de las técnicas de inoculación, pero la variabilidad de los resultados fue más grande de lo conveniente y resulta inaceptable en relación con ciertas técnicas de inoculación. Todavía no se ha creado una técnica de inoculación que produzca un nivel constantemente alto y uniforme de infección en los granos y de concentración de aflatoxinas sin dañar los granos.*

Although the principles of screening maize (*Zea mays* L.) for resistance to *Aspergillus flavus* Link ex Fries and/or aflatoxin production are not necessarily different than for other diseases, many rather distinctive problems have been encountered. There are a number of difficulties associated with obtaining high uniform levels of kernel infection with *A. flavus*:

- *Aspergillus flavus* is a weak pathogen;

- The fungus may or may not produce aflatoxin; and
- Infection of kernels must occur when they are on the ear.

There are also problems associated with the evaluation of genotypes for response to infection, including:

- Variation among samples;
- Evaluation procedures (both for kernel infection and aflatoxin determination) are time consuming and costly;
- The resistance level must be high because the acceptable tolerance level of aflatoxin in grain is very low;
- Kernels are the only portion of the plant that can be evaluated but different tissues in the kernel have different genotypes; and
- Infection cannot be determined just by looking at the ear.

Because the evolution of genotypes is time consuming and costly, the results from one year's testing may not be available to plant for the following year's experiments.

This paper reviews studies published on the reaction of two or more maize genotypes to *A. flavus*, even though the main purpose of some of the studies may have been other than screening for resistance among maize genotypes. Data obtained when screening genotypes for response to a pathogen are often not appropriate for publication as journal articles. At best, they are often presented at annual commodity meetings and appear in abstracts, i.e., King and Scott (12). However, because of the importance of aflatoxin contamination, some screening experiments have been published as journal articles. King and Wallin (13) have previously reviewed the literature on methods for screening maize for resistance to *A. flavus* and/or aflatoxin production.

Rambo *et al.* (30) evaluated four hybrids for incidence of visible *A. flavus* growth and bright greenish-yellow fluorescence (BGYF). They inoculated by spraying the silks with a spore suspension, by injecting the ear through the husk and inserting a swab dusted with spores into a hole drilled into the cob. They did not report whether or not the differences among hybrids were significant.

Five South Carolina single-cross hybrids and one open-husk short-season maize cultivar were evaluated by LaPrade and Manwiller (14). Four methods of inoculation were used: inoculum applied to the silks, inoculum applied to the surface of uninjured tip seeds of each ear following hand removal of the husks, inoculum injected forcibly through the husk into a single seed and inoculum injected forcibly through the husk into three seeds per ear. They found significant differences among hybrids for aflatoxin concentration.

Lillehoj *et al.* (23) evaluated opaque-2 and normal endosperm types of two different hybrids. They relied on natural inoculation, ear inoculation with hypodermic syringe and ear damage without inoculation by stapling through the husk. Aflatoxin concentration, averaged over four locations, differed significantly between the two hybrids. Differences between endosperm types were not significant.

In 1974 Lillehoj *et al.* (22) evaluated six maize hybrids at Florence, South Carolina, and Gainesville, Florida. Hybrids included five South Carolina experimental single crosses and a commercial single-cross hybrid adapted to the Corn Belt. Treatments included inoculating silks with spores of *A. flavus*, mechanical damage to uninoculated ears, an untreated control and applying an insecticide to untreated plants. The ears inoculated with *A. flavus* had a higher incidence of BGYF and higher aflatoxin levels than the controls. Significant differences among hybrids for BGYF and aflatoxin concentration were observed, with the highest amounts being found in the hybrid adapted to the Corn Belt.

In a study by Zuber *et al.* (42), no significant differences were found in aflatoxin concentration between two single crosses (one with a thick pericarp and the other with a thin pericarp) grown at 22 locations under conditions of natural infection. Using six different kernel injury inoculation techniques, Calvert *et al.* (2) compared aflatoxin amounts in kernels with a thick pericarp to amounts in kernels from a thin pericarp single cross. The thin pericarp single cross had significantly higher levels of aflatoxin.

LaPrade and Manwiller (15) inoculated nine short-season hybrids with an *A. flavus* spore suspension at three sites on each ear using needle inoculation through the husk. The aflatoxin concentration varied from 88 to 145 μg/kg, and differences among hybrids were significant. They also evaluated 27 long-season hybrids and found that, although differences for aflatoxin concentration ranged from only 0 to 46 μg/kg, differences among hybrids were significant.

Ten full-season and 10 short- to midseason hybrids were tested under natural infection by Widstrom *et al.* (40). No significant differences among hybrids were detected when toxin concentrations were averaged over all locations. However, at the Tifton location, significant differences occurred among hybrids, in tests with and without corn earworm [*Heliothis zea* (Boddie)] infestation. They also reported results from 30 dent and 15 sweetcorn three-way crosses from plots infested with 30 corn earworm eggs and inoculated with *A. flavus* spores injected into the silk mass. Later each ear was also inoculated through the husk at three places. Genotypic differences for aflatoxin concentration were not detected in either of the three-way tests.

Zuber *et al.* (43) evaluated 28 single crosses of an 8-parent diallel for aflatoxin. In this study, the husks were pulled back and the kernels injured with a pinboard. The ears were then inoculated with a spore suspension of *A. flavus*, after which the husks were repositioned over the ear and secured with rubber bands. Significant differences for aflatoxin concentration were found among the crosses.

In a study designed to develop a system to estimate preharvest aflatoxin contamination, Lillehoj *et al.* (24) tested two commercial hybrids at nine locations under natural *A. flavus* infection. They found no difference between hybrids for toxin occurrence.

Lillehoj *et al.* (17), in an attempt to evaluate aflatoxin concentration in blends of inoculated and noninoculated kernels, inoculated two maize hybrids with *A. flavus* using a pinboard device. No significant differences for aflatoxin concentration occurred between the two single crosses, B73 X Mo17 and N28 X Mo17.

Jones and Wallin (7) and Wallin *et al.* (34) used a decapped kernel method to screen maize for response to *A. flavus*. Decapped kernels were placed cut-side down in petri dishes on moist paper containing *A. flavus* conidia. After inoculation, kernels were evaluated for visible *A. flavus* growth, BGYF and aflatoxin content. A few of the genotypes had significantly lower levels of *A. flavus* growth and BGYF emission.

Lillehoj *et al.* (27) evaluated 12 commercial and experimental single and three-way cross hybrids at 12 locations. Treatments included a control, inoculation into the extended silk channel with *A. flavus* and damage of kernels with a pinboard pressed through the husk. Significant differences were found among hybrids for aflatoxin concentration in the inoculated samples.

Lillehoj *et al.* (25) tested four hybrids (two developed for the central Corn Belt and two developed for the southern region) with three planting dates at nine locations under conditions of natural inoculation with *A. flavus*. They reported significant differences among hybrids at some of the locations. However, the relative response of each hybrid over locations was not as consistent as desired.

In another study, Lillehoj *et al.* (20) grew 12 hybrids in Georgia, Florida and Tennessee. Treatment involved a control (natural *A. flavus* infection) and inoculation into the silk channel. Aflatoxin varied among hybrids at a location, but uniform susceptibility of individual hybrids at all locations was not observed.

McMillian *et al.* (29) tested mature ears of two hybrids using maize weevils (*Sitophilus zeamais* Motschulsky), some of which had been dusted with spores of *A. flavus*. They found significant differences among hybrids for percentage of ear area with visible *A. flavus*.

Two genotypes were evaluated by Thompson *et al.* (31) in controlled environment chambers. Ears were inoculated by peeling back the husk on one side, applying a pinboard to damage two adjacent rows of kernels and spraying with an *A. flavus* spore suspension. Husks were then repositioned, secured with a rubber band and covered with a paper bag. No significant difference between maize genotypes for aflatoxin concentration was reported.

Wilson *et al.* (41) monitored the incidence of *A. flavus* on insect-damaged maize ears, BGYF of kernels and aflatoxin concentration of kernels. They used treatment combinations of silk inoculation with *A. flavus* and infestation with corn earworms,

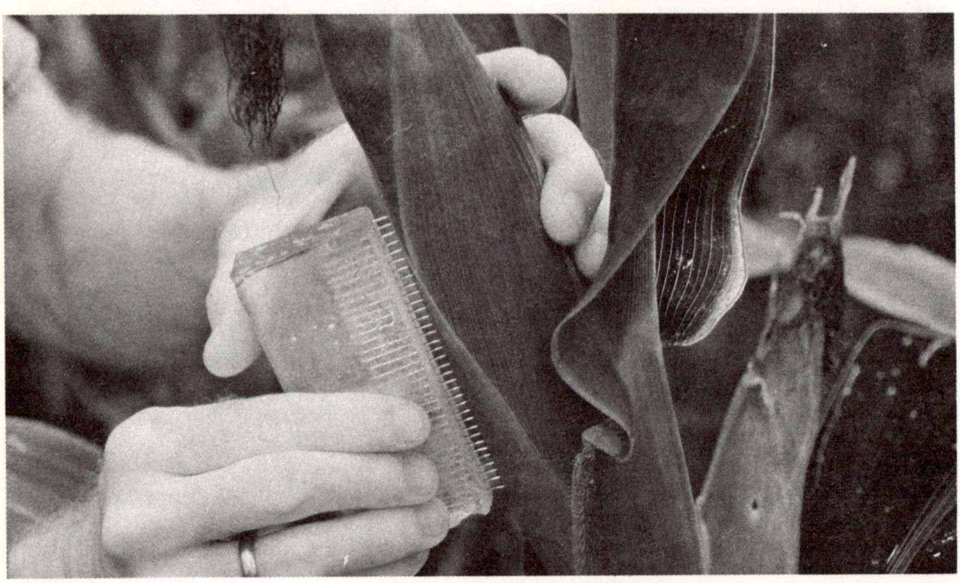

Fungal growth on one row as a result of pinbar inoculation

including infested-inoculated, infested only, inoculated only and the control. Aflatoxin levels were found to be significantly different among hybrids in 1975 but not in 1976.

Using the pinboard technique, Jones and Duncan (8) inoculated ears of two hybrids with *A. flavus* in a study to evaluate cultivar, planting date, harvest date and nitrogen fertilization. No significant differences among maize genotypes were found.

Widstrom *et al.* (36) inoculated four hybrids with *A. flavus* using a needle through the husks, a knife, and multiple-puncture injury to the kernels. They measured aflatoxin concentration, insect damage to the ear and percentage of ears with visible *A. flavus*. Differences among hybrids were not significant for any of the three characteristics when the averages for two years were compared. However, differences among hybrids were significant in the heavily contaminated 1977 test.

Lauver and Calvert (16) inoculated surface-sterilized kernels with conidia of *A. flavus* and tumbled them in water to increase their moisture content to 25 to 28%. The kernels were then incubated for 14 days and assayed for frequency of infection and aflatoxin content. No differences were detected among the three lines tested.

Ten commercial maize hybrids were evaluated by King and Scott (11) using four inoculation techniques and a noninoculated control. No significant differences among genotypes with natural infection or with silk inoculation were detected. Exposed kernel inoculation (the husk on one side of the ear was peeled back, the ear inoculated with *A. flavus* spore suspension, and the husk repositioned) had statistically significant differences among hybrids for each of the two years, but agreement between years was not good. Significant differences among hybrids for percent of kernels infected with *A. flavus* were detected

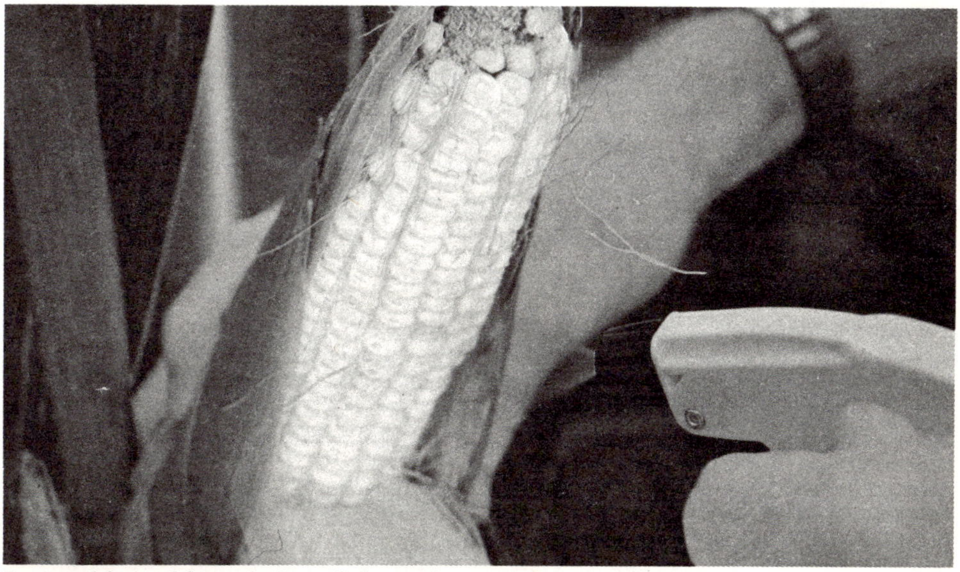

Exposed kernel inoculation of the maize ears

with single kernel infection (evaluating the first, second, third and fourth kernels from the inoculated kernel) and with pinbar inoculation (evaluating the kernels on the first and second row from the inoculated row).

McMillian *et al.* (28) evaluated 17 popcorn genotypes for three years. In the first two years, ears were inoculated 20 days after full silk by injecting an *A. flavus* spore suspension into the top, middle and base of the ear with a hypodermic syringe. In 1980, ears were inoculated by the knife inoculation technique. Over the three-year evaluation, the geometric mean for aflatoxin concentration ranged from 31 to 320 ppb. Differences among genotypes were significant.

In a study to determine the interactions among the European corn borer, three fungal species and developing kernels of two maize genotypes, Lillehoj *et al.* (18) reported results from three locations. No significant differences were observed between hybrids. Widstrom *et al.* (37) grew two commercial maize hybrids in 1978 and 1980 to evaluate 13 sampling methods (grain-sampling-inoculation combinations) for *A. flavus* concentration in the grain. The two inoculation techniques used were knife inoculation and needle inoculation of the silk masses at the ear tips. Significant differences between the two hybrids, averaged over the 13 sampling methods, were detected for aflatoxin concentration.

Gardner (6) evaluated three inbred lines with three *A. flavus* inoculation techniques. Inoculation techniques used were pinboard, silk and toothpick-cob. Differences in aflatoxin production were not detected among the three inbred lines. In 1980, Lillehoj *et al.* (19) evaluated 26 hybrids grown at Florence, South Carolina, in yield trials. These hybrids had received only natural infection with *A. flavus*. The range in aflatoxin concentration was from 1 to 614 μg/kg. Differences among hybrids were statistically significant, but primarily because of one hybrid that had the highest level of aflatoxin. More levels of significance were obtained among the hybrids for BGYF.

Twenty hybrids of different endosperm types were tested by Lillehoj *et al.* (26). Hybrids were subjected to natural infection with *A. flavus*. Significant differences among hybrids were detected for the mean aflatoxin concentration over locations. Zuber *et al.* (44) evaluated four adapted maize hybrids and eight open-pollinated varieties under conditions of natural *A. flavus* infection. One open-pollinated variety, Huffman, had significantly higher levels of aflatoxin concentration than the other genotypes evaluated.

King and Scott (10) employed a laboratory technique based on the extent of growth of *A. flavus* on kernels in petri dishes to differentiate among genotypes. Using the same technique, Adams *et al.* (1) inoculated kernels with a spore suspension of *A. flavus* and incubated the kernels to allow the fungus to grow and develop. A rating scale of 1 to 8 was used to indicate the amount of growth and development of *A. flavus* on the maize kernels. Significant differences among 10 single crosses and four commercial hybrids were found.

Differences in *A. flavus* growth on inoculated kernels were also found among four inbreds and 16 single crosses. Lillehoj *et al.* (21) evaluated eight treatments of two hybrids for reaction to *A. flavus*/aflatoxin. The treatments were: control; *A. flavus* (silk) inoculation; corn earworm (CEW); fungicide; *A. flavus* + CEW; *A. flavus* + fungicide; CEW + fungicide; and *A. flavus* + CEW + fungicide. No differences between the hybrids for aflatoxin concentration were found.

Eight hybrids were grown by Thompson *et al.* (32) in the phytotron and in the field, and kernels were inoculated with *A. flavus* by peeling back the husk on one side of the ear, applying a pinbar to damage adjacent rows and then spraying with *A. flavus* spore suspension. Husks were repositioned, secured and covered with plastic and paper bags for three days to maintain humidity. Ears were shelled at maturity and assayed for aflatoxin concentration. Significant differences for aflatoxin concentration were detected among hybrids.

Widstrom *et al.* (38) evaluated the 36 possible single crosses of a nine-parent diallel of dent inbreds and the 28 possible single crosses of an eight-parent diallel of sweetcorn inbreds. Ears were inoculated with a spore suspension of *A. flavus* in the tip, middle and base with a hypodermic syringe or with the knife inoculation technique. Significant differences for aflatoxin concentration were detected. Widstrom *et al.* (39) inoculated eight commercial hybrids in the field using a hypodermic needle through the husks. They found differences for aflatoxin concentration among hybrids. Grain from other commercial hybrids was also infused in agar, placed in petri plates and inoculated with *A. flavus*. Colony diameter was a better criterion than sporulation characteristics for detecting differences among hybrids.

Fortnum and Manwiller (5) evaluated 15 commercial hybrids for field aflatoxin concentration. Natural infection produced significant differences for aflatoxin concentration among some hybrids, but hybrid response was not consistent. When damaged and intact kernels were inoculated with a spore suspension of *A. flavus*, differences were obtained among hybrids, although again the consistency over years was not high.

Tucker *et al.* (33) evaluated four single crosses using the pinbar, knife, exposed kernel and silk inoculation techniques, as well as natural infection. They found significant differences among hybrids for aflatoxin concentration with each of the inoculation techniques, but the aflatoxin concentration with silk and natural inoculation were very low on all hybrids. For *A. flavus* kernel infection, differences among hybrids were detected with all inoculation techniques. However, again, the silk inoculated and noninoculated material had a very low percentage of kernels infected. The percentage of kernels with BGYF differed among hybrids only with pinbar inoculation.

Davis *et al.* (3) presented results from aflatoxin analysis of 215 maize hybrids grown at one or more of 12 locations in Alabama in one or more years during the period 1976-1981. All hybrids were subjected to natural infection. The authors concluded that there was no resistance in the hybrids tested. However, they did not present any statistical analysis of their data.

Zummo and Scott (45) evaluated maize genotypes for *A. flavus* kernel infection by a number of techniques (pinbar, needle application of inoculum through the silk channel directly onto the ear, infested toothpick in the ear, infested toothpick in the silk channel and infested string around the silks). Kernel infection percentages were higher for the pinbar and needle inoculation, and both techniques showed significant differences among genotypes tested. Needle inoculation did not produce more kernel infection than the pinbar technique, but it did not cause kernel damage, was easier to apply and did not require hand shelling of inoculated ears.

The crosses of a ten-parent diallel have been evaluated (unpublished data) for percentage of *A. flavus* kernel infection for two years (Table 1). Based on previous single-cross data, five inbreds were designated as resistant and five inbreds as susceptible. Single crosses were inoculated with a pinbar, and kernels from the adjacent row were evaluated for percentage of infected kernels. Significant differences among hybrids were detected each year. In 1983, the results from resistant x resistant, resistant x susceptible and susceptible x susceptible crosses fairly well verified the previous classification of the parents as resistant or susceptible. However, with the lower infection level in 1984, differences based on previous parental classification were not obvious.

As can be seen by the information presented in this paper, a number of inoculation techniques have been used in an attempt to identify maize hybrids resistant to infection by *A. flavus* and/or aflatoxin production. Natural infection has been effective in detecting significant differences among genotypes in some studies (5,19,25,26,33,44) but not in others (3,11,24,42). Natural infection often produces relatively low aflatoxin concentrations. Needle inoculation of one or more kernels through the husks at a few positions on the ear has been effective in detecting significant differences among genotypes in some cases (14,15,22,23,28,37,39), but not in others (36,40). Inoculation of the silks has also been successful in some cases (20,37,33,41) and not in others (6,11,14,21,40). Of all inoculation techniques evaluated, the pinboard inoculation technique resulted in the highest aflatoxin concentration in the ear. However, differences among genotypes based on pinboard inoculation are inconsistent (2,32,43 and 6,8,17,31,36). Inoculation with *A. flavus* through the husk using a

Table 1. Percentage of kernels infected with *Aspergillus flavus* in row adjacent to pinbar-inoculated row in resistant x resistant, resistant x susceptible and susceptible x susceptible crosses, Mississippi, 1983 and 1984

Common parent	1983 Crossed with		1984 Crossed with	
	Resistant	Susceptible	Resistant	Susceptible
Resistant				
Mp337	15	21	10	6
GA209	13	14	11	11
Mp317	14	11	11	8
Tx601	13	19	7	10
T232	11	15	11	12
Mean	13	16	10	9
Susceptible				
Mp307	15	29	9	8
Mp428	16	27	8	6
Mp486	20	33	8	9
SC212M	19	20	11	9
SC376	14	27	10	8
Mean	17	27	8	8

knife was effective in a number of cases (28,33,37), but in another study was not (36). The pinbar inoculation technique usually shows differences among the genotypes being evaluated (11,15,33). Needle inoculation, in which inoculum is applied down the silk channel and onto the ear, has been effective in only one reported case (45).

The inability to consistently detect significant differences among genotypes being tested can be the result of the facts that the genotypes do not differ, that inoculation techniques result in infection rates that are too low or too variable, or that sampling and/or evaluation procedures do not give a reasonably good estimate of the genotypic mean for the field plot from which the sample was harvested. A major constraint has been that highly resistant and susceptible genotypes have not been identified for use in comparing various inoculation and evaluation techniques. Without known resistant and susceptible genotypes, the merits of the inoculation and evaluation techniques cannot be evaluated, and without the proper inoculation and evaluation techniques, the genotypes for resistance and susceptiblity cannot be identified.

It seems self-evident that screening maize genotypes for infection by *A. flavus* or aflatoxin production may not necessarily show differences among them. Of course, with larger numbers being evaluated, the probability is greater that some will have a higher level of resistance than others.

The inoculation and evaluation techniques currently used may contribute to the inability to precisely measure differences among genotypes. What is needed is a representative sample of the response of a genotype to *A. flavus* and/or aflatoxin production. Thus, it also becomes a sampling problem. For instance, if the inoculation technique involves kernel damage, will the best results be obtained by evaluating only damaged kernels, only intact kernels or all kernels? If only damaged kernels are assessed, the aflatoxin concentration will probably be high and a possible source of resistance may be missed because its kernels have been damaged mechanically. If only intact kernels are evaluated, the spread of *A. flavus* from the damaged kernels is being determined and the samples will probably vary because the kernels adjacent to the damaged kernels will have a higher *A. flavus* infection and/or aflatoxin concentration than kernels further removed from them. If all kernels are examined, then the sample evaluation may more accurately reflect the number of damaged kernels in the sample than the genotypic response to infection by *A. flavus* or aflatoxin concentration. Dickens and Whitaker (4) and Whitaker and Dickens (35) have discussed the problems of sampling. Even inoculation techniques that do not damage the kernel pose some problems in evaluation; this is primarily because of insect damage and the relative number of insect-damaged kernels that go into the sample being evaluated.

A better understanding of the infection process of *A. flavus* in maize ears and kernels is needed. Although the organism is mainly considered a saprophyte, an understanding of the factors involved in the shift to parasitic metabolism is critical in establishing a base of knowledge of the fungus in developing maize.

Screening maize genotypes for kernel infection with *A. flavus* and/or aflatoxin production is not an impossible task. There is no question that it offers a challenge. For screening, it would be desirable to have elevated pathogenicity of *A. lavus*, but such increased disease potential is undesirable in the agronomic context. Some progress has been made in screening for resistance, and because genotypes have been identified that differ in response to *A. flavus* infection, the rate of progress in detecting higher levels of resistance should increase.

References

1. Adams, N.J., G.E. Scott and S.B. King. 1984. Quantification of *Aspergillus flavus* growth on inoculated excised kernels of corn genotypes. Agronomy Journal 76:98-102.

2. Calvert, O.H., E.B. Lillehoj, W.F. Kwolek and M.S. Zuber. 1978. Aflatoxin B_1 and G_1 production in developing *Zea mays* kernels from mixed inocula of *Aspergillus flavus* and *A. parasiticus*. Phytopathology 68:501-506.

3. Davis, N.D., C.G. Currier and U.L. Diener. 1985. Response of corn hybrids to aflatoxin formation by *Aspergillus flavus*. Alabama Agricultural Experiment Station Bulletin 575. Auburn, Alabama, USA.

4. Dickens, J.W., and T.B. Whitaker. 1983. Aflatoxin testing procedures for corn. *In* Aflatoxin and *Aspergillus flavus* in Corn, U.L. Diener, R.L. Asquith and J.W. Dickens, eds. Southern Cooperative Series Bulletin 279. Alabama Agricultural Experiment Station, Auburn, Alabama, USA. Pp. 35-37.

5. Fortnum, B.A., and A. Manwiller. 1985. Effects of irrigation and kernel injury on aflatoxin B_1 production in selected maize hybrids. Plant Disease 69:262-265.

6. Gardner, C.A. 1982. Studies relating to aflatoxin production in maize. PhD Thesis. University of Missouri, Columbia, Missouri, USA.

7. Jones, D.L., and J.R. Wallin. 1978. Evaluating resistance of corn strains to *Aspergillus flavus* and aflatoxin contamination. Transactions of the Missouri Academy of Science 12:154. Abstract.

8. Jones, R.K., and H.E. Duncan. 1981. Effect of nitrogen fertilizer, planting date, and harvest date on aflatoxin production in corn inoculated with *Aspergillus flavus*. Plant Disease 65:741-744.

9. Jones, R.K., H.E. Duncan, and P.B. Hamilton. 1981. Planting date, harvest date, and irrigation effects on infection and aflatoxin production by *Aspergillus flavus* in field corn. Phytopathology 71:810-816.

10. King, S.B., and G.E. Scott. 1981. Screening maize for resistance to kernel infection by *Aspergillus flavus*. Phytopathology 71:231. Abstract.

11. King, S.B., and G.E. Scott. 1982. Field inoculation techniques to evaluate maize for reaction to kernel infection by *Aspergillus flavus*. Phytopathology 72:782-785.

12. King, S.B., and G.E. Scott. 1982. Screening maize single crosses for resistance to preharvest infection of kernels by *Aspergillus flavus*. Phytopathology 72:942. Abstract.

13. King, S.B., and J.R. Wallin. 1983. Methods for screening corn for resistance to kernel infection and

aflatoxin production by *Aspergillus flavus*. *In* Aflatoxin and *Aspergillus Flavus* in Corn, U.L. Diener, R.L. Asquith and J.W. Dickens, eds. Southern Cooperative Series Bulletin 279. Alabama Agricultural Experiment Station, Auburn, Alabama, USA.

14. LaPrade, J.C., and A. Manwiller. 1976. Aflatoxin production and fungal growth on single cross corn hybrids inoculated with *Aspergillus flavus*. Phytopathology 66:675-677.

15. LaPrade, J.C., and A. Manwiller. 1977. Relation of insect damage, vector, and hybrid reaction to aflatoxin B_1 recovery from field corn. Phytopathology 67:544-547.

16. Lauver, E.L., and O.H. Calvert. 1981. A laboratory screening procedure for resistance in corn aflatoxin production. Phytopathology 71:888. Abstract.

17. Lillehoj, E.B., O.H. Calvert, W.F. Kwolek and M.S. Zuber. 1979. Aflatoxin variation among corn samples with varying ratios of *Aspergillus flavus*-inoculated/ noninoculated kernels. Journal of the Association of Official Analytical Chemists 62(5):1083-1086.

18. Lillehoj, E.B., W.F. Kwolek, W.D. Guthrie, D. Barry, W.W. McMillian and N.W. Widstrom. 1982. Aflatoxin accumulation in preharvest maize kernels: Interaction of three fungal species, European corn borer and two hybrids. Plant and Soil 65:95-102.

19. Lillehoj, E.B., A. Manwiller, T.B. Whitaker and M.S. Zuber. 1983. Hybrid differences in estimation of preharvest occurrence of bright greenish-yellow fluorescence and aflatoxin in corn. Journal of Environmental Quality 12:216-218.

20. Lillehoj, E.B., W.F. Kwolek, E.S. Horner, N.W. Widstrom, L.M. Josephson, A.O. Franz and E.A. Catalano. 1980. Aflatoxin contamination of preharvest corn: Role of *Aspergillus flavus* inoculum and insect damage. Cereal Chemistry 57:255-257.

21. Lillehoj, E.B., W.W. McMillian, N.W. Widstrom, W.D. Guthrie, J.L. Jarvis, D. Barry and W.F. Kwolek. 1984. Aflatoxin contamination of maize kernels before harvest. Mycopathologia 85:77-81.

22. Lillehoj, E.B., W.F. Kwolek, A. Manwiller, J.A. DuRant, J.C. LaPrade, E.S. Horner, J. Reid and M.S. Zuber. 1976. Aflatoxin production in several corn hybrids grown in South Carolina and Florida. Crop Science 16:483-485.

23. Lillehoj, E.B., W.F. Kwolek, E.E. Vandegraft, M.S. Zuber, O.H. Calvert, N.W. Widstrom, M.C. Futrell and A.J. Bockholt. 1975. Aflatoxin production in *Aspergillus flavus* inoculated ears of corn grown at diverse locations. Crop Science 15:267-270.

24. Lillehoj, E.B., W.F. Kwolek, M.S. Zuber, O.H. Calvert, E.S. Horner, N.W. Widstrom, W.D. Guthrie, G.E. Scott, D.L. Thompson, W.R. Findley and A.J. Bockholt. 1978. Aflatoxin contamination of field corn: Evaluation of regional test plots for early detection. Cereal Chemistry 55:1007-1013.

25. Lillehoj, E.B., W.F. Kwolek, M.S. Zuber, E.S. Horner, N.W. Widstrom, W.D. Guthrie, M. Turner, D.B. Sauer, W.R. Findley, A. Manwiller and L.M. Josephson. 1980. Aflatoxin contamination caused by natural fungal infection of preharvest corn: Variation between hybrids developed for specific regions but grown at diverse locations. Plant and Soil Science 54:469-475.

26. Lillehoj, E.B., M.S. Zuber, L.L. Darrah, W.F. Kwolek, W.R. Findley, E.S. Horner, G.E. Scott, A. Manwiller, D.B. Sauer, D. Thompson, H. Warren, D.R. West and N.W. Widstrom. 1983. Aflatoxin occurrence and levels in preharvest corn kernels with varied endosperm characteristics grown at diverse locations. Crop Science 23:1181-1184.

27. Lillehoj, E.B., W.F. Kwolek, M.S. Zuber, A.J. Bockholt, O.H. Calvert, W.R. Findley, W.D. Guthrie, E.S. Horner, L.M. Josephson, S.B. King, A. Manwiller, D.B. Sauer, D. Thompson, M. Turner and N.W. Widstrom. 1980. Aflatoxin in corn before harvest: Interaction of hybrids and locations. Crop Science 20:731-734.

28. McMillian, W.W., N.W. Widstrom and D.M. Wilson. 1982. Aflatoxin production on various popcorn genotypes. Agronomy Journal 74:156-157.

29. McMillian, W.W., N.W. Widstrom, D.M. Wilson and R.A. Hill. 1980. Transmission by maize weevils of *Aspergillus flavus* and its survival on selected corn hybrids. Journal of Economic Entomology 73:793-794.

30. Rambo, G.W., J. Tuite and P. Crane. 1973. Preharvest inoculation and infection of dent corn ears with *Aspergillus flavus* and *A. parasiticus*. Phytopathology 64:797-800.

31. Thompson, D.L., E.B. Lillehoj, K.J. Leonard, W.F. Kwolek and M.S. Zuber. 1980. Aflatoxin concentration in corn as influenced by kernel development stage and postinoculation temperature in controlled environments. Crop Science 20:609-612.

32. Thompson, D.L., J.O. Rawlings, M.S. Zuber, G.A. Payne and E.B. Lillehoj. 1984. Aflatoxin accumulation in developing kernels of eight maize single crosses after inoculation with *Aspergillus flavus*. Plant Disease 68:465-467.

33. Tucker, D.H. Jr., L.E. Trevathan, S.B. King and G.E. Scott. 1986. Infection and aflatoxin contamination of resistant and susceptible corn hybrids inoculated with *Aspergillus flavus* using four inoculation techniques. Phytopathology 76:290-293.

34. Wallin, J.R., M.S. Zuber, D.V. Loonan and C.A.C. Gardner. 1980. The response of inbred lines of maize challenged with *Aspergillus flavus*. Transactions of the Missouri Academy of Science 14:174. Abstract.

35. Whitaker, T.B., and J.W. Dickens. 1983. Evaluation of a testing program for aflatoxin in corn. Journal of the Association of Official Analytical Chemists 66:1055-1058.

36. Widstrom, N.W., D.M. Wilson and W.W. McMillian. 1981. Aflatoxin contamination of preharvest corn as influenced by timing and method of inoculation. Applied and Environmental Microbiology 42:249-251.

37. Widstrom, N.W., D.M. Wilson and W.W. McMillian. 1982. Evaluation of sampling methods for detecting aflatoxin contamination in small test plots of maize inoculated with *Aspergillus flavus*. Journal of Environmental Quality 11:655-657.

38. Widstrom, N.W., D.M. Wilson and W.W. McMillian. 1984. Ear resistance of maize inbreds to field aflatoxin contamination. Crop Science 24:1155-1157.

39. Widstrom, N.W., W.W. McMillian, D.M. Wilson, D.L. Garwood and D.V. Glover. 1984. Growth characteristics of *Aspergillus flavus* on agar infused with maize kernel homogenates and aflatoxin contamination of whole kernel samples. Phytopathology 74:887-890.

40. Widstrom, N.W., B.R. Wiseman, W.W. McMillian, W.F. Kwolek, E.B. Lillehoj, M.D. Jellum and J.H. Massey. 1978. Evaluation of commercial and experimental three-way corn hybrids for aflatoxin B_1 production potential. Agronomy Journal 70:986-988.

41. Wilson, D.M., N.W. Widstrom, L.R. Marti and B.D. Evans. 1981. *Aspergillus flavus* group, aflatoxin, and bright greenish yellow fluorescence in insect-damaged corn in Georgia. Cereal Chemistry 58:40-42.

42. Zuber, M.S., O.H. Calvert, E.B. Lillehoj and W.F. Kwolek. 1976. Preharvest development of aflatoxin B_1 in corn in the United States. Phytopathology 66:1120-1121.

43. Zuber, M.S., O.H. Calvert, W.F. Kwolek, E.B. Lillehoj and M.S. Kang. 1978. Aflatoxin B_1 production in an eight-line diallel of *Zea mays* infected with *Aspergillus flavus*. Phytopathology 68:1346-1349.

44. Zuber, M.S., L.L. Darrah, E.B. Lillehoj, L.M. Josephson, A. Manwiller, G.E. Scott, R.T. Gudauskas, E.S. Horner, N.W. Widstrom, D.L. Thompson, A.J. Bockholt and J.L. Brewbaker. 1983. Comparison of open-pollinated maize varieties and hybrids for preharvest aflatoxin contamination in the southern United States. Plant Disease 67:185-187.

45. Zummo, N., and G.E. Scott. 1985. Evaluation of field inoculation techniques for screening corn genotypes against kernel infection by *Aspergillus flavus* in Mississippi. Phytopathology 75:1330.

Regionally Adapted Maize Hybrids with Respect to Levels of Aflatoxin Contamination

A. Manwiller, USAID, Nairobi, Kenya

When it was first generally recognized that aflatoxin could be a preharvest problem in maize (*Zea mays* L.), it seemed practical to evaluate commercial hybrids in actual production. In 1977, Manwiller (3) sampled 70 hybrids grown under severe drought stress from two South Carolina coastal plain locations. Thirteen hybrids had aflatoxin levels of 20 ng/g^{-1} or less, 43 yielded 21 to 100 ng/g^{-1} and 15 exceeded 100 ng/g^{-1}; the worst had 280 ng/g^{-1}. For the most part, the entries with 20 ng/g^{-1} or less were statistically better than those with over 100 ng/g^{-1}. The full-season types (1200 AES maturity) with long, tight husks had a better record than the semident Corn Belt group with shorter, looser husks. Also, the grain quality ratings (ear rots and insect damage) were reasonably well correlated with aflatoxin levels.

Therefore, it seemed that some agronomically satisfactory hybrids in each maturity group were safe for human food, while many others could be fed to the more resistant types of livestock. The marks of excellence were a long tight husk and resistance to rots, molds and the major insect pests.

The following year Manwiller and Fortnum (4) repeated the experiment and found poor correlation between aflatoxin levels of individual hybrids over the two years. Only a few cases were sufficiently consistent to be classified as high or low for both years. These were among those used by Lillehoj et al. (1,2) and Zuber et al. (5) in artificial inoculation experiments, insect evaluations, etc., in which the same material was planted in several Corn Belt states, together with a number of southern states, and the data were pooled. Since uninoculated checks were used routinely, it was possible to follow natural contamination patterns over a wide range of environments. The overall toxin averages were often reassuring, but these were misleading as the hazardous results were diluted with those from areas where little or no aflatoxin was produced. In every case examined, one or more locations yielded unacceptable amounts of aflatoxin for each hybrid entry.

The aflatoxin levels under artificial inoculation were, on occasion, extraordinarily high and showed the true potential for aflatoxin production, given their optimum conditions of kernel injury, fungal presence and weather stress. While the problem is probably under complex genetic control, it is a classic case of overkill; the susceptible inoculated samples yield 800 ng/g^{-1} and the resistant ones 25% as much (either level can kill).

The huge breeding effort to date has netted a very small positive result. This is not to say that a "safe" hybrid cannot be produced, but present evidence is still very sketchy. Perhaps the gene splicers will be able to add something from the soybean (*Glycine max* L. Merril) that will be able to break through the barrier.

References

1. Lillehoj, E.B., W.K. Kwolek, M.S. Zuber, E.S. Horner, N.W. Widstrom, W.D. Guthrie, M. Turner, D.B. Sauer, W.R. Findley, A. Manwiller and L.M. Josephson. 1980. Aflatoxin contamination caused by natural fungal infection of preharvest corn. Plant and Soil 54:475.

2. Lillehoj, E.B., W.K. Kwolek, M.S. Zuber, A.J. Bockholt, O.J. Calvert, W.R. Findley, W.D. Guthrie, E.S. Horner, L.M. Josephson, S. King, A. Manwiller, D.B. Sauer, D. Thompson, M. Turner and N.W. Widstrom. 1980. Aflatoxin in corn before harvest: Interaction of hybrids and locations. Crop Science 20:731-734.

3. Manwiller, A. 1978. Aflatoxin in preharvest corn from the South Carolina coastal plain. South Carolina Agricultural Experiment Station Agronomy and Soils Research Series 98. Clemson, South Carolina, USA.

4. Manwiller, A., and B. Fortnum. 1979. A comparison of aflatoxin levels in preharvest corn from the South Carolina coastal plain for 1977-78. South Carolina Agricultural Experiment Station Agronomy and Soils Research Series 101. Clemson, South Carolina, USA.

5. Zuber, M.S., L.L. Darrah, E.B. Lillehoj, L.M. Josephson, A. Manwiller, G.E. Scott, R.T. Gudauskas, E.S. Horner, N.W. Widstrom, D.L. Thompson, A.J. Bockholt and J.L. Brewbaker. 1983. Comparison of open-pollinated maize varieties and hybrids for preharvest aflatoxin contamination in the southern United States. Plant Disease 67:185-187.

Yield and the Genetic Control of Aflatoxin in Maize

L.L. Darrah, Agricultural Research Service, US Department of Agriculture, M.S. Zuber, Department of Agronomy, both of the University of Missouri, Columbia, Missouri, and E.B. Lillehoj, Agricultural Research Service, US Department of Agriculture, New Orleans, Louisiana, USA

Abstract

*Germplasm substantially reducing the production of aflatoxin B_1 in maize (*Zea mays *L.) kernels infected with* Aspergillus flavus *Link ex Fries has not been identified. Despite several studies that indicate different types of genetic control of aflatoxin production in US maize, progress in breeding has been lacking. To broaden the search for resistance to naturally occurring contamination, 12 CIMMYT germplasm pools and 17 advanced populations were evaluated at seven sites in the USA and Mexico. Because stress caused by lack of adaptation may enhance the production of aflatoxin, the pools and populations were topcrossed onto the inbred lines B73 and Mo17, which represent the dominant heterotic groups in US maize production. Entries were evaluated for yield to obtain information on combining ability and for aflatoxin B_1. Differences among entries for yield were significant at all sites. Population 29 yielded significantly more when crossed to B73 than when crossed to Mo17. Three genotypes, Pool 26 and Populations 22 and 43, appeared to cross well with both testers but did not fit into either heterotic group. Differences among entries for aflatoxin B_1 were significant only in South Carolina and in the combined analysis. Three entries, all pools or populations per se, had significantly more aflatoxin than the mean of all entries and none had less aflatoxin. Correlations between yield and aflatoxin were not significant. Grouping the pools and their topcrosses by flint and dent endosperm types showed a significant difference in the pools, where the dent endosperm types had significantly more aflatoxin B_1. Differences in the B73 and Mo17 topcrosses exhibited similar trends, but the variation was not significant. Effects of the flint or dent endosperm may have been diluted in the testcrosses.*

Resumen

*No se ha identificado ningún germoplasma que reduzca en forma sustancial la producción de aflatoxina B1 en los granos de maíz (*Zea mays *L.) infectados con esporas de* Aspergillus flavus *Link ex Fries. A pesar de los diversos estudios efectuados que indican diferentes tipos de control genético de la producción de aflatoxinas en el maíz estadounidense, no se ha realizado ningún progeso en el mejoramiento. A fin de ampliar la búsqueda de la obtención de resistencia a la contaminación que se produce en forma natural, se evaluaron 12 complejos germoplásmicos y 17 poblaciones avanzadas del CIMMYT en siete localidades de Estados Unidos y México. Dado que el estrés causado por la falta de adaptación suele incrementar la producción de aflatoxinas, se efectuaron cruzas triples entre los complejos y las poblaciones y las líneas endogámicas B73 y Mo17, que representan a los grupos heterocigóticos dominantes en la producción estadounidense de maíz. Las entradas se evaluaron respecto al rendimiento a fin de obtener información sobre la capacidad de combinación y respecto a la aflatoxina B_1. En todas las localidades, las diferencias entre las entradas en cuanto al rendimiento fueron significativas. La*

Población 29 tuvo un rendimiento mucho mayor cuando se la cruzó con B73 que cuando se la cruzó con Mo17. Tres genotipos, el Complejo 26 y las Poblaciones 22 y 43, parecieron cruzarse bien con los dos probadores, pero no se adaptaban a ninguno de los dos grupos heterocigóticos. Las diferencias de las entradas en cuanto a la aflatoxina B_1 sólo fueron significativas en Carolina del Sur y en el análisis combinado. Tres entradas, que eran complejos o poblaciones, presentaron un nivel significativamente mayor de aflatoxinas que la media de todas las entradas y ninguna presentó niveles menores de aflatoxina. No fueron significativas las correlaciones entre el rendimiento y el nivel de aflatoxinas. La clasificación de los complejos y sus cruzas triples de acuerdo con el tipo de endosperma (cristalino o dentado) mostró una diferencia significativa en los complejos, ya que los tipos de endosperma dentado presentaron una cantidad mucho mayor de aflatoxina B1. Las diferencias en las cruzas triples con B73 y Mo17 presentaron tendencias semejantes, aunque la variación no fue significativa. Es posible que los efectos del endosperma cristalino o dentado se hayan perdido en las cruzas de prueba.

Production of aflatoxin B_1 in maize (*Zea mays* L.) kernels infected with *Aspergillus flavus* Link ex Fries appears to be controlled genetically. Zuber *et al.* (4), using the pinboard inoculation method, found significant differences for aflatoxin B_1 content among the 28 possible F_1 crosses of eight maize inbred lines; two replications of a 13-plant plot were used. Estimates of general combining ability were highly significant, and estimates of specific combining ability were nonsignificant. General combining ability effects and specific combining ability effects for aflatoxin levels reported in the study may have been affected by ear size, because kernels from whole ears, rather than only infected kernels, were ground and analyzed.

A similar diallel, including seven of the eight lines used by Zuber *et al.* (4), was grown by Gardner (2), who also used the pinboard inoculation method; in this study only inoculated kernels were analyzed. Genotypic differences were significant for both aflatoxin B_1 and B_2, as was the variance associated with general and specific effects. Specific combining ability sum of square estimates accounted for approximately 65% of the genotype sum of squares. Coefficients of variation were high, 92% for B_1 and 90% for B_2. In spite of the differences in the studies, the estimates of general combining ability effects and rankings of aflatoxin levels from crosses and parental line means were in general agreement with Zuber *et al.* (4).

Results from other, unpublished, genetic studies have shown large error variances and poor separation of means. These results have been attributed to very large genotype x environment interactions and the lack of suitable methods of artificial inoculation. The pinboard method defeats any protection offered by the husk, silk channel, pericarp and/or aleurone layer. The importance of these barriers is indicated by the significant rank correlation between insect damage rating and aflatoxin levels from natural contamination found in a study of 12 genotypes at eight locations over two years (3).

The question arose as to whether results from artificial inoculation studies would be the same as those from studies conducted under naturally occurring aflatoxin levels. In 1981, 21 F_1 crosses from a diallel mating of the same parents as used by Gardner (2) were grown in six environments with six replications each (Darrah, personal communication). Silks were sprayed with a spore suspension of *A. flavus* approximately five days after full extrusion, and the ears were covered for three days to produce conditions of high humidity. Sites in Florida, Georgia, Mississippi, North Carolina, South Carolina and Tennessee were chosen because of the high levels of natural aflatoxin incidence found in previous studies (no data were obtained from Florida).

Aflatoxin levels differed significantly among genotypes and environments. Mean aflatoxin B_1 levels ranged from a low of 54 ng/g^{-1} at Knoxville, Tennessee, to 1788 ng/g^{-1} at Mississippi State, Mississippi. General and specific combining ability mean squares were highly significant. Comparison of aflatoxin levels by F_1 crosses, or estimates of general combining ability effects under natural contamination versus results of artificial inoculation, showed negative relationships in this study. Results indicated that genotypic information on genetic control of aflatoxin contamination obtained by artificial inoculation with *A. flavus* was not the same as that provided by natural contamination.

Naturally occurring aflatoxin contamination was compared in a two-year study in the southern USA by Zuber *et al.* (3). Four widely grown commercial hybrids and eight open-pollinated varieties adapted to the region were grown with two replications at eight locations in Alabama, Florida, Georgia, Mississippi, North Carolina, South Carolina, Texas and Hawaii. No attempt was made to control insects or inoculate with *Aspergillus* species. Aflatoxin levels differed significantly among entries. The open-pollinated variety Huffman had significantly more aflatoxin than any other entry, and it also had the highest insect-damage rating. The mean of the four commercial hybrids was significantly lower than the mean of the eight open-pollinated varieties. Contributing to this difference may be greater tolerance to stress in the commercial hybrids as compared to the older, open-pollinated varieties.

No single source of resistance has been identified for incorporating resistance into breeding material. Artificial inoculation techniques do not mimic natural infestation adequately; therefore, the best current procedure appears to be the evaluation of naturally occurring contamination. Evaluation should be conducted in geographic areas where aflatoxin occurs with demonstrated frequency.

Tropical maize germplasm is often not adapted to the USA and could be considered to be under stress when grown there. Since resistance to aflatoxin contamination in US germplasm appeared to be limited, several of the CIMMYT germplasm pools and populations were considered. To reduce the lack of adaptation, the pools and populations were topcrossed onto the inbred lines B73 and Mo17, representing the major heterotic groups in US maize production. The objective of the study was to evaluate the combining ability of the pools and populations and to screen for possible resistance to contamination by naturally occurring aflatoxin.

Materials and Methods

Twelve CIMMYT pools and 17 populations were chosen for topcross evaluation (Table 1). The pools included all combinations of the tropical lowland classification of early, intermediate or late maturities, white or yellow kernel color and dent or flint endosperm (1). The population represented a range of material included in the CIMMYT international maize testing program.

Topcrosses were made in the winter of 1982 in Maui, Hawaii, by DeKalb-Pfizer Genetics. Except for Population 28 (Amarillo Dentado), topcrosses were successfully made to both B73 and Mo17 in sufficient quantity for the evaluation. Included in the evaluation were the pools and populations, the topcrosses to B73 and Mo17 and checks to bring the total number of entries to 90.

Ten environments were sampled for the evaluation, ranging from Iowa to Mexico; of these, sites in Iowa, Missouri and Costa Rica did not produce data. The US sites were all planted in April or May, 1983, and the two sites in Mexico, at Poza Rica and Tlaltizapan, were planted in December, 1983. The experimental design was a quadruple rectangular lattice with an experimental unit of one row of 20 seeds. Analyses of variance were made assuming that the entries were fixed variables and that the combinations of locations and year were random environments.

Table 1. Tropical lowland pools and populations included in the topcross evaluation, USDA-University of Missouri study, 1982 and 1983

Pool or population	Name	Characteristic
Pool 15		Early white flint
Pool 16		Early white dent
Pool 17		Early yellow flint
Pool 18		Early yellow dent
Pool 19		Intermediate white flint
Pool 20		Intermediate white dent
Pool 21		Intermediate yellow flint
Pool 22		Intermediate yellow dent
Pool 23		Late white flint
Pool 24		Late white dent
Pool 25		Late yellow flint
Pool 26		Late yellow dent
Pop. 21	Tuxpeño 1	
Pop. 22	Mezcla Tropical Blanco	
Pop. 23	Blanco Cristalino-1	
Pop. 24	Antigua Veracruz 181	
Pop. 26	Mezcla Amarilla	
Pop. 27	Amarillo Cristalino-1	
Pop. 28	Amarillo Dentado	
Pop. 29	Tuxpeño Caribe	
Pop. 30	Blanco Cristalino-2	
Pop. 31	Amarillo Cristalino-2	
Pop. 35	Antigua Republica Dominicana	
Pop. 36	Cogollero	
Pop. 38	Poza Dulce (MS) 6 QPM[a]	
Pop. 39	Yellow QPM[a]	
Pop. 40	White QPM[a]	
Pop. 43	La Posta	
Pop. 49	Blanco Dentado-2	

[a] QPM refers to quality protein maize.

For aflatoxin analyses, 10 ears were harvested after physiological maturity and immediately dried at 60°C for three to four days to stop further fungal growth. After drying, ears were shelled and the grain samples sent to the US Department of Agriculture's Southern Regional Research Center in New Orleans, Louisiana, for aflatoxin analyses. Although results were provided for aflatoxin B_1 and B_2, the occurrence of aflatoxin B_2 was sporadic, and significant differences among entries were found only at Winterville, North Carolina. The combined analysis of variance for aflatoxin B_2 did not show significant differences among entries.

Results and Discussion

Data were received from sites in Georgia, Mississippi, North Carolina, South Carolina, Texas and Mexico (Poza Rica and Tlaltizapan) for aflatoxin and from all but Mississippi for yield. Yields ranged from 62.7 quintals/ha^{-1} (1 quintal = 100 kg) in Georgia down to 25.4 q/ha^{-1} in North Carolina (Table 2). The US growing season in 1983 was characterized by a prolonged drought during the latter part of July and early August, accompanied by higher than normal temperatures; crop failure occurred in Iowa and Missouri. Differences among entries were significant at each site where yield was measured. Coefficients of variation had a not unexpected spread of values, with the Mexican trials at Poza Rica and Tlaltizapan having the lowest coefficients of variation (CVs) of 12.9 and 12.8%, respectively. The relatively higher CVs noted for the US sites may be largely a reflection of low mean yields.

Differences among entries for aflatoxin B_1 were significant in South Carolina and in the combined analysis. Amounts of aflatoxin B_1 ranged from 1.6 ng/g^{-1} at Tlatizapan to 162.7 ng/g^{-1} in North Carolina (Table 2); South Carolina averaged 20.7 ng/g^{-1}. The CVs differed considerably. North Carolina had a CV of 157.2%, whereas Poza Rica had 803.1% and Tlaltizapan, 967.0%. As with yield, the low mean incidence of natural contamination at the Mexican sites contributed to the relatively high CVs. The combined analysis CV, based on the genotype x site interaction, was 273.4%. The range of CVs observed substantiates the need for adequate replication and for choosing sites where incidence of natural contamination is high.

Mean yields varied from 4.0 q/ha^{-1} for the inbred Mo17 to 67.1 q/ha^{-1} for the topcross of Population 29 to B73 (Table 3). Seventeen entries had yields significantly exceeding the mean for all entries (48.4 q/ha^{-1}). Of these, 13 involved topcrosses to B73 and the

Table 2. Location and combined statistics for yield (q/ha^{-1}) and aflatoxin B_1 (ng/g^{-1}), USDA-University of Missouri study, 1982 and 1983

	USA					Mexico		
Character	Georgia	Mississippi	North Carolina	South Carolina	Texas	Poza Rica	Tlaltizapan	Combined
Yield								
Mean	62.7	—	25.4	37.1	46.9	58.2	60.0	48.4
LSD 0.05	22.0	—	12.5	11.0	13.6	10.6	10.8	11.6
CV %	24.9	—	34.7	21.0	20.4	12.9	12.8	41.5
Aflatoxin B_1								
Mean	55.9	3.7	162.7	20.7	38.7	9.2	1.6	43.2
LSD 0.05	ns	ns	ns	92.2	ns	ns	ns	67.6
CV %	191.7	716.5	157.2	320.5	192.4	803.1	967.0	273.4

remaining four were topcrosses to Mo17. Over half of the pools and populations yielded significantly less than the mean of all entries and 60% or less than the mean for the topcrosses.

The topcross yields were examined to determine if any heterotic patterns might be found. Only Population 29 had topcrosses to B73 and Mo17 that differed significantly, with the cross to B73 yielding 67.1 q/ha^{-1} and the topcross to Mo17 yielding 54.5 q/ha^{-1}. Pools or populations that favored neither heterotic group, but performed well with both testers, were Pool 26 and Populations 22 and 43. All of the topcrosses for these three sources yielded significantly above the mean of all entries. Adaptation of the topcrosses was adequate in most environments, as reflected by the average or better yields obtained. The pools and populations, however, were probably under some degree of adaptive stress as evidenced by their relatively low yield.

Significant differences among entries for aflatoxin B_1 were found in the combined analysis of variance. Three

Table 3. Combined yield and aflatoxin data from sites growing the 1983 topcross evaluation of CIMMYT pools and populations, USDA-University of Missouri, study, 1982 and 1983

Entry	No.	Yield (q/ha^{-1})	Aflatoxin B_1 (ng/g^{-1})
Pool 15 (TEWF)	1	31.3	113.5
Pool 16 (TEWD)	2	34.2	55.5
Pool 17 (TEYF)	3	31.3	29.2
Pool 18 (TEYD)	4	33.4	35.4
Pool 19 (TIWF)	5	36.1	66.0
Pool 20 (TIWD)	6	43.2	146.3
Pool 21 (TIYF)	7	43.5	35.9
Pool 22 (TIYD)	8	41.2	98.9
Pool 23 (TLWF)	9	40.9	33.4
Pool 24 (TLWD)	10	39.8	73.0
Pool 25 (TLYF)	11	40.1	20.2
Pool 26 (TLYD)	12	36.1	65.0
Popn. 21 (Tuxpeño-1)	13	29.8	53.4
Popn. 22 (Mezcla Tropical Blanco)	14	40.9	46.5
Pop. 23 (Blanco Cristalino-1)	15	37.9	61.3
Pop. 24 (Antigua Veracruz 181)	16	36.0	70.9
Pop. 26 (Mezcla Amarilla)	17	36.9	30.8
Pop. 27 (Amarillo Cristalino-1)	18	36.5	122.2
Pop. 28 (Amarillo Dentado)	19	32.2	76.2
Pop. 29 (Tuxpeño Caribe)	20	37.6	40.3
Pop. 30 (Blanco Cristalino-2)	21	38.0	17.6
Pop. 31 (Amarillo Cristalino-2	22	33.3	25.7
Pop. 35 (Antigua Republica Dominicana)	23	36.3	47.9
Pop. 36 (Cogollero)	24	39.3	43.6
Pop. 38 (Poza Dulce (MS) 6 QPM)	25	28.3	76.7

Table 3.(continued)

Entry	No.	Yield (g/ha^{-1})	Aflatoxin B_1 (ng/g^{-1})
Pop. 39 (Yellow QPM)	26	34.3	27.0
Pop. 40 (White QPM)	27	35.9	37.9
Pop. 43 (La Posta)	28	34.3	69.8
Pop. 49 (Blanco Dentado-2)	29	30.2	83.6
Pool 15 x B73	30	50.3	10.9
Pool 16 x B73	31	54.6	16.6
Pool 17 x B73	32	45.6	9.8
Pool 18 x B73	33	49.2	11.4
Pool 19 x B73	34	61.3	28.1
Pool 20 x B73	35	62.2	32.9
Pool 21 x B73	36	56.4	15.8
Pool 22 x B73	37	63.0	16.9
Pool 23 x B73	38	61.9	46.3
Pool 24 x B73	39	64.3	67.1
Pool 25 x B73	40	62.0	7.0
Pool 26 x B73	41	60.3	25.2
Pop. 21 x B73	42	64.4	66.5
Pop. 22 x B73	43	63.7	41.9
Pop. 23 x B73	44	60.3	16.0
Pop. 24 x B73	45	61.8	34.3
Pop. 26 x B73	46	53.4	30.6
Pop. 27 x B73	47	59.4	38.6
Filler (B73 x Mo17)	48	48.0	55.7
Pop. 29 x B73	49	67.1	37.2
Pop. 30 x B73	50	49.7	13.9
Pop. 31 x B73	51	51.9	3.6
Pop. 35 x B73	52	59.0	93.1
Pop. 36 x B73	53	59.1	85.4
Pop. 38 x B73	54	56.5	28.0
Pop. 39 x B73	55	50.4	30.4
Pop. 40 x B73	56	55.5	55.1
Pop. 43 x B73	57	64.3	62.9
Pop. 49 x B73	58	59.9	43.8
Pool 15 x Mo17	59	47.5	14.1
Pool 16 x Mo17	60	52.0	22.8
Pool 17 x Mo17	61	47.3	29.4
Pool 18 x Mo17	62	49.3	64.6
Pool 19 x Mo17	63	52.6	79.1
Pool 20 x Mo17	64	52.3	51.8
Pool 21 x Mo17	65	54.1	29.8

Table 3. (continued)

Entry	No.	Yield (g/ha^{-1})	Aflatoxin B_1 (ng/g^{-1})
Pool 22 x Mo17	66	57.9	22.9
Pool 23 x Mo17	67	54.4	31.5
Pool 24 x Mo17	68	57.5	16.3
Pool 25 x Mo17	69	57.1	21.6
Pool 26 x Mo17	70	61.1	47.4
Pop. 21 x Mo17	71	59.8	62.2
Pop. 22 x Mo17	72	60.1	55.5
Pop. 23 x Mo17	73	51.9	14.5
Pop. 24 x Mo17	74	59.6	23.1
Pop. 26 x Mo17	75	57.8	12.2
Pop. 27 x Mo17	76	54.1	47.5
Filler (B73 x Mo17)	77	45.5	15.5
Pop. 29 x Mo17	78	54.5	11.8
Pop. 30 x Mo17	79	54.7	27.2
Pop. 31 x Mo17	80	41.3	14.2
Pop. 35 x Mo17	81	52.8	12.8
Pop. 36 x Mo17	82	51.2	36.0
Pop. 38 x Mo17	83	51.3	16.2
Pop. 39 x Mo17	84	51.1	33.3
Pop. 40 x Mo17	85	60.9	17.7
Pop. 43 x Mo17	86	63.2	54.7
Pop. 49 x Mo17	87	57.1	65.3
B73 x Mo17	88	46.5	26.9
B73	89	8.9	37.4
Mo17	90	4.0	18.4
Mean		48.4	43.2
LSD 0.05		11.6	67.6
CV% (based on site x entry mean square)		41.5	273.4
Number of sites		6	7

entries had significantly more aflatoxin than the mean for all entries, Pool 15 (113.5 ng/g^{-1}), Pool 20 (146.3 ng/g^{-1}) and Population 27 (122.2 ng/g^{-1}). No entry was significantly less than the mean of all entries, because the LSD 0.05 was greater than the mean. However, Population 31 topcrosses to B73 had only 3.6 ng/g^{-1} aflatoxin B_1. Whether that low value was due to inherent resistance or one of the barriers of husk protection, thick pericarp or a tight silk channel, is not known.

Differences among groups of genotypes were further examined by averaging similar entries. For instance, the four early maturity pools had an average yield of 32.6 q/ha^{-1}, with an average of 113.5 ng/g^{-1} of aflatoxin B_1 (Table 4). The early maturity pools averaged significantly lower yield than the intermediate or late pools. Similarly, the topcrosses of the early pools to B73 yielded significantly less than topcrosses of the intermediate and the late pools. For the Mo17 pool topcrosses, only the averages of the early pools and late pools differed significantly for yield. The populations yielded significantly less than their topcrosses to both B73 and Mo17.

Aflatoxin B_1 amounts found in the early, intermediate and late pools differed significantly, with more aflatoxin found in the earlier maturity entries. It is possible that this response represents a lack of adaptation through maturation before other maize in the vicinity. Usually, maize that matures early relative to other maize is more prone to bird damage and subsequent infestation with *A. flavus*. The populations had significantly more aflatoxin than their topcrosses to Mo17 (54.8 versus 31.5 ng/g^{-1}, respectively).

Correlations of yield and aflatoxin B_1 were calculated on the combined means for each grouping (Table 4) and for all entries. The correlation coefficient for all entries was r = -0.19, which was not significant at P = 0.05. None of the correlation coefficients for

Table 4. Summary of combined yield and aflatoxin B_1 data averaged across pool maturities and populations by testers, USDA-University of Missouri study, 1982 and 1983

Groups	No. of entries	Yield (q/ha^{-1})	Aflatoxin B_1 (ng/g^{-1})	r[a]
Early maturity pools	4	32.6	113.5	-0.33
Intermediate maturity pools	4	41.0	86.8	0.22
Late maturity pools	4	39.2	47.9	-0.54
Populations	17	35.2	54.8	-0.31
Early pools x B73	4	49.9	12.2	0.92
Intermediate pools x B73	4	60.7	23.4	0.46
Late pools x B73	4	62.1	36.4	0.69
Populations x B73	16	58.5	42.6	0.43
Early pools x Mo17	4	49.0	32.7	0.09
Intermediate pools x Mo17	4	54.2	45.9	-0.76
Late pools x Mo17	4	57.5	29.2	0.54
Populations x Mo17	16	55.1	31.5	0.46
B73 x Mo17 check hybrid	3	46.7	32.7	0.99
B73	1	8.9	37.4	—
Mo17	1	4.0	18.4	—
LSD 0.05 for comparing average of 4 entries		5.8	33.8	
LSD 0.05 for comparing average of 4 and 16/17 entries		4.6	26.7	
LSD 0.05 for comparing average of 16/17 entries		2.9	16.9	

[a] Correlation of yield and aflatoxin B_1 in each group; none of the coefficients were significant at P=0.05

the individual groupings were significant at P = 0.05. Signs and magnitudes of the coefficients suggested random variation. However, with only four entries in most of the groupings, significance was difficult to obtain. Yield was not significantly related to the levels of aflatoxin observed.

A question that is often asked is whether endosperm type has any relation to susceptibility to aflatoxin contamination. Means were calculated for the flint and dent endosperm types for all maturities of the pools and their topcrosses to B73 and Mo17 (Table 5). Mean yields did not differ significantly for any comparison of the endosperm types. However, the dent pools had significantly more aflatoxin B_1 than the flint pools (79.0 versus 49.7 ng/g^{-1}, respectively). Differences between aflatoxin levels for the flint and dent endosperm topcrosses to B73 and to Mo17 were not significant. The differences were in the same direction as the pools, suggesting that a dilution effect of the tester may have occurred. Characterization of the flint and dent pools for husk coverage and pericarp thickness might explain a part of the observed difference, but it is likely that the major effect is of endosperm vitreousness and decreased ability of the fungus to infect the vitreous kernels.

References

1. CIMMYT. 1982. CIMMYT Review. El Batan, Mexico.

2. Gardner, C.A.C. 1982. Studies relating to aflatoxin production in maize. PhD Dissertation. University of Missouri, Columbia, Missouri, USA.

3. Zuber, M.S., L.L. Darrah, E.B. Lillehoj, L.M. Josephson, A. Manwiller, G.E. Scott, R.T. Gudauskas, E. S. Horner, N.W. Widstrom, D. L. Thompson, A.J. Bockholt and J.L. Brewbaker. 1983. Comparison of open-pollinated maize varieties and hybrids for preharvest aflatoxin contamination in the southern United States. Plant Disease 67:185-187.

4. Zuber, M.S., O.H. Calvert, W.F. Kwolek, E.B. Lillehoj and M.S. Kang. 1978. Aflatoxin production in an eight line diallel of *Zea mays* L. infected with *Aspergillus flavus*. Phytopathology 68:1346-1349.

Table 5. Comparison of six entries of flint and dent endosperm maize types and their topcrosses to B73 and Mo17 for yield and aflatoxin B_1 using combined means, USDA-University of Missouri study, 1982 and 1983

Groups	Yield (q/ha^{-1})	Aflatoxin B_1 (ng/g^{-1})
Flint pools	37.2	49.7
Dent pools	38.0	79.0
Flint pools x B73	56.3	19.7
Dent pools x B73	58.9	28.3
Flint pools x Mo17	52.2	34.3
Dent pools x Mo17	55.0	37.6
LSD 0.05 for comparing groups	4.7	27.6

Potential Involvement of Plant Metabolites in Maize Resistance to Aflatoxin Contamination

D.M. Wilson, Department of Plant Pathology, Coastal Plain Experiment Station, University of Georgia, Tifton, Georgia, USA

Abstract

Limited information is available on the effects of volatile and nonvolatile metabolites of maize on Aspergillus flavus *and* A. parasiticus *and their aflatoxin production. The active volatile aldehydes are generally more inhibitory than active ketones or the corresponding alcohols. Beta-ionene has an unusual effect on growth and morphology of the* A. flavus *group. Some extracts of maize containing nonvolatile metabolites may contain unidentified compounds that will suppress growth and/or aflatoxin production by the* A. flavus *group. Breeding programs may be possible which manipulate inhibitory maize metabolites for resistance to aflatoxin contamination.*

Resumen

Se cuenta con muy poca información acerca de los efectos de los metabolitos volátiles y no volátiles del maíz sobre Aspergillus flavus *y* A. parasiticus *y su producción de aflatoxinas. En general, los aldehídos volátiles activos son más inhibitorios que las cetonas activas o los alcoholes correspondientes. La beta-ciclocitrilidenacetona ejerce un efecto desusado en el crecimiento y morfología del grupo* A. flavus. *Ciertos extractos del maíz que contienen metabolitos no volátiles suelen presentar compuestos no identificados que suprimen el crecimiento y/o la producción de aflatoxinas por parte del grupo* A. flavus. *Quizá sea posible crear programas de mejoramiento que manipulen los metabolitos inhibitorios del maíz para obtener resistencia a la contaminación por aflatoxinas.*

Aflatoxin contamination of maize, *Zea mays* L., follows the successful germination, colonization and growth of *Aspergillus flavus* Link ex Fries or *A. parasiticus* Speare. Aflatoxin contamination cannot occur without prior growth of a toxigenic member of the *A. flavus* group. However, maize colonized by a member of the *A. flavus* group is not necessarily colonized by an aflatoxin-producing isolate (6). Experimental approaches to aflatoxin control can emphasize either the control of fungal growth or inhibition of secondary aflatoxin production.

Plant metabolites that affect *A. flavus* group growth or aflatoxin production are infrequently reported in the literature (15). R. Gueldner at the US Department of Agriculture (USDA) Richard Russell Laboratory, Athens, Georgia, and N.W. Widstrom and W.W. McMillian at the USDA laboratory at Tifton, Georgia, have encouraging preliminary results in a cooperative project designed to identify chemicals in maize that affect the growth and sporulation of the *A. flavus* group.

Some maize metabolites that affect the *A. flavus* group and aflatoxin contamination have been reported by Nagarajan and Bhat (12). They found that a 5% aqueous NaCl extract of opaque-2 *Zea mays* L. inhibited toxin production, probably due to a low molecular weight protein. Priyadarshini and Tulpule (13) did not find a correlation between fungal growth and aflatoxin production and suggested that genotypic differences were due to nutritional factors. Wilson and co-workers (14) reported that beta-ionone affected both growth and aflatoxin synthesis. Other work in Georgia (9) demonstrated that several volatile maize metabolites inhibited *A. flavus* group growth.

Lillehoj, Garcia and Lambrow (11) suggested that mineral nutrition of maize, especially zinc, could influence aflatoxin contamination. They postulated that metal complexes with naturally occurring chelating agents, such as phytic acid, could be important in controlling aflatoxin contamination of maize. Gupta, Maggon and Venkitasubramanian (8) suggested that zinc-related glycolysis may be involved in aflatoxin biosynthesis.

Extracts from several plants and spices have been found to inhibit aflatoxin production, and quite often *A. parasiticus* has been more sensitive to inhibition than *A. flavus* (10). Buchanan and Shepard (2) demonstrated that thymol was generally more inhibitory to growth of the fungus than to toxin production.

Carrot root extracts affect growth and development of *A. parasiticus* and inhibit aflatoxin production (1). Raw carrot roots do not support growth of *A. parasiticus*, but cooked carrot roots support growth and aflatoxin production. Batt, Solberg and Ceponis (1) extracted carrot roots and found that chloroform extracts contained substances that inhibited sporulation and aflatoxin production by *A. parasiticus*; the extracts had little effect on its growth as measured by dry weight. Buttery and co-workers (3,5) identified several volatile compounds from carrot roots.

Plant volatiles have been implicated in regulation of fungal development (15). Buttery, Ling and Chan (4) used a steam distillation technique to identify 56 compounds in maize husk volatile oil and 34 compounds in maize kernel volatile oil. Flath and coworkers (7) identified 63 components in the volatile oil of maize silks. Wilson and co-workers (14) tested several of the volatile metabolites of maize in culture using a nontoxic *A. flavus* isolate. The aldehydes 2,4-hexadienal, trans-2-hexenal and trans-trans-2,4-decadienal all inhibited *A. flavus* growth. Several alcohols, including 1-heptanol, 2-nonanol, 1-nonanol, geraniol, 2-octanol and 2-decanol slightly inhibited *A. flavus* growth. Gueldner, Wilson and Heidt (9) found that furfural also inhibited growth of *A. flavus*. When natural and synthetic active compounds were tested, aldehydes were the most inhibitory, followed by ketones; the corresponding alcohols were the least inhibitory to growth.

The active aldehydes and alcohols inhibited growth and sporulation of *A. flavus* but did not alter conidiophore morphology. Beta-ionone severely restricted both growth and sporulation in culture. Direct contact with beta-ionone applied next to the agar surface resulted in very restricted growth, little sporulation and arrested asexual reproductive development. Contact with beta-ionone only in the vapor state resulted in slow growth and a typical conidial development, such as branching and unusual conidiophore morphology. In shake culture, *A. flavus* grew more slowly with 250 μl of beta-ionone per liter of liquid medium than with 50 μl of beta-ionone per liter of liquid medium. Concentrations of 100 μl/l and above of beta-ionone in the culture medium inhibited aflatoxin accumulation (14,15). It is interesting to note that beta-ionone also occurs in carrot roots (5). Perhaps beta-ionone and related carotenes contribute to the inhibition of *A. parasiticus* germination and growth in raw carrot roots. The loss of volatile compounds during cooking may be one reason that cooked carrots support *A. parasiticus* growth and aflatoxin production.

The manipulation of volatile or other maize metabolites in maize breeding programs for aflatoxin resistance seems possible. This experimental approach is not yet very advanced, but the published and preliminary work by the USDA in Georgia is encouraging. When optimum mixtures of maize ear volatile and nonvolatile compounds are identified, breeders may be able to develop tolerant genotypes based on the fungistatic properties of these chemicals.

References

1. Batt, C., M. Solberg and M. Ceponis. 1980. Inhibition of aflatoxin production by carrot root extract. Journal of Food Science 45:1210-1213.

2. Buchanan, R.L., and A.J. Sheperd. 1981. Inhibition of *Aspergillus parasiticus* by thymol. Journal of Food Science 46:976-977.

3. Buttery, R.G., D.R. Black, W.F. Haddon, L.C. Ling and R. Teranishi. 1979. Identification of additional volatile constituents of carrot roots. Journal of Agricultural and Food Chemistry 27:1-3.

4. Buttery, R.G., L.C. Ling and B.G. Chan. 1978. Volatiles of corn kernels and husks: possible corn ear worm attractants. Journal of Agricultural and Food Chemistry 26:866-869.

5. Buttery, R.G., R.M. Seifert, D.G. Guadagni, D.R. Black and L.C. Ling. 1968. Characterization of some volatile constituents of carrots. Journal of Agricultural and Food Chemistry 16:1009-1015.

6. Davis, N.D., U.L. Diener and V.P. Agnihotri. 1967. Production of aflatoxins B_1 and G_1 in chemically defined medium. Mycopathologia et Mycologia Applicata 31:251-256.

7. Flath, R.A., R.R. Forrey, J.O. John and B.G. Chan. 1978. Volatile components of corn silks (*Zea mays* L.): Possible *Heliothis zea* (Boddie) attractants. Journal of Agricultural and Food Chemistry 26:1290-1293.

8. Gupta, S.K., K.K. Maggon and T.A. Venkitasubramanian. 1977. Effect of zinc on tricarboxylic acid cycle intermediates and enzymes in relation to aflatoxin biosynthesis. Journal of General Microbiology 99:43-48.

9. Gueldner, R.C., D.M. Wilson and A.R. Heidt. 1985. Volatile compounds inhibiting *Aspergillus flavus*. Journal of Agricultural Chemistry 33:411-413.

10. Hitokoto, H., S. Morozumi, T. Wauke, S. Sakai and I. Ueno. 1978. Inhibitory effects of condiments and herbal drugs on the growth and toxin production of toxigenic fungi. Mycopathologia 66:161-167.

11. Lillehoj, E.B., W.J. Garcia and M. Lambrow. 1974. *Aspergillus flavus* infection and aflatoxin production in corn: Influence of trace elements. Applied Microbiology 28:763-767.

12. Nagarajan, V., and R.V. Bhat. 1972. Factor responsible for varietal differences in aflatoxin production in maize. Journal of Agricultural and Food Chemistry 20:911-914.

13. Priyadarshini, E., and P.G. Tulpule. 1978. Relationship between fungal growth and aflatoxin production in varieties of maize and groundnut. Journal of Agricultural and Food Chemistry 26:249-252.

14. Wilson, D.M., R.C. Gueldner, J.M. McKinney, R.H. Lievsay, B.D. Evans and R.A. Hill. 1981. Effect of β-ionone on *Aspergillus flavus* and *Aspergillus parasiticus* growth, sporulation, morphology and aflatoxin production. Journal of the American Oil Chemists' Society 58:959A-961A.

15. Wilson, D.M., R.C. Gueldner and R.A. Hill. 1983. Effects of plant metabolites on the *Aspergillus flavus* group and aflatoxin production. *In* Aflatoxin and *Aspergillus flavus* in Corn, U.L. Diener, R.L. Asquith and J.W. Dickens, eds. Southern Cooperative Series Bulletin 279. Alabama Agricultural Experiment Station, Auburn, Alabama, USA.

Maize Plant Resistance to Insect Damage and Associated Aflatoxin Development

W.W. McMillian, Insect Biology and Population Management Research Laboratory, Agricultural Research Service, US Department of Agriculture, Tifton, Georgia, USA

Abstract
*Field tests conducted at Tifton, Georgia, indicate that maize (*Zea mays *L.) hybrids adapted to the southern USA average less insect damage and aflatoxin contamination than hybrids adapted to the Corn Belt. The differences are apparently due primarily to the complete husk cover on the southern ear. Other evaluations of maize germplasm have revealed that both chemical and physical attributes of the silk and kernel may influence insect damage,* Aspergillus flavus *Link ex Fries growth and/or aflatoxin development.*

Resumen
*Pruebas de campo efectuadas en Tifton, Georgia, indican que los híbridos de maíz (*Zea mays *L.) adaptados a la zona sur de Estados Unidos presentan, en promedio, menores daños causados por insectos y menor contaminación por aflatoxinas que los híbridos adaptados a la faja maicera. Al parecer, las diferencias se deben fundamentalmente a la cobertura completa que presentan las mazorcas de la zona sur. Otras evaluaciones del germoplasma de maíz indican que las propiedades físicas y químicas de los granos y estigmas pueden influir en el daño ocasionado por los insectos, el crecimiento de las esporas de* Aspergillus flavus *Link ex Fries y/o el desarrollo de aflatoxinas.*

In 1920, J.J. Taubenhaus (4) stated in the bulletin, A Study of the Black and Yellow Molds of Ear Corn [*Zea mays* L.], "To control yellow mold [presumably *Aspergillus flavus*], one has only to plant those varieties of corn that have ears that hang down. To do so is especially important in localities with heavy rainfall." Passing years have shown the problem to be a little more involved than Taubenhaus once suggested.

A large portion of the maize research effort at the Insect Biology Population Management Research Laboratory in Tifton, Georgia, involves cooperative plant resistance studies. The purpose of this paper is to briefly review selected research presently conducted there by cooperating entomologists, plant geneticists and chemists.

On several occasions, mature ears that were either upright or hanging tip-down on the stalk have been harvested and compared. To date, no significant differences have been detected in either insect damage or aflatoxin accumulation as a result of ear posture.

A number of researchers (3,8,19,20) have reported that lower levels of overall insect damage and/or aflatoxin accumulation are sustained by maize hybrids adapted to the southern USA than those adapted to the Corn Belt. The differences in susceptibility are apparently due primarily to the complete husk cover of the southern ear (5,9,12,13). For example, in a test conducted at Tifton over the last six years, a representative southern-adapted hybrid with complete husk cover averaged only about one-sixth the amount of aflatoxin contamination compared to an open, loose-husk Corn Belt hybrid. Insect damage—mainly corn earworm, *Heliothis zea* (Boddie)—to these same hybrids averaged 3.2 cm of feeding on the southern type compared to 6.1 cm of feeding on the Corn Belt type.

Wiseman *et al.* (14,16,18) reported that silks of some maize germplasm, such as Zapalote Chico, express the resistance mechanisms of nonpreference and antibiosis to corn earworm larvae, resulting in significantly less ear damage, smaller larvae and fewer surviving earworms. It is suggested that maize hybrids that possess the Zapalote Chico type of resistance could be used as an integral part of a pest-management program. Studies by Widstrom *et al.* (10) have suggested that maize plants with fewer trichomes are least preferred by the adult earworm for oviposition. Further studies by Wiseman *et al.* (15) have demonstrated ear resistance to earworms, due to a long, tight silk-channel filled with a large silk mass that allows small larvae to complete development before reaching the kernel area and inflicting damage.

The fall armyworm, *Spodoptera frugiperda* (J.E. Smith), often inflicts severe feeding damage on whorl stage as well as mature maize plants. Field studies over three years (17) demonstrated that maize genotype Antigua 2D-118 was not favored by fall armyworm larvae, resulting in fewer larvae per plant and half the damage sustained by the susceptible check Cacahuacintle X.

Field experiments in Georgia by Widstrom *et al.* (11) have demonstrated that damage by the European corn borer larvae, *Ostrinia nubilalis* (Hübner), enhances aflatoxin contamination in preharvest maize. An interregional field test by Guthrie *et al.* (1) demonstrated that a higher incidence of *A. flavus* group isolates occurred in corn borer larvae collected from Oh43 x W182E, a susceptible hybrid, than from B86 x SC213, a resistant hybrid. In this evaluation, however, there appeared to be no association between hybrid resistance to corn borer damage and aflatoxin contamination levels.

Studies by Widstrom *et al.* (6,7) have demonstrated that maternal tissue in the maize kernel is important in maize weevil (*Sitophilus zeamais* Motschulsky) resistance. Husk cover and kernel hardness are probably the principal contributors to this resistance.

Other evaluations of maize germplasm have revealed that both physical and chemical attributes of the kernel may influence insect population build-up and aflatoxin development (2). Maize weevils exposed to spores of *A. flavus* or *A. parasiticus* and reared on dent, flint, waxy and high-amylose versions of the maize hybrid B37 x C103 produced progenies of 445, 438, 456 and 557 weevils, respectively. The dent version of B37 x C103 contained the highest level of aflatoxin (2,622 ng/g^{-1}), followed by flint (2,088 ng/g^{-1}), waxy (981 ng/g^{-1}) and high amylose (391 ng/g^{-1}) versions. Fungal growth appeared to be closely associated with the area damaged by maize weevils.

Recently in the field in Tifton, an ear of maize from an unidentified hybrid was observed to be segregating for visible *A. flavus* sporulation on the kernels. The ear was harvested and the kernels were separated into *A. flavus* resistant and susceptible categories. For the past five years, each population was increased and evaluated. The kernels that were free of visible *A. flavus* on the original ear have produced progeny that have consistently sustained lower aflatoxin contamination levels than the infected kernels, even under field-inoculated conditions (W.W. McMillian, unpublished data).

Recent laboratory bioassays of crude plant extracts have indicated that some maize genotypes possess chemicals that inhibit *A. flavus* growth. When a low concentration of a methanol extract of the resistant genotypes was applied to paper discs, air dried and then placed on *A. flavus*-inoculated agar medium, *A. flavus* growth on and around the discs was inhibited for several days (W.W. McMillian, unpublished data). Although these plant chemical evaluations are very preliminary, they suggest possibilities for development of chemically-based resistance in maize to insects and/or fungi.

References

1. Guthrie, W.D., E.B. Lillehoj, W.W. McMillian, D. Barry, W.F. Kwolek, A.O. Franz, E.A. Catalano, W.A. Russell and N.W. Widstrom. 1981. Effect of hybrids with different levels of susceptibility to second generation European corn borers on aflatoxin contamination in corn. Journal of Agricultural and Food Chemistry 29:1170-1172.

2. McMillian, W.W., N.W. Widstrom and D.M. Wilson. 1981. Rearing the maize weevil on maize genotypes when aflatoxin-producing *Aspergillus flavus* and *A. parasiticus* isolates were present. Environmental Entomology 10:760-762.

3. McMillian, W.W., N.W. Widstrom and B.R. Wiseman. 1982. Pink scavenger caterpillar resistance among selected dent corn hybrids. Journal of the Georgia Entomological Society 17:93-96.

4. Taubenhaus, J.J. 1920. A study of the black and yellow molds of ear corn. Texas Agricultural Experiment Station Bulletin 270. College Station, Texas.

5. Widstrom, N.W. 1974. Selection indexes for resistance to corn earworm based on realized gains in corn. Crop Science 14:673-675.

6. Widstrom, N.W., W.D. Hanson and L.M. Redlinger. 1975. Inheritance of maize weevil resistance to maize. Crop Science 15:467-470.

7. Widstrom, N.W., W.W. McMillian, L.M. Redlinger and W.J. Wiser. 1983. Dent corn inbred sources of resistance to the maize weevil (Coleoptera: Curculionidae). Journal of Economic Entomology 76:31-33.

8. Widstrom, N.W., W.W. McMillian and D.M. Wilson. 1984. Contamination of preharvest corn by aflatoxin. *In* Proceedings of the 39th Annual Corn and Sorghum Industry Research Conference. Pp. 68-83.

9. Widstrom, N.W., W.W. McMillian and B.R. Wiseman. 1970. Resistance in corn to the corn earworm and the fall armyworm. IV. Earworm injury to corn inbreds related to climatic conditions and plant characteristics. Journal of Economic Entomology 63:803-808.

10. Widstrom, N.W., W.W. McMillian and B.R. Wiseman. 1979. Ovipositional preference of the corn earworm and the development of trichomes on two exotic corn selections. Environmental Entomology 8:833-839.

11. Widstrom, N.W., A.N. Sparks, E.B. Lillehoj and W.F. Kwolek. 1975. Aflatoxin production and lepidopteran insect injury on corn in Georgia. Journal of Economic Entomology 68:855-856.

12. Widstrom, N.W., and M.S. Zuber. 1983. Prevention and control of aflatoxin in corn: Sources and mechanisms of genetic control in the plant. *In* Aflatoxin and *Aspergillus flavus* in Corn, U.L. Diener, R.L. Asquith and J.W. Dickens, eds. Southern Cooperative Series Bulletin 279. Alabama Agricultural Experiment Station, Auburn, Alabama, USA.

13. Wiseman, B.R., W.W. McMillian and N.W. Widstrom. 1970. Husk and kernel resistance among maize hybrids to an insect complex. Journal of Economic Entomology 63:1260-1262.

14. Wiseman, B.R., W.W. McMillian and N.W. Widstrom. 1978. Potential of resistant corn to reduce corn earworm production. Florida Entomologist 61:92.

15. Wiseman, B.R., N.W. Widstrom and W.W. McMillian. 1977. Ear characteristics and mechanisms of resistance among selected corns to corn earworm. Florida Entomologist 60:97-103.

16. Wiseman, B.R., N.W. Widstrom and W.W. McMillian. 1981. Influence of corn silks on corn earworm feeding response. Florida Entomologist 64:395-399.

17. Wiseman, B.R., N.W. Widstrom and W.W. McMillian. 1981. Effects of 'Antigua 2D-118' resistant corn on fall armyworm feeding and survival. Florida Entomologist 64:516-519.

18. Wiseman, B.R., N.W. Widstrom and W.W. McMillian. 1983. Influence of resistant and susceptible corn silks on selected developmental parameters of corn earworm (Lepidoptera: Noctuidae) larvae. Journal of Economic Entomology 76:1288-1290.

19. Wiseman, B.R., N.W. Widstrom and W.W. McMillian. 1984. Increased seasonal losses in field corn to corn earworm. Journal of the Georgia Entomological Society 19:41-43.

20. Zuber, M.S., and E.B. Lillehoj. 1979. Status of the aflatoxin problem in corn. Journal of Environmental Quality 8:1-5.

The Laboratory Response of Two Maize Varieties and Three Hybrids to Two Types of Aflatoxin-Producing *Aspergillus* Strains

Anguiano, R.G.L., and D. Guzman de Peña, Center for Research and Advanced Studies, National Polytechnic Institute, Irapuato, Guanajuato, Mexico

Abstract

One of the most recent approaches for controlling aflatoxin contamination of maize is the use of resistant maize hybrids. This study investigated the susceptibility in the laboratory of three hybrids and two varieties of maize grown in the Bajio region of Mexico to two aflatoxin-producing strains of Aspergillus. *These strains were found in samples of maize in the region; their toxin-producing capacity had been recognized in previous tests. Fifty-gram samples of the maize genotypes were inoculated with a determined number of spores suspended in distilled water, and were incubated at room temperature (25° to 27°C) for seven days. The methods used for the analysis and quantification of aflatoxin were those of Shotwell* et al. *(11), Nabney and Nesbitt (9) and a modified CB test (1). Despite the small number of materials studied, it was possible to separate the maize genotypes into three groups according to their response, susceptible (good substrates), resistant (poor substrates) and ambiguous, depending on the strain used as the inoculum.*

Resumen

Uno de los enfoques más recientes en el control de la contaminación del maíz con aflatoxinas es el uso de híbridos resistentes. El objeto del presente estudio fue investigar la sensibilidad en el laboratorio de tres híbridos y dos variedades de maíz que se cultivan en la región del Bajío en México a dos cepas toxigénicas de Aspergillus. *Estas cepas se encontraron en muestras provenientes de la región; su capacidad para producir toxinas había sido identificada en ensayos anteriores. Muestras de 50 g de los genotipos de maíz fueron inoculadas con un número constante de esporas suspendidas en agua destilada e incubadas a la temperatura ambiente (25° a 27° C) durante siete días. Los métodos utilizados para el análisis y cuantificación de las aflatoxinas fueron los de Shotwell* et al. *(11), Nabney y Nesbitt (9) y una prueba CB-modificada (1). A pesar de limitado número de materiales estudiados, fue posible separar los genotipos de maíz en tres grupos según su respuesta: sensible (buenos substratos), resistente (malos substratos) y ambigua, dependiendo de la cepa que se utilizó como inóculo.*

Aflatoxin production in cereals and oilseeds is related to a number of complex factors, such as the presence of strains of toxin-producing fungi, substrate susceptibility and environmental factors, the most significant of which are temperature and moisture. These factors affect the behavior of both the plant and the microorganism. The toxicity of certain species of *Aspergillus* has been recognized as a strain-related trait, a fact which allows for great variability in toxin-producing capacity (4). Leaich *et al.* (5) have reported that *Aspergillus flavus* mutants can be classified as having high, medium or low aflatoxin-producing capacity. Similar results were observed in a study of *Aspergillus* strains in maize from the Bajio region of the state of Guanajuato on the central Mexican plateau (G.L. Anguiano, unpublished data, 1984).

A review of the literature shows that maize genotypes exhibit differing levels of susceptibility to the attacks of storage fungi; this suggests the possibility of a similar response in the case of aflatoxin production (2,7). Priyadarshini *et al.* observed differences in the responses of maize varieties to both fungal growth and aflatoxin accumulation. These differences have been associated with qualitative changes in the characteristics of genotypes (10). Zuber *et al.* demonstrated that aflatoxin B_1 was produced in low concentrations in maize hybrids and in high concentrations in crosses of the maize variety OH-545 with hybrids (14).

Studies of environmental factors affecting aflatoxin production have indicated that such conditions as high temperatures (35°C), drought and high moisture levels during storage favor aflatoxin production. Other determining factors are the differences in the level of susceptibility of certain plants (for example, maize and peanuts) to the growth and/or sporulation of the main groups of storage fungi (2,6). This study was conducted at the Center for Research and Advanced Studies of the National Polytechnic Institute at Irapuato, Mexico, in order to evaluate the laboratory production of aflatoxins in three maize hybrids and two varieties in the Bajio in the presence of strains of varying toxin-producing capacity.

Materials and Methods

Microorganisms

A good toxin-producing strain, *A. flavus* UI-1, and a nonproducing strain that was thought to be *A. restrictus* UI, were used in the study. Both samples were morphologically classified according to M.C. Christensen (3) but their classification must be confirmed by comparison with an internationally accepted microbial collection. The toxigenic capability of the strains was analyzed in the laboratory on the maize variety VS-373, using both intact and damaged kernels. The strains were kept in inclined tubes with potato-dextrose agar. Sterile, distilled water (10 ml) was added to each tube to obtain spore suspensions for use as inocula.

Inoculation and incubation

Five types of maize were used in this study, the opaque-2 double-cross hybrid H-332-OP, the inbred line x variety hybrid HB-15 and the hybrid H-363-R, the synthetic dwarf inbred line VS-361 and the improved variety CO-82. They were analyzed by method I of the Association of Official Analytical Chemists (AOAC) (1) to ascertain that they were free of aflatoxins. Fifty grams of damaged kernels were sterilized in 500-ml Erlenmeyer flasks at 15 pounds pressure for 15 minutes. Each sample was inoculated with the corresponding spore suspension, at concentrations of 2200 spores of *A. flavus* UI-1 and 9000 spores of the nonproducing strain.

To provide sufficient moisture in the substrate, which had been dried to a 12% moisture level, 3 ml of sterile, distilled water was added to each flask after three days and 8 ml of water the fourth day. Five repetitions of each treatment of maize genotype x fungal strain were performed. The flasks were then incubated in the dark at 25° to 27°C for seven days.

Extraction and quantification of aflatoxins

The procedure used to extract and quantify the aflatoxin consisted of a combination of methods. For extraction of the toxin, the method of Shotwell *et al.* (11) was used. Maize samples were sterilized at 15 pounds pressure for 15 minutes, after which 500 ml of distilled water was added to each of the flasks; they were then shaken for five minutes in a Lab-Line multiwrist shaker. Five hundred ml of chloroform was added to each flask, and flasks were shaken again for 30 minutes. This mixture was filtered through cotton in a funnel and the two layers so obtained separated in a separation funnel. The chloroform layer was collected, and the aqueous layer was again extracted. Another 200 ml of chloroform was added. The chloroform layers were mixed together and put into a 1.01 Erlenmeyer flask and 25 g of anhydric sodium sulfate added. The flask was hand shaken and the mixture filtered through Whatman No. 4 paper. The fraction so obtained was evaporated to 10 ml and purified in a column of silica gel, acording to method I (CB) of the AOAC (1). The fraction obtained from the column was concentrated to 500 μl, and was analyzed by thin-layer chromatography by the method described by the AOAC for visual quantification (1). After the extracts were purified on preparatoy plates, a spectrophotometric determination was made using the Nabney and Nesbitt method (9).

Statistical analysis

Statistical analysis was performed, using the criteria of Whitaker and Dickens (13) for determining the total error associated with each reading. The differences of the means were calculated by the Tukey method and the Student (LSD) method. In both cases, a 5% level of significance was used (13).

Results

In the laboratory, the maize varieties studied showed various levels of susceptibility to aflatoxin accumulation. The analysis of their responses in the presence of the highly productive strain showed two hybrids (H-332-OP and H-363-R) and one improved variety (CO-82) as susceptible (good substrates for aflatoxin production, 1555 μg/kg). The hybrid HB-15 and the variety VS-361 were resistant (poor substrates for toxin production, <86 μg/kg).

This response is clearly seen when aflatoxin production is measured by spectrophotometry, as shown in Table 1. When production is determined by visual comparison with a standard in thin-layer chromatography (Table 2), the differences are less pronounced but still detectable. Nevertheless, statistical analysis reveals a significant difference in the response of maize genotypes to *A. flavus* UI-1.

When the different varieties of maize were inoculated with the strain that seemed to be *A. restrictus*, it was observed that this fungus (currently considered nontoxigenic) was capable

Table 1. Production of aflatoxin B_1 in maize by *Aspergillus flavus* UI-1, Irapuato, Mexico [a/]

	Aflatoxin (μg/kg)[b/]	
Genotype	**Mean**	**Standard deviation**
H-332-OP	3923	1346
HB-15	780	408
H-363-R	1555	1289
VS-361	859	239
CO-82	2559	1041

a/ Determined by spectrophotometry
b/ Average of 5 repetitions

Table 2. Production of aflatoxin B_1 in maize by *Aspergillus flavus* UI-1, Irapuato, Mexico [a/]

	Aflatoxin (μg/kg)[b/]	
Genotype	**Mean**	**Standard deviation**
H-332-OP	2380	887
HB-15	1421	917
H-363-R	2083	1235
VS-361	1388	362
CO-82	2809	1522

a/ Determined by thin-layer chromatography
b/ Average of 5 repetitions

of producing low levels of aflatoxin B_1 (Tables 3 and 4). The hybrid H-363-R continued to be a good substrate for aflatoxin accumulation with this strain, whereas HB-15, H-332-OP and VS-361 were poor substrates; the amount of aflatoxin accumulation was very low. The variety CO-82 proved to be the most toxin-resistant substrate. The statistical analysis of these results shows significant differences in the response of the various maize hybrids and varieties to the strain that is thought to be *A. restrictus* (Table 5).

Based on these results, maize response can be classified in three categories, susceptible (good substrate), H-363-R; resistant or partly resistant (poor substrate), HB-15 and VS-361; and ambiguous, CO-82 and H-332-OP.

Table 3. Production of aflatoxin B_1 in maize by *Aspergillus restrictus*, Irapuato, Mexico [a]

Genotype	Aflatoxin (μg/kg)[b] Mean	Standard deviation
H-332-OP	53	32
HB-15	38	84
H-363-R	567	599
VS-361	73	103
CO-82	21	47

a/ Determined by spectrophotometry
b/ Average of 5 repetitions

Table 4. Production of aflatoxin B_1 in maize by *Aspergillus restrictus*, Irapuato, Mexico [a]

Genotype	Aflatoxin (μg/kg)[b] Mean	Standard deviation
H-332-OP	99	57
HB-15	46	86
H-363-R	147	141
VS-361	26	43
CO-82	0	0

a/ Determined by thin-layer chromatography
b/ Average of 5 repetitions

Discussion

In this study it was observed that the ability to sustain aflatoxin production depended on the interaction between the characteristics of the maize and the fungal inocula. The maize hybrid H-363-R showed a uniform response of susceptibility to both species of *Aspergillus*. The HB-15 and VS-361 genotypes were equally stable in their response for resistance. The CO-82 and H-332-OP varieties were susceptible only for the highly productive strain and resistant for the strain of low toxin production. These responses indicate that some maize varieties have no inhibitors to aflatoxin production; the presence of these inhibitors in maize has been suggested by various authors (8,10,12).

The majority of studies reported in literature concerning differences in the capability of maize genotypes to sustain aflatoxin production have been performed with *A. flavus* and *A. parasiticus*. Both strains are recognized as outstanding aflatoxin producers (5,10,14). However, it was

Table 5. Interaction between fungal strain and maize genotype, Irapuato, Mexico

Genotype	*A. flavus* (μg/kg)[a]	*A. restrictus* (μg/kg)[a]
H-332-OP	55.08850 a [b]	8.41507 b
HB-15	31.84130 c	4.07580 b, c
H-363-R	40.09050 b	15.08550 a
VS-361	33.06695 c	4.89376 b, c
CO-82	50.78285 a	1.93441 c
LSD 0.05 = 6.22		

a/ Averages converted according to Whitaker and Dickens (13)
b/ Numbers followed by the same letter not significantly different

found in this study that when the strain thought to be *A. restrictus*, considered in the literature as nontoxigenic, was present in high levels of inoculum with various varieties of maize, it produced low levels of aflatoxin B_1 in most materials. The fact that some genotypes vary their resistant/susceptible response in the presence of this strain could lead to more rigorous selection. Two of the varieties maintained their resistant response through all repetitions and with both aflatoxin-producing strains.

The results obtained again demonstrate that maize varieties differ in their levels of susceptibility to aflatoxin accumulation, as reported by Zuber and Priyadarshini (10,14). The results also suggest that aflatoxin-resistant varieties have inhibitors that are not present in varieties showing a susceptible or ambiguous response. The presence of such inhibitors as Beta-ionone in certain materials has been reported by Wilson *et al.* (12). Future plans should include the measurement of the level of this inhibiting substance in resistant maize varieties.

References

1. Association of Official Analytical Chemists. 1980. Official Methods of Analysis, 11th ed. P. 436.

2. Cantone, F.A., J. Tuite, L.A. Bauman and R. Stroshine. 1983. Genotypic differences in reaction of stored corn kernels to attack by selected *Aspergillus* and *Penicillium* spp. Phytopathology 73:1250-1255.

3. Christensen, M.C., and H.J. Kaufman. 1969. Grain Storage: The Role of Fungi in Quality Loss. University of Minnesota Press, Minneapolis, Minnesota, USA.

4. Hesseltine, C.W., W.G. Sorenson and M. Smith. 1970. Taxonomic studies of the aflatoxin producing strains in the *Aspergillus flavus* group. Mycologia 62:123-132.

5. Leaich, L.L., and K.E. Papa. 1974. Aflatoxins in mutants of *Aspergillus flavus*. Mycopathologia et Mycologia Applicata 52:223-229.

6. Mixon, A.C. 1981. Reducing aflatoxin contamination in peanut genotypes by selection and breeding. Memorial Symposium on Mycotoxins. Journal of the American Oil Chemists' Society 58:961.

7. Moreno-Martínez, E., R.M. Reza and R.G. Lombardo. 1978. Diferencias entre lineas, cruzas simples y dobles de maíz en su susceptibilidad al daño por condiciones adversas de almacenamiento. Turrialba 28:233-237.

8. Nagarajan, V. 1972. Factor responsible for varietal differences in aflatoxin production in maize. Journal of Agricultural and Food Chemistry 20:911-914.

9. Nabney, J., and B.F. Nesbitt. 1965. A spectrophotometric method for determining aflatoxin. The Analyst 90:155-160.

10. Priyadarshini, E., and P.G. Tulpule. 1978. Relationship between fungal growth and aflatoxin production in varieties of maize and groundnut. Journal of Agricultural and Food Chemistry 26:249-252.

11. Shotwell, O.L., G.W. Hesseltine, R.D. Stubblefield and W.G. Sorenson. 1966. Production of aflatoxin in rice. Applied Microbiology 14:425-428.

12. Wilson, D.M., R.C. Gueldner and R.A. Hill. 1981. Effects of plant metabolites on the *A. flavus* group and aflatoxin production. *In*: Aflatoxin and *Aspergillus flavus* in Corn, U.L. Diener, R.L. Asquith and J.W. Dickens, eds. Southern Cooperative Series Bulletin 279. Alabama Agricultural Experiment Station, Auburn, Alabama, USA. Pp. 13-15.

13. Whitaker, T.B., and J.W. Dickens. 1979. Variability associated with testing corn for aflatoxin. Journal of the American Oil Chemists' Society 56:789-794.

14. Zuber, M.S., O.H. Calvert, W.F. Kwolek, E.B. Lillehoj and M.S. Kang. 1978. Aflatoxin B production in an eight line diallel of *Zea mays* infected with *Aspergillus flavus*. Phytopathology 68:1346-1349.

Decontamination of Aflatoxin-Contaminated Maize Grain

E.B. Lillehoj and J.H. Wall, Southern Regional Research Center, Agricultural Research Service, US Department of Agriculture, New Orleans, Louisiana, USA

Abstract

Widespread outbreaks of aflatoxin contamination of agricultural commodities have stimulated research interest in prevention and detoxification. The discovery of widespread preharvest contamination of maize by aflatoxin demonstrated that prevention approaches will not provide rapid solutions. Researchers have focused on developing procedures for detoxifying certain target commodities to provide feed-grade materials, in an effort to avoid massive economic sanctions. This review outlines the fundamental techniques that have been developed for physical separation of aflatoxin-contaminated seed, solvent extraction of aflatoxin and chemical inactivation of the toxin. Information is presented on the methods utilized for decontamination of peanuts and cottonseed because these procedures are relevant to the evolution of effective and practical techniques for detoxification of aflatoxin-contaminated maize.

Resumen

La gran cantidad de brotes de contaminación de productos agrícolas con aflatoxinas despertaron el interés de la investigación en la prevención y destoxificación. El descubrimiento de una gran contaminación del maíz con aflatoxinas durante la fase de precosecha demostró que los métodos de prevención no ofrecen una solución inmediata al problema. Los investigadores han concentrado su atención en la creación de precedimientos para eliminar la contaminación de ciertos productos con el fin de proporcionar materiales aptos para el consumo animal y tratar de evitar sanciones económicas masivas. En este trabajo se describen las principales técnicas creadas para separar físicamente las semillas contaminadas con aflatoxinas, extraer las aflatoxinas mediante solventes e inactivar la toxina por medios químicos. Se presenta información acerca de los métodos utilizados para descontaminar los cacahuates y la semilla de algodón porque estos procedimientos son importantes para el desarrollo de técnicas prácticas y eficaces para la eliminación de toxinas del maíz contaminado con aflatoxinas.

The most effective strategy for controlling aflatoxin contamination of food and feed commodities is to reduce or eliminate the initial toxin production process. However, extended observations have demonstrated that control procedures may be only partially effective, and that such unanticipated environmental factors as prolonged droughts during crop development can introduce high, widespread levels of aflatoxin. The ongoing nature of aflatoxin contamination has underscored the need for techniques to identify commodities at high risk of contamination. Relatively rapid quantitative methods for detecting aflatoxin have been developed (90), but technical expertise is required to carry out the assays.

A simple, presumptive method has been developed that is based on the presence of bright greenish-yellow (BGY) fluorescence in aflatoxin-contaminated maize (50,92). Presence of the BGY fluorescence in a commodity is clear indication of the activity of *Aspergillus flavus* Link ex

Fries and/or *A. parasiticus* Speare, but the test is not an absolute measure of the qualitative occurrence of aflatoxin. The test is used widely at rural grain elevators in the USA to detect a potential aflatoxin problem in maize. Unfortunately, a certain amount of spurious price manipulation has evolved from the BGY test. Furthermore, the fate of the devalued, BGY-positive material is not clear. An equitable management procedure would include a concise method for pricing BGY-positive maize and a mechanism for directing the material to an appropriate destination.

Physical methods to identify and separate aflatoxin-contaminated seed from larger lots have been developed. These procedures are based on the heterogeneous distribution of contaminated seed within a parent lot. Only a few seeds are generally responsible for contamination of a large lot. In the peanut industry, kernels are traditionally examined at shelling to remove discolored and immature seed by hand sorting and/or mechanical screening (43). In addition, electronic sorters have been developed to individually examine each kernel and reject discolored seed, resulting in a concomitant reduction of aflatoxin in the parent lots (29). A similar procedure has been developed to eliminate fungal-infected tree nuts and cottonseed by electro-optical sorting for BGY fluorescence (10,37,86). Most BGY fluorescence in *A. flavus*-infected maize kernels is within apparently sound kernels; therefore, the BGY fluorescence test requires damaged kernels (57). Since maize kernel fragments are not suitable for electronic sorting, BGY fluorescence tests are incompatible with electron sorting of contaminated maize.

A number of heat treatments have been tested in aflatoxin-detoxification studies (80,89). Steaming under pressure and other cooking techniques often reduce aflatoxin levels but do not eliminate the toxin (25,27,87). Dry roasting of peanuts and maize kernels also reduces contaminant concentrations (11,87), but the procedures do not routinely lower the levels to acceptable concentrations.

The US Food and Drug Administration (FDA) identified a guideline of 20 ppb total aflatoxin level, with some variance for animal feeds (82). The concept of guidelines and tolerances is currently being reviewed by the US Supreme Court (6). Although 20 ppb has also been considered acceptable for feed, FDA regulations in recent years have concentrated on aflatoxin and aflatoxin derivatives in tissue residues of milk, meat and eggs (78). With an expected ratio of 100:1 feed to tissue distribution of aflatoxin, levels of toxin in the feed could be at the 50 ppb level and still not violate FDA action levels of 0.5 ppb residue in tissues (83).

Under appropriate conditions aflatoxin is degraded by light, particularly ultraviolet radiation. Sunlight has been utilized to reduce aflatoxin levels in contaminated peanut cake and vegetable oils (84,91). Ultraviolet light has been used to degrade aflatoxin M_1 in milk; the process removed all of the toxin in milk more efficiently in the presence of small quantities of hydrogen peroxide (103). Ultraviolet irradiation of oilseed meals to destroy aflatoxin has been patented (9). Simple, light-mediated aflatoxin decontamination techniques are available and are practical in certain instances for use in developing countries (70).

Aflatoxin is also degraded by gamma radiation (100). However, gamma treatment of aflatoxin-contaminated commodities has provided mixed results. Feuell (38) observed no reduction in toxicity, whereas Aibara and Miyaki (1) and Ogbadu and Bassir (69) detected decreased toxicities of contaminated feeds after gamma radiation. A study by Temcharoen and

Thilly (97) identified a reduction in toxicity of gamma-irradiated, aflatoxin-contaminated peanut meal, but Ames tests demonstrated the presence of a mutagen in the treated meal. Although physical processes for removal or detoxification of aflatoxin-contaminated materials ameliorate problems in peanuts, evaluation of the literature suggests that the most economical processes for contaminated maize will be linked to chemical procedures.

Solvent Extractions

Solvent extraction of aflatoxin from contaminated commodities has a number of desirable aspects:

- Potential for quantitative removal of the toxin;
- Technologies that are available on an industrial scale for recovery of vegetable oils;
- Simultaneous extraction of oils and toxin from seed fractions can be followed by established refining procedures using bleaching clays that remove the toxin from the oil fraction (45,72); and
- Selective extraction of the aflatoxin and free fatty acids with negligible oil and protein solubility, leaving a full fat commodity for conventional oil recovery.

One negative feature of solvent extractions is the removal of desired portions of the seed and reduction of the net value of the commodity.

Feuell (38) tested several ordinary organic solvents for detoxification of toxin-contaminated peanut meal. He found that toxin extractability from meal is not the same as anticipated from solubility data of the toxin alone. He concluded that in the meal aflatoxin is bound to polar macromolecules. A system capable of removing aflatoxin from contaminated meal without extracting large quantities of solids, using a tertiary system of acetone-hexane-water, was developed and patented (44). Gardner *et al.* (41) evaluated a tertiary solvent and a binary aqeuous acetone for aflatoxin removal from contaminated oilseed meal. Both procedures offered an economical technique for removing aflatoxin to the 30 ppb level in the extracted material.

Although polar solvents such as methanol and ethanol effectively reduce aflatoxin levels in contaminated materials, they also extract significant quantities of solids (28,38,43). Rayner *et al.* (79) reported that 80% aqueous isopropanol completely removed aflatoxin in cottonseed and peanut meal, but it also removed 8.7% and 9.5%, respectively, of the meal solids. Cottonseed extracted with aqueous (25 to 30%) acetone has been reported to remove most aflatoxin with only negligible removal of neutral oils and protein (43). The aflatoxin-free material is a full-fat product that is ideally suited for subsequent processing. Sreenivasamurthy *et al.* (94) described an aqueous calcium chloride solution that extracted most of the aflatoxin from peanut meal, with only 8% removel of the protein. Calcium chloride solutions added in the preparation of protein isolates prevented toxin from associating with the precipitated protein. Aibara and Yano (3) described a procedure for methoxymethane extraction of aflatoxin from contaminated peanut meals that effectively removed oil and aflatoxin in a single step with no solvent residues in dried meals. An intriguing new process, described by Hron and Koltun (47), utilizes an aqueous ethanol for cottonseed oil extraction with subsequent phase separation of oil fractions from solids. The system would provide a unique method for solvent recovery of aflatoxin from contaminated meal; the method is currently being assessed for its efficiency in extracting aflatoxin (Hron, personal communication).

Chemical Inactivation

Chemical inactivation of aflatoxin *in situ* appears to offer the most promising potential for effective control of contaminated agricultural commodities. The two most vulnerable sites on the aflatoxin molecule for chemical reaction are the coumarin moiety and the double bond of the terminal furan (28).

Oxidation/reduction agents

Trager and Stoloff (98) examined a number of chemical reactions including oxidations as potential detoxifying procedures. Benzoyl peroxide and osmium tetroxide reacted with aflatoxin B_1 and G_1 through the olefinic double bond, whereas NaOCl, KMn_4O, $NaBO_3$, $Ce(NH_4)_2(SO_4)_3$ and 3% H_2O_2:$NaBO_2$ (1:1) reacted with all four toxins (B_1, B_2, G_1 and G_2). Bioassay demonstrated the ability of gaseous chloride, chlorine dioxide, nitrogen dioxide and 5% NaOCl to detoxify contaminated meal. Feuell (38) identified the capacity of sulfur dioxide to decontaminate meals. The initial studies identified oxidative cleavage as the reason for the vulnerability of the double bond in the terminal furan (B_1, G_1 and M_1) in toxin molecules. The mechanism for the observed oxidative degradation of the toxin structures with no olefinic furan bond (B_2, G_2 and M_2) has not been established.

Hydrogen peroxide has been utilized to develop practical techniques for oxidative decontamination of aflatoxin in tainted meals (28,43). Sreenivasmurthy *et al.* (95) described an effective method that involved addition of an equal weight of a 6% H_2O_2 to a 10% solids suspension followed by incubation for 0.5 hours at 80°C; a similar procedure has been patented for peanut meal detoxification (95). Addition of basic compounds to the reaction facilitated the oxidative degradation of the toxin. Apparently, opening of the lactone ring facilitates peroxide cleavage of the toxin molecule. A 3% H_2O_2 technique has been developed for maize that reduces the aflatoxin level from 397 ppb to below 20 ppb, with less than 0.6% loss in proteins and lipids (20). Spraying peanuts with H_2O_2 after removal of skins also reduced aflatoxin levels (73). Applebaum and Marth (7) developed an innovative method to inactivate aflatoxin M_1 in milk by adding 1% H_2O_2 plus 0.5 mM riboflavin to contaminated milk at 30°C for 30 minutes, followed by heating at 63°C for 30 minutes. In an earlier study, they had also utilized sulphite or bentonite to remove aflatoxin M_1 from naturally contaminated milk (8).

Sodium hypochlorite has been widely utilized as a decontamination agent in laboratories to degrade glass-bound toxin (28). Castegnaro *et al.* (18) demonstrated that excessive aflatoxin:NaOCl ratios can produce the aflatoxin B_1-2,3-dicholoride. Early work identified the dihalide toxin derivative as an effective carcinogen (96). Therefore, the use of hypochlorite to effect oxidations was questioned. A treatment developed to eliminate the halaform reaction included adding hypochlorite to an aqueous aflatoxin solution, diluting it to a final concentration of 1:1.5% of the oxidizing solution, incubating the solution at ambient conditions for 30 minutes, and adding it to a final 5% acetone:95% water solution (v:v). Although formation of the dihalide derivative of aflatoxin raises serious questions about the practical value of the technique for detoxification of commodities, it has been used to eliminate aflatoxin with no residual toxicity in the treated material (33, 102).

Natarajan *et al.* (64) developed a method for the NaOCl-mediated oxidation of aflatoxin-contaminated peanut protein isolate and defatted meal. Results of the studies identified concentration of the oxidizing agent and pH as the critical components of the toxin-degradation system. At pH 8.0, 0.4% NaOCl reduced aflatoxin B_1 from 725 ppb to trace levels, whereas at pH 9.0 only 0.3% NaOCl was required. Similar observations were made in tests of defatted meal.

Oxidative destruction of aflatoxin by ozone has been considered as a practical method for decontamination of oilseed meal (34). After two hours of incubation with ozone at 100°C in meal containing 22 to 30% moisture, all of the aflatoxin B_1 was inactivated, whereas 78 to 90% of the other toxins (B_2 and G_1) was degraded. Dollear *et al.* (31) compared the ability of ozone, oxygen and air to reduce aflatoxin levels in contaminated cottonseed and peanut meal. Heating meal at 22% moisture for two hours at 100°C in air and oxygen effected a 67 to 76% reduction in toxin levels, whereas ozonized air completely destroyed aflatoxin B_1 and G_1 in one hour. Ozone in water has also been utilized for decontamination (19); the technique could be useful for industrial-scale detoxification of the water extracts from wet maize milling, since the toxin accumulates in the steep water fraction (101). However, ozone preference for the olefinic bonds of B_1 and G_1, with less effect on B_2 and G_2, reduced its usefulness. Although the aggressive oxidizing capacities of substances such as H_2O_2 and ozone have an inherent ability to degrade the aflatoxin molecule, they also react with other entities in contaminated commodities and often reduce the value of the treated material.

Bisulfite is a highly reactive compound that is often used in wines, fruit juices, jams, dried fuits and other foods to inhibit enzymatic and nonenzymatic browning and growth of microbes by reductive processes (81). Since bisulfite is an acceptable food additive, Doyle and Marth (32) tested its ability to degrade aflatoxin. Bisulfite reacted with aflatoxins B_1 and G_1 and reduced the fluorescence of solutions of the toxins. The results suggested that treating food with 50 to 500 ppm SO_2 would not effectively degrade aflatoxin, but treatment with 2000 ppm SO_2 at higher temperatures would distinctly reduce aflatoxin concentrations. Moerck *et al.* (63) observed extensive destruction of aflatoxin after 24 hours at ambient temperatures in 0.5 to 2.0% bisulfite. Hagler *et al.* (46) examined the effect of sodium bisulfite on maize grain containing 2350 ppb aflatoxin B_1 and 45 ppb B_2. Under appropriate conditions all of the B_1 was destroyed by bisulfite, whereas B_2 was relatively recalcitrant. Moisture, bisulfite levels, time and temperature had significant effects on aflatoxin degradation. The most effective procedure involved soaking maize kernels in 10% sodium bisulfite for 72 hours, removing the solvent, placing the kernels in plastic bags and incubating at 50°C for 21 days. Sodium bisulfite exhibited antimicrobial activity similar to propionic acid and was considered an effective antimicrobial agent in high-moisture stored maize.

Acids

Aflatoxin B_1 in acids is hydrated to hydroxidihydro aflatoxin B_1 (aflatoxin B_{2a}). The hydrated toxin is significantly less toxic than the parent compound (56), it is somewhat unstable and is converted to aphenolic dialdehyde under alkaline conditions (74). Acidification of free fatty acids in commercial soapstocks has real potential for toxin removal during the refining process (75). Kinetic studies of the acid-mediated hydration showed that heating at 100°C for 10 minutes at pH 1 converted 95% of the B_1 to B_{2a}, whereas at pH 3 and 100°C, seven hours were required to achieve the same conversion. Since relatively drastic conditions are required to induce satisfactory conversions, it is unlikely that aqueous acid solutions will be utilized on a broad scale in aflatoxin decontamination of maize.

Bases

Inorganic and organic bases have been widely utilized in food processing to achieve a number of objectives. Vegetable oil refining represents a good example of the industrial use of bases; crude oils are washed with sodium hydroxide solutions to remove gums, free fatty acids and base-soluble pigments (4). After water washing, the treated oils are routinely bleached with special clays to remove other pigments. Although the procedures are utilized to achieve desired chemical properties of the oil, the technique also removes most of the aflatoxin present in the crude oil; however, a signicant portion of the toxin is retained in the meal.

Initial consideration of chemical treatments to destroy aflatoxin in meal routinely included bases such as sodium hydroxide (28,38,43). A cooking procedure with aflatoxin-contaminated peanut meal was developed that utilized 2% sodium hydroxide and 30% moisture for two hours at 100°C. The technique reduced total aflatoxin to less than 5 ppb. However, the procedure significantly reduced lysine so that amendments were needed before treated material could be used as complete feed rations. Spent grains from ethanol fermentations of aflatoxin-contaminated maize were decontaminated by addition of bases such as NaOh, with subsequent incubation of the hot stillage (58).

Dollear *et al.* (31) tested methylamine for degradation of aflatoxin. A 2% level of the base added to peanut meal with 8.9% moisture reduced the toxin concentration to less than 5 ppb when the treatment included heating at 100°C for 90 minutes. A reduction in lysine levels similar to the NaOH treatments was observed. In a later study, Mann *et al.* (59) increased the efficiency of aflatoxin degradation in cottonseed meal by simultaneous incorporation of 1% NaOH and 2% methylamine. The treatment did not reduce the amino acid availability of the meal, but in animal-feeding trials the methylamine-treated meal produced liver abnormalities. Brandt *et al.* (14) patented a detoxification process based on use of alkali and an organic amine.

The Use of Ammonia for Decontamination

Aqueous or gaseous ammonia with or without elevated temperature and pressure appears to be the most efficient approach to decontaminating commodities containing aflatoxin (5,21,45). Masri *et al.* (61) patented a process for ammonia detoxification of aflatoxin-contaminated peanut meal, in which moistened meal was incubated at 94°C for 60 minutes at 20 psig anhydrous ammonia pressure. Similar procedures, developed by Dollear *et al.* (31), were effective at somewhat milder conditions of temperature and time (Table 1). Although ammonia treatments increased the overall nitrogen levels of meal, some reductions in lysine levels were noted.

For large-scale runs, Gardner *et al.* (40) developed optimum techniques to decontaminate ton lots of oilseed meal by ammoniation. The most effective conditions included adjustment of moisture levels to 12.5%, 114° to 122°C, 50 psig ammonia and 60 minutes incubation time. Monocalcium phosphate was added to the treated material to absorb residual ammonia, since in the initial studies a relatively significant amount of ammonia was released at the end of the trials.

A new procedure for enzymatic (urease) release of ammonia from urea has been developed that effectively detoxifies aflatoxin in contaminated meal (85). Although the initial studies showed a real potential for industrial-scale use, a number of problems were identified. Base-mediated reactions with aflatoxin involve opening the lactone ring with a concomitant loss of the fluorescent chromophore. Parker and Melnick (72) identified this loss under basic conditions, and the reappearance of the intact aflatoxin molecule when the solution was acidified. Since the initial *in situ* ammonia procedures did not identify the fate of the aflatoxin moleucles, the possibility existed that the toxin would be reconstituted intact in animal digestive systems. Uncertainty about the safety of the products made a subsequent biological assay necessary. The ammoniated cottonseed meal prepared by Gardner *et al.* (40) was utilized in dairy feeds, and no aflatoxin M_1 was found in milk (62).

The success of pressure ammonia treatments in reducing aflatoxin in contaminated oilseed materials increased interest in developing similar procedures in maize. To determine the on-farm practicality of a procedure, studies were carried out at ambient pressures. Brekke *et al.* (17) recirculated NH_3-air mixtures through a column of shelled maize at 25°C and 17.6% moisture until the NH_3 was uniformly distributed. After ammoniation, the columns were sealed and incubated at 25°C for two weeks, resulting in a reduction of initial aflatoxin levels from 100 ppb to 10 ppb (Table 2). The procedure was expanded to larger lots (1000 bushels or about 40 metric tons), using a standard grain bin with a perforated floor and a fan for recirculating the atmosphere above the grain through the floor (5,17). Grain was adjusted to 15 to 22% moisture, with a target range of 18 to 19% after equilibration (11,67). Gaseous ammonia was introduced to provide a level of 0.5 to 15% (dry weight basis) and the closed atmosphere was recirculated for 24 hours. The treated grain was then incubated at ambient conditions for about two weeks. The procedure effectively reduced aflatoxin from levels exceeding 100 ppb to less than 20 ppb. In a subsequent study, Nofsinger and Anderson (65) stored ammonia-treated maize in a bin during a winter in the midwestern USA with average temperatures during the six-month period of -5° to 16°C. With 2.5% ammonia-treated grain, original aflatoxin levels of 226 ppb were reduced to 15 ppb.

Table 1. Ammoniation of aflatoxin-contaminated peanut meal under different conditions of pressure, temperature and time

NH_3 pressure (psig)	Temperature (°C)	Time (min)	Aflatoxin total (μg/kg)
0	-	-	550
20	64	10	81
20	63	20	29
45	46	15	15
45	57	20	5
45	86	45	0
30	64	30	0

Source: Dollear *et al.* (31)

Table 2. Effectiveness of gaseous and aqueous ammonia for inactivation of aflatoxin in maize[a/]

NH_3 (g/100g maize)	NH_3 form	Maize moisture (% wet wt. basis)	Reaction time (days)	Residual aflatoxin B (μg/kg)
0				1000
0.5	gas	16.7	14	23
0.5	aqueous	17.5	14	30
1.0	gas	19.7	14	8
1.0	aqueous	20.0	8	7
1.5	gas	16.7	15	8
1.5	aqueous	17.5	8	15
2.0	gas	16.7	15	11
2.0	aqueous	17.5	8	11

a/ At 25°C
Source: Brekke *et al.* (17)

Utilization of anhydrous, gaseous ammonia as a decontaminating agent in maize containing aflatoxin has distinct advantages:

- Anhydrous ammonia is widely used in agriculture as a source of nitrogen fertilizer and is readily available;
- Widespread use of anhydrous ammonia has led to its economical production and distribution; and
- Gaseous ammonia is very penetrating and distributes rapidly within a storage bin and into stored kernels.

The disadvantages of anhydrous ammonia for detoxification are:

- Release of noxious vapors;
- Discoloration of the maize grain; and
- Need for careful monitoring of gas release and uniform distribution.

The latter problem can be corrected by introducing NH_3 in aqueous solution. Bothast *et al.* (13) described a method in which a 22% aqueous solution of ammonia was used to obtain an average 0.48% ammonia concentration in the maize. The liquid ammonia was applied as a coarse spray to maize grain being augered into a bin. The equipment required for this procedure was relatively simple compared to the metering device required for introducing gaseous ammonia. In addition to ease of application, with this procedure the moisture levels needed to expedite the detoxification process could be achieved at the same time. Ammonia can also be used in high-moisture maize grain to control microbial development (13,36). Ammonia (0.5 to 1.2%) restricted fungal development on maize stored at 27% moisture, but after one month *Scopulariopsis brevicaulis* Saccando & Brainier became the predominant microbe on the stored commodity.

Ammoniated maize grain has a very strong ammonia odor. Although this odor was not identified as a significant rejection factor in ruminant feeding trials, poultry and swine reduced their intake of rations containing maize ammoniated with 1.5 to 2.0% NH_3 (5,15). Drying techniques were developed to deodorize the ammonia-treated maize. Reducing moisture to 10% distinctly reduced the ammonia odor of treated grain (Table 3), and incorporating ammonia-treated, dried, ground maize with other ingredients in a mixed feed provided a satisfactory ration for poultry and swine.

Another objection to ammonia-treated maize is kernel discoloration. White maize turns yellow to tan and yellow kernels become dark tan to mahogany (15). The ammonia-treated kernels represent some potential problems for the processing industry since

Table 3. Drying ammonia-treated maize to reduce ammonia odor[a/]

Drying time (hr)	Moisture (%)	NH_3 (%)	NH_3 odor
0	18.5	0.50	Very Strong
1.5	13.1	0.36	Very Strong
4.2	10.3	0.32	Moderate
9.5	10.1	0.29	Slight

a/ Maize treated with 1% NH_3 for 23 days at 24°-32°C and dried with circulating air at 51°C

Source: Brekke *et al.*(15)

incorporating some treated kernels into large lots through blending would make it difficult to detect treated material. However, the residual ammonia in the starch fraction could become widely distributed in a processing procedure and introduce an organoleptically undesirable element into food commodities. Certain safety precautions are also required in dealing with large quantities of maize during amomonia treatment. Adequate ventilation is required to avoid exposing workers to ammonia concentrations exceeding safety limits specified by such regulatory agencies as the US Environmental Protection Agency.

A number of studies have been carried out to evaluate the feed efficiency of ammonia-treated maize grain (5,38,43). Brekke *et al.* (15) examined the ability of ammonia-treated, aflatoxin-contaminated maize to reduce carcinogenicity in trout. During a 12-month feeding period, untreated maize contaminated with 180 ppb aflatoxin produced a 96 to 98% incidence of tumors in trout, whereas the same quantities of ammonia-treated maize produced less than a 3% occurrence of tumors. In a study comparing aflatoxin-free with ammonia-detoxified maize fed to chickens, Hughes *et al.* (48) recorded that the ammonia-treated feed had no deleterious effect on production, egg quality, feed consumed per dozen eggs or mortality. In somewhat more sophisticated studies, Norred (68) examined the effect of ammonia-treated, aflatoxin-contaminated maize on liver function in rats. The changes in liver function observed when the rats were fed aflatoxin-contaminated maize were not observed in livers of rats fed ammonia-detoxified maize. Similar studies by Southern and Clawson (93) and Norred (66) support the contention that ammoniation of maize contaminated with aflatoxin provides a practical and economical method for detoxification.

Ammonia Detoxification: Industrial Processes

In response to the studies of ambient ammonia detoxification procedures developed for maize, scientists in Arizona developed similar techniques for cottonseed. Price *et al.* (78) utilized cottonseed samples contaminated with 400 to 7000 ppb of aflatoxin in multi-ton lots. The contaminated cottonseed was treated with 1½% ammonia, packed into polyethylene bags (3 x 30M) and held for 21 days. Ammoniated cottonseed was fed to lactating dairy cattle at a level of 3.5 kg per cow per day for 19 days. Less than 0.1 ppb of aflatoxin M_1 was detected in the bulk milk during the trial. Ammoniation of whole cottonseed reduced the level of aflatoxin M_1 in milk by approximately 90%.

Several pilot-scale processes developed in Europe provide new and imaginative approaches to using ammonia to detoxify aflatoxin-contaminated commodities (23,42,76). In Senegal, a processing plant (handling 150 tons/day) using a detoxification process based on that developed by Lesieur Cotelle et Associes has been built by Draiswerke of Mannheim for the Societé Nationale de Commercialisation des Oleagineux du Sénégal. The procedure involves charging a reactor with peanut cake (12 to 15% moisture) and subsequent treatment with steam and ammonia. The treatment is carried out at 2 to 3 bars pressure for 16 to 30 minutes at 90°C with intermittent agitation (23). Residual ammonia is removed by forced ventilation. Nutritional studies of the treated material indicate effective detoxification without significant loss of nutritive components. A pilot-scale plant (6.25 tons/day) is being built at Port Sudan by Extraktionstechnik (Extechnik) of Hamburg. The plant is under construction, and no information is available on the efficiency of the process.

The third design involves a collaborative project between the Tropical Development and Research Institute (TDRI) and the Ministry of Agriculture, Fisheries and Food (MAFF) in the UK. The moisture content of peanut cake is raised by steam to 15 to 20% moisture, and gaseous ammonia is circulated through the hot cake at pressures not exceeding 1 bar. The ammonia is left in contact with the hot cake for one hour and then is slowly reduced in pressure to ambient. Excess ammonia is purged form the system with steam. The procedure reduces aflatoxin about 95% with slight losses of the amino acids cystine and lysine (23).

Since the process of *tortilla* production uses lime, which elevates pH levels, that method has been considered for detoxification of aflatoxin-contaminated maize. A preliminary study by Ulloa-Sosa and Schroeder (99) demonstrated that boiling the maize in limewater, to facilitate peeling and softening of kernels before grinding them to produce *masa*, also significantly reduced aflatoxin levels, from 49.1 ppb in the starting material to 15.5 ppb in the *tortillas*. In a subsequent study, Price and Jorgensen (77) examined maize naturally contaminated with 127 ppb aflatoxin (Table 4). Treatments varied in percent $Ca(OH)_2$, boiling and holding time; aflatoxin determinations plus Ames tests (bacterial mutagenesis) were carried out at various stages of the procedure. Toxin and mutagen levels were also assesssed in reacidified samples. Although all of the treatments caused a decrease in aflatoxin, acidification of the samples prior to assay caused most of the original aflatoxin to re-form. Treatments also decreased the level of mutagenic potential of samples with the exception of the acidified *tortillas*; the latter contained more mutagens than the original maize. The studies elucidated flaws in earlier work on aflatoxin decontamination during *tortilla* production, and provided evidence for the inability of traditional methods used in *tortilla* manufacture to effectively reduce aflatoxin levels in contaminated maize products.

The active base ingredient of limewater, $Ca(OH)_2$, has also been used to degrade aflatoxin in contaminated meal (22). At 25% moisture, 2.0% $Ca(OH)_2$ and 1.0% formaldehyde, meal containing 600 ppb aflatoxin was reduced in one hour at 115° to 117°C to less than 5 ppb of toxin. Formol has also been incorporated into procedures with ammonia to increase the rate of aflatoxin degradation (39).

Ammonia-Degraded Aflatoxin Products

Since the fate of ammonia-degraded aflatoxin will determine the final utility of the process, a number of important chemical studies have been carried out to define the pertinent mechanisms. In a model system, reaction of aflatoxin B_1 with NH_4OH at 100°C in a Parr bomb produced two products (286 MW and 206 MW) (26,54). The proposed reaction mechanism involved initial opening of the lactone ring that occurs under mild alkaline conditions (25). Subsequently, the acid undergoes decarboxylation to form aflatoxin D_1 (286 MW) or the dihydro-4-hydroxy-6-methoxyfuro-(2,3,6) benzofuran (206 MW). The studies provided convincing evidence of the chemical degradation of the aflatoxin molecule during pressure ammonia treatment.

Ammonia treatments of contaminated cottonseed meal verified the conversion of B_1 to D_1 under the conditions of nominal detoxification (52). Ames tests for the presence of mutagenic substances in extracts of ammoniated materials containing aflatoxin demonstrated only slight mutagenic activity (51,53,71). Schroeder *et al.* (87) determined the distribution of C-14 label in ammonia-treated cottonseed meal after spiking with C-14 ring-labelled aflatoxin. Twenty-five percent of the label was recovered by methylene chloride extraction of the meal, 5% was recovered by methanol extraction, 3% was released by weak acid treatment of meal and 19% of the label was released by pronase digestion. The results verify the binding of aflatoxin derivatives to protein structures in ammonia-treated meals.

Examination of the radio-labeled, aflatoxin-binding properties of ammonia-treated derivatives in commodities demonstrated that ammoniating maize meal at 25° to 50°C for periods of 3 to 30 days induced two types of aflatoxin binding to meal macromolecules (12). A reversible association of the toxin resulted from the opening of the lactone ring and electrostatic and/or hydrogen bonding with meal components. The irreversible binding was facilitated by covalent bonding through the dihydrofuran ring of aflatoxin. The nature of the covalent binding of aflatoxin derivatives remains somewhat obscure, but the overwhelming body of information on the reduced toxicity in ammonia-treated commodities provides assurance that the bound toxin does not contain the intact aflatoxin structure.

Summary

The intrinsic hazards associated with ingesting aflatoxin-contaminated food or feed have served as the impetus for an impressive research effort to develop effective and economical methods for detoxification. A number of physical procedures for identifying contaminated grain have emerged, based on the detection and elimination of the limited number of contaminated kernels in a large lot. Electronic sorters are widely used in the peanut industry and, in conjunction with screening and hand-sorting, they reduce aflatoxin levels in the commercial commodity. The extensive effort to develop solvent extraction techniques to chemically remove aflatoxin from food and feed have provided several important, efficacious procedures. Although the

Table 4. Effect of *tortilla* preparation process on aflatoxin and mutagen levels in naturally contaminated maize

	Aflatoxin Total (μg/kg)			*Salmonella typhimurium* revertant number		
Treatment	**Untreated**	***Tortilla* (alkaline)**	***Tortilla* (acid)**	**Untreated**	***Tortilla* (alkaline)**	***Tortilla* (acid)**
Cooked 20 min. in 0.33 lime-water, soaked 15 hr.	135	62	93	815	408	1248
Cooked 1 hr. in 0.25% lime-water, soaked 24 hr.	145	67	84	1002	932	1101
Soaked 15 hr. in 0.33% lime-water, no cooking	142	38	110	862	400	1320

Source: Price and Jorgensen (77)

methods may be limited by economic considerations, they remain an important repository of technology that might be used in the future. The procedures that currently appear to offer the most effective and economic detoxification are based on chemical inactivation of the toxin *in situ*. Gaseous and aqueous solutions of ammonia provide the best, most practical inactivation methods. However, some constraints in the techniques are linked to the inability to define the fate of the ammonia-degraded toxin. Efforts to identify the degradation products of the treated toxin have begun, but important questions remain that can only be answered by extensive feeding studies. Results from an array of feeding studies conducted on a range of animal species will provide the best basis for assessing whether ammonia-treated, aflatoxin-contaminated commodities are safe sources of feed for mature domestic animals.

References

1. Aibara, K., and K. Miyaki. 1970. Aflatoxin and its radiosensitivity. *In* Radiation Sensitivity of Toxins and Animal Poisons, Proceedings of the IAEA Panel, Bangkok, Thailand, May 19-22, 1969. IAEA, Vienna, Austria. Pp. 41-62.

2. Aibara, K., and S. Yamagishi. 1970. Effects of ultraviolet irradiation on the destruction of aflatoxin B_1. *In* Toxic Microorganisms, M. Herzberg, ed. UJNR and Department of the Interior, US Government Printing Office, Washington, D.C., USA. Pp. 211-221.

3. Aibara, K., and N. Yano. 1977. New approach to aflatoxin removal. *In* Mycotoxins in Human and Animal Health, J.V. Rodricks, C.W. Hesseltine and M.A. Mehlman, eds. Pathotox, Park Forest South, Illinois, USA. Pp. 151-172.

4. Andersen, A.J.C. 1962. Refining of Oils and Fats for Edible Purposes, 2nd ed., P.N. Williams, ed. Pergamon Press, New York, N.Y., USA.

5. Anderson, R.A. 1983. Detoxification of aflatoxin-contaminated corn. *In* Aflatoxin and *Aspergillus flavus* in Corn, U.L. Diener, R.L. Asquith and J.W. Dickens, eds. Southern Cooperative Series Bulletin 279, Alabama Experiment Station, Auburn, Alabama, USA. Pp. 87-89.

6. Anonymous. 1985. Court invalidates aflatoxin action levels. Food Chemistry News 1:1-6.

7. Applebaum, R., and E.H. Marth. 1982. Inactivation of aflatoxin M_1 using hydrogen peroxide and hydrogen peroxide plus riboflavin or lactoperoxidase. Journal of Food Protection 45:557-560.

8. Applebaum, R., and E.H. Marth. 1982. Use of sulphite or bentonite to eliminate aflatoxin M_1 from naturally contaminated raw whole milk. Zeitschrift für Lebensmittel Untersuchung und Forschung 174:303-305.

9. Arthur, J.C., Jr., and J.A. Robertson, Jr. 1970. Photochemical process for detoxification of fungal metabolites. US Patent 3,506,452.

10. Ashworth, L.J., Jr., J.L. McMeans, J.L. Pyle, C.M. Brown, J.W. Osgood and R.E. Ponton. 1968. Aflatoxin in cottonseed: Influence of weathering on toxin content of seeds and on a method for mechanically sorting seed lots. Phytopathology 58:102-107.

11. Bagley, E.B. 1979. Decontamination of corn containing aflatoxin by treatment with ammonia. Journal of the American Oil Chemists' Society 56:808-811.

12. Beckwith, A.C., R.F. Vesonder and A. Ciegler. 1975. Action of weak bases upon aflatoxin B_1 in contact with macro-molecular reactants. Journal of Agricultural and Food Chemistry 23:582-587.

13. Bothast, R.J., G.H. Adams, E.E. Hatfield and E.B. Lancaster. 1975. Preservation of high-moisture corn: A microbiological evaluation. Journal of Dairy Science 58: 386-391.

14. Brandt, J., C. Giddey and G. Butner. 1975. Method of reducing the aflatoxin content of oilseed meal products. US Patent 3,890,452.

15. Brekke, O.L., A.J. Peplinski and E.B. Lancaster. 1977. Aflatoxin inactivation in corn by aqua ammonia. Transactions of the American Society of Agricultural Engineers 20:1160-1168.

16. Brekke, O.L., R.O. Sinnhuber, A.J. Peplinski, J.H. Wales, G.B. Putnam, D.J. Lee and A. Ciegler. 1977. Aflatoxin in corn: Ammonia inactivation and bioassay with rainbow trout. Applied Environmental Microbiology 34:34-37.

17. Brekke, O.L., A.C. Stringfellow and A.J. Peplinski. 1978. Aflatoxin inactivation in corn by ammonia gas: laboratory trials. Journal of Agricultural and Food Chemistry 26:1383-1389.

18. Castegnaro, M., M. Friesen, J. Michelon and E.A. Walker. 1981. Problems related to the use of sodium hypochlorite in the detoxification of aflatoxin B_1. Journal of the American Industrial Hygiene Association 42:389-401.

19. Caulfield, M.J., G.R. Burleson and M. Pollard. 1979. Ozonation of mutagenic and carcinogenic alkylating agents, pesticides aflatoxin B_1 and benzidine in water. Cancer Research 39:2155-2159.

20. Chakrabarti, A.G. 1981. Detoxification of corn. Journal of Food Protection 44:591-592.

21. Ciegler, A. 1976. Detoxification of aflatoxin-contaminated agricultural commodities. *In* Toxins: Animal, Plant and Microbial, P. Rosenberg, ed. Toxicon Supplement 1, Pergamon Press, Oxford, England. Pp. 729-738.

22. Codifer, F.G., Jr., G.E. Mann and F.G. Dollear. 1976. Aflatoxin inactivation: Treatment of peanut meal with formaldehyde and calcium hydroxide. Journal of the American Oil Chemists' Society 53:204-206.

23. Coker, R.D., K. Jewers and B.D. Jones. 1985. The destruction of aflatoxin by ammonia: Practical possibilities. Tropical Science 25:139-154.

24. Conway, H.F., R.A. Anderson and E.B. Bagley. 1978. Detoxification of aflatoxin-contaminated corn by roasting. Cereal Chemistry 55:115-117.

25. Coomes, T.J., P.C. Crowther, A.J. Feuell and B.J. Francis. 1966. Experimental detoxification of groundnut meals containing aflatoxin. Nature 209:406-407.

26. Cucullu, A.F., L.S. Lee, W.A. Pons, Jr., and J.B. Stanley. 1976. Ammoniation of aflatoxin B_1: Isolation and characterization of a product with molecular weight 206. Journal of Agricultural and Food Chemistry 24:408-410.

27. Dada, L.O., and H.G. Muller. 1983. The fate of aflatoxin B_1 in the production of *ogi*, a Nigerian fermented sorghum porridge. Journal of Cereal Science 1:63-70.

28. Detroy, R.W., E.B. Lillehoj and A. Ciegler. 1971. Aflatoxin and related compounds. *In* Microbial Toxins, vol. 8, A. Ciegler, S. Kadis and S.J. Ajl, eds. Academic Press, New York, N.Y., USA. Pp. 4-178.

29. Dickens, J.W., and T.B. Whitaker. 1975. Efficacy of electronic color sorting and hand picking to remove aflatoxin contaminated kernels from commercial lots of shelled peanuts. Peanut Science 2:45-50.

30. Dollear, F.G. 1969. Detoxification of aflatoxins in foods and feeds. *In* Aflatoxin, L.A. Goldblatt, ed. Academic Press, New York, N.Y., USA. Pp. 359-391.

31. Dollear, F.G., G.E. Mann, L.P. Codifer, Jr., H.K. Gardner, Jr., S.P. Koltun and H.L.E. Vix. 1968. Elimination of aflatoxins from peanut meal. Journal of the American Oil Chemists' Society 45:862-865.

32. Doyle, M.P., and E.H. Marth. 1978. Bisulfite degrades aflatoxin: Effect of temperature and concentration of bisulfite. Journal of Food Protection 41:774-780.

33. Draughon, F.A., and E.A. Childs. 1982. Chemical and biological evaluation of aflatoxin after treatment with sodium hypochlorite, sodium hydroxide and ammonium hydroxide. Journal of Food Protection 45:703-706.

34. Dwarankanath, C.T., E.T. Rayner, G.E. Mann and F.G. Dollear. 1968. Reduction of aflatoxin levels in cottonseed and peanut meals by ozonization. Journal of the American Oil Chemists' Society. 45:93-95.

35. Espoy, H.M. 1972. Detoxification of oilseed meal with calcium hydroxide. US Patent 3,689,275.

36. Fandialan-Dalmaci, I., and D.Y.C. Fung. 1977. Microbiology of high-moisture corn treated with "cold flow" ammonia. Philippine Journal of Biology 6:203-216.

37. Farsaie, A., W.F. McClure and R.J. Monroe. 1981. Design and development of an automatic electro-optical sorter for removing BGY fluorescent pistachio nuts. Transactions of the American Society of Agricultural Engineers 24:1372-1375.

38. Feuell, A.J. 1966. Aflatoxin in groundnuts. IX. Problems of detoxification. Tropical Science 8:61-70.

39. Frayssinet, C., and C. LaFarge. 1972. Procédé pour éliminer les aflatoxines dans les tourteaux d'arachide. French Patent 2,098,711.

40. Gardner, H.K. Jr., S.P. Koltun, F.G. Dollear and E.T. Rayner. 1971. Inactivation of aflatoxins in peanut and cottonseed meals by ammoniation. Journal of the American Oil Chemists' Society 48:70-73.

41. Gardner, H.K. Jr., S.P. Koltun and H.L.E. Vix. 1968. Solvent extractions of aflatoxins from oilseed meals. Journal of Agricultural and Food Chemistry 16:990-993.

42. Giddy, C., J. Brandt and G. Bunter. 1977. The detoxification of oil-seed cakes polluted by aflatoxins: Research and development of an industrial process. Agricultural Technology Age 27:331-338.

43. Goldblatt, L.A. 1971. Control and removal of aflatoxin. Journal of the Americal Oil Chemists' Society 48:605-610.

44. Goldblatt, L.A., and R.A. Robertson, Jr. 1970. Process for removing aflatoxin from peanuts. US Patent 3,515,736.

45. Goldblatt, L.A., and F.G. Dollear. 1977. Detoxification of contaminated crops. *In* Mycotoxins in Human and Animal Health, J.V. Rodricks, C.W. Hesseltine and M.A. Mehlman, eds. Pathotox, Park Forest South, Illinois, USA. Pp. 139-150.

46. Hagler, W.M., J.E. Hutchins and P.B. Hamilton. 1982. Destruction of aflatoxin in corn with sodium bisulfite. Journal of Food Protection 45:1287-1291.

47. Hron, R.J., Jr., and S.P. Koltun. 1984. An aqueous ethanol extraction process for cottonseed oil. Journal of the American Oil Chemists' Society 61:1457-1460.

48. Hughes, B.L., B.D. Barnett, J.E. Jones and J.W. Dick. 1979. Safety of feeding aflatoxin-inactivated corn to white Leghorn layer-breeders. Poultry Science 58:1202-1209.

49. Jorgensen, K.V., and R.L. Price. 1981. Atmospheric pressure-ambient temperature reduction of aflatoxin B_1 in ammoniated cottonseed. Journal of Agricultural and Food Chemistry 29:555-558.

50. Kwolek, W.F., and O.L. Shotwell. 1979. Aflatoxin in white corn under loan. V. Aflatoxin prediction from weight percent of bright greenish-yellow fluorescent particles. Cereal Chemistry 56:342-345.

51. Lawlor, T.E., E. Zeiger, D.L. Park and L.S. Lee. 1985. Mutagenic potential of ammoniation-related aflatoxin reaction-products isolated from cottonseed meal. Journal of the American Oil Chemists' Society 62:1136-1138.

52. Lee, L.S., and A.F. Cucullu. 1978. Conversion of aflatoxin B_1 to aflatoxin D_1 in ammoniated peanut and cottonseed meals. Journal of Agricultural and Food Chemistry 26:881-884.

53. Lee, L.S., J.J. Dunn, A.J. DeLucca and A. Ciegler. 1981. Role of lactone ring of aflatoxin B_1 in toxicity and mutagenicity. Experientia 37:16-17.

54. Lee, L.S., S.P. Koltun and J.B. Stanley. 1984. Ammoniation of aflatoxin B_1 in a pressure chamber used to decontaminate toxin-containing cottonseed meal. Journal of the Americal Oil Chemists' Society 61:1607-1608.

55. Lesieur, B., and J. Colin. 1977. Mise au point d'un procédé industrial pour le traitement des tourteaux d'arachide contaminés par l'aflatoxine. Annals of Technology and Agriculture 27:367-371.

56. Lillehoj, E. B., and A. Ciegler. 1969. Biological activity of aflatoxin B_{2a}. Applied Microbiology 17:516-519.

57. Lillehoj, E.B., W.F. Kwolek, R.E. Peterson, O.L. Shotwell and C.W. Hesseltine. 1976. Aflatoxin contamination, fluorescence, and insect damage in corn infected with *Aspergillus flavus* before harvest. Cereal Chemistry 53:505-512.

58. Lillehoj, E.B., A. Lagoda and W.F. Maisch. 1979. The fate of aflatoxin in naturally contaminated corn during the ethanol fermentation. Canadian Journal of Microbiology 25:911-914.

59. Mann, G.E., H.K. Gardner, Jr., A.N. Booth and M.R. Gungmann. 1971. Aflatoxin inactivation: Chemical and biological properties of ammonia and methylamine treated cottonseed meal. Journal of Agricultural and Food Chemistry 19:1155-1158.

60. Martinez, R.R. 1979. Las aflatoxinas en las tortillas. Veterinaria México 10:37-39.

61. Masri, M.S., H.L.E. Vix and L.A. Goldblatt. 1969. Process for detoxifying substances contaminated with aflatoxin. US Patent 3,429,709.

62. McKinney, J.D., G.C. Cavanagh, J.T. Bell, A.S. Hoversland, D.M. Nelson, J. Pearson and R.J. Selkirk. 1973. Effects of ammoniation of aflatoxin in rations fed lactating cows. Journal of the American Oil Chemists' Society 50:79-84.

63. Moerck, K.E., P. McElfresh, A. Wohlman and B.W. Hilton. 1980. Aflatoxin destruction in corn using sodium bisulfite, sodium hydroxide and aqueous ammonia. Journal of Food Protection 43:571-574.

64. Natarajan, K.R., D.C. Rhee, C.M. Carter and K.F. Mattil. 1975. Destruction of aflatoxins

in peanut protein isolates by sodium hypochlorite. Journal of the American Oil Chemists' Society 52:160-163.

65. Nofsinger, G.W., and R.A. Anderson. 1979. Note on inactivation of aflatoxin in ammonia-treated shelled corn at low temperatures. Cereal Chemistry 56:120-121.

66. Norred, W.P. 1981. Effect of ammoniation on the toxicity of corn artificially contaminated with aflatoxin B_1. Applied Pharmacological Toxicology 51:411-416.

67. Norred, W.P. 1982. Ammonia treatment to destroy aflatoxins in corn. Journal of Food Protection 45:972-976.

68. Norred, W.P., and R.E. Morrissey. 1983. Effects of long-term feeding of ammoniated, aflatoxin-contaminated corn to Fischer 344 rats. Toxicology and Applied Pharmacology 70:96-104.

69. Ogbadu, G., and O. Bassir. 1979. Toxicological study of gamma-irradiated aflatoxins using the chicken embryo. Toxicology and Applied Pharmacology 51:379-382.

70. Okonkwo, P.O., and C. Nwokolo. 1975. Aflatoxin B_1: simple procedures to reduce levels in tropical food. Nutrition Reports International 17:387-395.

71. Park, D.L., L.S. Lee and S.A. Koltun. 1984. Distribution of ammonia-related aflatoxin reaction products in cottonseed meal. Journal of the American Oil Chemists' Society 61:1071-1074.

72. Parker, W.A., and D. Melnick. 1966. Absence of aflatoxin from refined vegetable oils. Journal of the American Oil Chemists' Society 43:635:638.

73. Paulsen, M.R., G.H. Brusewitz, B.L. Clary, G.V. Odell and J. Pominski. 1976. Aflatoxin content and skin removal by treatment with chemicals, water spray, heated air and nitrogen. Journal of Food Science 41:667-671.

74. Pohland, A.E., M.E. Cushmac and P.J. Andrellos. 1968. Aflatoxin B_1 hemiacetal. Journal of the Association of Official Analytical Chemists 51:907-910.

75. Pons, W.A., Jr., A.F. Cucullu, L.S. Lee, H.J. Janssen and L. A. Goldblatt. 1972. Kinetic study of acid catalyzed conversion of aflatoxin B_1 and G_1 to B_{2a} and G_{2a}. Journal of the American Oil Chemists' Society 49:124-128.

76. Prevot, A. 1974. Evolution et methodes d'élimination des aflatoxines dans les produits oléagineux (huiles et tourteaux). Revue Française des Corps Gras 21:91-103.

77. Price, R.L., and K.V. Jorgensen. 1985. Effects of processing on aflatoxin levels and on mutagenic potential of tortillas made from naturally contaminated corn. Journal of Food Science 50:347-349.

78. Price, R.L., O.G. Lough and W.H. Brown. 1982. Ammoniation of whole cottonseed at atmospheric pressure and ambient temperature to reduce aflatoxin M_1 in milk. Journal of Food Protection 45:341-344.

79. Rayner, E.T., S.P. Koltun and F.G. Dollear. 1977. Solvent extraction of aflatoxins from contaminated agricultural products. Journal of the American Oil Chemists' Society 54:242A-244A.

80. Rehana, F., S.C. Basappa and V.S. Murthy. 1979. Destruction of aflatoxin in rice by different cooking methods. Journal of Food Science and Technology 16:111-112.

81. Roberts, A.C., and D. J. McWeeny. 1972. Use of sulfur dioxide in the food industry. Journal of Food Technology 7:221-238.

82. Rodricks, J.V., and H.R. Roberts. 1977. Mycotoxin regulation in the United States. *In* Mycotoxins in Human and Animal Health, J.V. Rodricks, C.W. Hesseltine and M.A. Mehlman, eds. Pathotox, Park Forest South, Illinois, USA. Pp. 753-758.

83. Rodricks, J.V., and L. Stoloff. 1977. Aflatoxin residues from contaminated feed in edible tissues of food-producing animals. *In* Mycotoxins in Human and Animal Health, J.V. Rodricks, C.W. Hesseltine and M.A. Mehlman, eds. Pathotox, Park Forest South, Illinois, USA. Pp. 67-80.

84. Samarajeewa, U., S.N. Arseculerante and C.H.S.R. Bandunatha. 1977. Degradation of aflatoxin in coconut oil and copra meal (*poonac*). Journal of the National Science Council (Sri Lanka) 5:1-12.

85. Santi, E., P. Amedeo and G. Piva. 1983. Detoxification of aflatoxin B_1 in peanut meal by urea treatment. Annali di Chimica 22:299-302.

86. Schade, J.E., K. McGreevey, A.D. King, Jr., and G. Fullwe. 1975. Incidence of aflatoxin in California almonds. Applied Microbiology 29:48-59.

87. Schroeder, T., U. Zweifel, P. Sagelsdorff, U. Friederich, J. Luthy and C. Schlatter. 1985. Ammoniation of aflatoxin-containing corn: Distribution, *in vivo* covalent deoxyribonucleic acid binding, and mutagenicity of ration products. Journal of Agricultural and Food Chemistry 33:311-316.

88. Scott, P.M. 1984. Effects of food processing on mycotoxins. Journal of Food Protection 47:489-499.

89. Seenapa, M. and I.K. Nyagahungu. 1982. Retention of aflatoxin in *ugali* and bread made from contaminated flour. Journal of Food Science and Technology 19:64-65.

90. Shanon, G.M., and O.L. Shotwell. 1979. Minicolumn detection methods for aflatoxin in yellow corn. Journal of the Association of Official Analytical Chemists 62:1070-1075.

91. Shantha, T., and V. Sreenivasa Murthy. 1981. Use of sunlight to partially detoxify groundnut (peanut) cake flour and casein contaminated with aflatoxin B_1. Journal of the Association of Official Analytical Chemists 64:291-293.

92. Shotwell, O.L., M.L. Goulden and C.W. Hesseltine. 1972. Aflatoxin contamination: Association with foreign material and characteristic fluorescence in damaged corn kernels. Cereal Chemistry 52:458-465.

93. Southern, L.L., and A.J. Clawson. 1980. Ammoniation of corn contaminated with aflatoxin and its effects on growing rats. Journal of Animal Science 50:459-466.

94. Sreenivasamurthy, J., H.A.B. Parpia, S. Srikanta and A.S. Murti. 1967. Detoxification of aflatoxin in peanut meal by hydrogen peroxide. Journal of the Association of Official Analytical Chemists 50:350-354.

95. Sreenivasamurthy, V., S. Srikanta and H.A.B. Parpia. 1971. Removing aflatoxin from peanut seed meals. Indian Patent 120,257.

96. Swenson, D.H., J.S. Miller and E.C. Miller. 1975. The reactivity and carcinogenicity of aflatoxin B_1-2,3 dichloride, a model for the putative 2,3 oxide metabolite of aflatoxin B_1. Cancer Research 35:3811-3823.

97. Temcharoen, P., and W.G. Thilly. 1982. Removal of aflatoxin B_1 toxicity but not mutagenicity by 1 megarad gamma radiation of peanut meal. Journal of Food Safety 4:199-205.

98. Trager, W., and L. Stoloff. 1967. Possible reactions for aflatoxin detoxification. Journal of Agricultural and Food Chemistry 15:679-681.

99. Ulloa-Sosa, M., and H.W. Schroeder. 1969. Note on aflatoxin decomposition in the process of making tortillas from corn. Cereal Chemistry 46:397-400.

100. Van Dyck, P.J., P. Tobback, M. Feys and H. van de Voorde. 1982. Sensitivity of aflatoxin B_1 to ionizing radiation. Applied Environmental Microbioloy 43:1317-1319.

101. Yahl, K.R., S.A. Watson, R.J. Smith and R. Barabolak. 1971. Laboratory wet-milling of corn containing high levels of aflatoxin and a survey of commercial wet-milling products. Cereal Chemistry 48:385-391.

102. Yang, C.Y. 1972. Comparative studies on the detoxification of aflatoxins by sodium hypochlorite and commercial bleaches. Applied Microbiology 24:885-890.

103. Yousef, A.E., and E.H. Marth. 1985. Degradation of aflatoxin M_1 in milk in ultraviolet energy. Journal of Food Protection 48:697-698.

The Use of Urea as a Control of Aflatoxin in Maize

A. Gonzalez P., Quality Control Laboratory, CONASUPO, Mexico, D.F., Mexico

Abstract

Urea at levels of 0.1, 0.3 and 0.5% was evaluated to determine its effectiveness in decontaminating cereals for human and animal consumption. Sets consisted of maize with pulverized urea; maize inoculated with A. flavus *spore solution; maize inoculated with* A. flavus *and treated with urea; and a control (uninoculated, untreated maize) kept at relative humidities of 70%, 85% and 95%. Sets were incubated at 22°C, and fungal growth and aflatoxin contamination were monitored for 12 weeks. At 70% relative humidity, urea inhibited fungal growth until the seventh week and aflatoxin until the twelfth week. When relative humidity increased, fungal growth and aflatoxin were delayed by higher concentrations of urea. Under extreme relative humidity (95%) urea inhibited generation of aflatoxins for up to six weeks in inoculated grain and eight weeks in grain naturally infected by the atmosphere.*

Resumen

Se evaluaron niveles de urea del 0.1, 0.3 y 0.5% con el fin de determinar su eficacia en detoxificar cereales para el consumo humano y animal. Se incubaron a una temperatura de 22 C un lote de maíz con urea pulverizada, otro de maíz inoculado con una solución de esporas de A. flavus, *un tercero de maíz inoculado con* A. flavus *y tratado con urea, y un lote testigo (maíz no tratado ni inoculado) a una humedad relativa de 70%, 85% y 95%. Se observó el crecimiento fungoso y la contaminación con aflatoxinas durante 12 semanas. Al 70% de humedad relativa, la urea inhibió el crecimiento fungoso hasta la séptima semana y las aflatoxinas hasta la duodécima semana. Al incrementar la humedad relativa, el crecimiento fungoso y las aflatoxinas fueron inhibidas con concentraciones más elevadas de urea. A una humedad relativa extrema (95%), la urea inhibió la generación de aflatoxinas durante un máximo de seis semanas en grano inoculado y ocho semanas en grano infectado por su propia flora natural.*

A number of methods have been tested for the decontamination of cereals and oil seeds infected with mycotoxins (5); several substances have been shown to inhibit or destroy aflatoxins (7). Any such agent must reduce mycotoxin levels to acceptable limits, leave little toxic residue and not decrease the nutritive value of the treated commodity. When the economics of detoxification are also considered, few substances have proven effective.

Detoxifying agents may be classified as alkalis, acids or oxidants. Of these, the alkalis have been shown to be the most efficient and relatively inexpensive for the control of aflatoxins. Alkalis are widely used to decontaminate oil seeds, through washing with NaOH 0.3-8.3 N. The process, which eliminates gums, free acids and soluble pigments, lowers aflatoxin contamination to less than 1 ppb (1).

Calcium hydroxide effectively controls aflatoxins in copra, peanuts, and cottonseed meal (6), when used in particles of 50 μ or less.

When aflatoxin is randomly located in cereals, the use of volatile alkalis, such as methylamine and ammonia, has been especially successful. Ammonia has been found to deactivate 95 to 98% of the aflatoxins in contaminated materials (5).

CONASUPO, the National Basic Commodities Supply Agency, buys and sells all kinds of foods as a means of raising the nutritional level of the lower-income groups in the country. The organization buys a great deal of maize grain, which must be handled and stored, often under suboptimal conditions which lead to fungal growth and aflatoxin development. The objective of the research reported here was the evaluation of the effectiveness of the ammonia-generating compound, urea, for decontaminating cereals for human and animal consumption.

Materials and Methods

Experimental materials included 27 chambers with a capacity of five liters with controlled relative humidity, and 108 perforated 200-ml containers for maize. United States Pharmacopoeia (USP)-grade urea was used, as well as solvents for performing mycotoxin quantification according to the method of Stoloff (2). Number 2 grade maize and *Aspergillus flavus* isolates were used.

The chambers were divided into three sets of nine each. Each of these was further subdivided into three chambers each, with four of the 200-ml containers for maize. The first container was used as the control (uninoculated, untreated maize), and the second contained maize with pulverized urea. The third container was inoculated with an *A. flavus* spore solution, and the fourth was inoculated with *A. flavus* and treated with urea. The first set had an internal relative humidity of 70%, the second 85% and the third 95%, attained by the use of saturated solutions of $CaCl_2$, $CuSO_4$ and NaCl, respectively. The sets were incubated at room temperature (22°C),

CONASUPO Maize Storage Conditions: humidity and grain damage lead to fungal growth

and fungal growth was checked weekly. Aflatoxin analyses were done every two weeks over the period of 12 weeks.

Results and Discussion

When the finely pulverized urea was mixed with maize, contact with moisture caused it to break down to ammonia and other trace elements. As shown in Table 1, at 70% relative humidity it inhibited fungal growth until the seventh week, and aflatoxin presence until the twelfth week. When the maize was inoculated with *A. flavus*, and urea was not used, aflatoxins were present from the first week.

When relative humidity was increased, fungal growth and aflatoxin development continued to be delayed by the use of a higher concentration of urea (Table 2). With a relative humidity of 95%, higher levels of aflatoxin were found, even though urea concentration was further increased (Table 3). However, even under this extreme condition of relative humidity, urea inhibits the generation of aflatoxins for up to six weeks in inoculated grain and eight weeks for grain naturally infected by the atmosphere.

It has been shown that ammonia levels of 0.5 to 2.0 g/kg of maize grain are effective for eliminating aflatoxin contamination (3). Urea decomposes slowly and therefore releases a constant supply of ammonia which has a continuous inhibiting effect on mold growth.

The toxic properties of urea in cereals for food and feed have been studied, and the levels utilized in these experiments (0.1, 0.3 and 0.5%) have been reported to be safe (5). In the case of ruminants, urea is beneficial; the animals are able to utilize the nitrogen generated. This is not true of other animals or of poultry.

The slow decomposition of urea lowers the actual concentration of ammonia present in cereals at any one time. Also important is the fact that for human consumption cereal grains are seldom eaten as they come from the

Table 1. Fungal growth and aflatoxin presence in maize grain treated with 0.1% urea and at 70% relative humidity CONASUPO urea study, Mexico

Week	Maize		Maize + urea		Maize + *A. flavus*		Maize + *A. flavus* + urea	
	A. flavus infection [a]	Aflatoxin (ppb) [b]	*A. flavus* infection	Aflatoxin (ppb)	*A. flavus* infection	Aflatoxin (ppb)	*A. flavus* infection	Aflatoxin (ppb)
1	-		-		+		-	
2	-	0	-	0	+	5	-	0
3	-		-		+		-	
4	+	0	-	0	+	5	-	0
5	+		-		+		+	
6	+	0	-	0	+	10	+	0
7	+		+		+		+	
8	+	5	+	0	+	10	+	0
9	+		+		+		+	
10	+	5	+	0	+	20	+	0
11	+		+		+		+	
12	+	10	+	5	+	20	+	5

a/ + = presence of infection, - = absence of infection
b/ Readings taken weeks 2, 4, 6, 8, 10, 12

field; they are processed into foodstuffs. For those reasons, even the highest recommended rate of urea is not dangerous to health.

Conclusions

The results of these experiments suggest that urea is an effective agent for inhibiting fungal growth and aflatoxin development in stored maize for up to eight weeks. It is recommended that further experiments evaluate other cereals and levels of relative humidity and temperatures. In this way, the effectiveness of urea for lowering aflatoxin contamination of various cereals under varying environmental conditions can be determined.

Table 2. Fungal growth and aflatoxin presence in maize grain treated with 0.3% urea and at 85% relative humidity CONASUPO urea study, Mexico

	Maize		Maize + urea		Maize + *A. flavus*		Maize + *A. flavus* + urea	
Week	*A. flavus* [a] infection	Aflatoxin [b] (ppb)	*A. flavus* infection	Aflatoxin (ppb)	*A. flavus* infection	Aflatoxin (ppb)	*A. flavus* infection	Aflatoxin (ppb)
1	-		-		+		-	
2	-	0	-	0	+	5	-	0
3	-		-		+		-	
4	+	0	-	0	+	10	+	0
5	+		-		+		+	
6	+	0	-	0	+	10	+	0
7	+		+		+		+	
8	+	5	+	0	+	15	+	0
9	+		+		+		+	
10	+	10	+	0	+	25	+	15
11	+		+		+		+	
12	+	10	+	15	+	25	+	15

a/ + = presence of infection, - = absence of infection
b/ Readings taken weeks 2, 4, 6, 8, 10, 12

Table 3. Fungal growth and aflatoxin presence in maize grain treated with 0.5% urea and at 95% relative humidity CONASUPO urea study, Mexico

	Maize		Maize + urea		Maize + *A. flavus*		Maize + *A. flavus* + urea	
Week	*A. flavus* [a] infection	Aflatoxin [b] (ppb)	*A. flavus* infection	Aflatoxin (ppb)	*A. flavus* infection	Aflatoxin (ppb)	*A. flavus* infection	Aflatoxin (ppb)
1	-		-		+		-	
2	+	0	-	0	+	15	-	0
3	+		-		+		+	
4	+	0	-	0	+	15	+	0
5	+		-		+		+	
6	+	10	+	0	+	25	+	15
7	+		+		+		+	
8	+	10	+	0	+	30	+	25
9	+		+		+		+	
10	+	15	+	15	+	50	+	25
11	+		+		+		+	
12	+	25	+	15	+	50	+	30

a/ + = presence of infection, - = absence of infection
b/ Readings taken weeks 2, 4, 6, 8, 10, 12

References

1. Andersen, A.J.C. 1962. Refining of Oils and Fats for Edible Purposes, 2nd ed., P.N. Williams, ed. Pergamon Press, New York, N.Y., USA.

2. Association of Official Analytical Chemists. 1980. *In* Official Methods of Analysis, 13th ed. Washington, D.C., USA.

3. Brekke, O.L., and E.B. Lancaster. 1973. Annual Report, Northern Regional Research Center, US Department of Agriculture, Peoria, Illinois, USA.

4. Dollear, F.G. 1969. Detoxification of aflatoxins in foods and feeds. *In* Aflatoxin: Scientific Background, Control and Implications, L.A. Goldblatt, ed. Academic Press, New York, N.Y., USA. Chapter XII.

5. Dollear, F.G., and E.T. Rayner. 1971. Inactivation of aflatoxins in peanuts and cottonseed meals by ammoniation. Journal of the American Oil Chemists' Society 48:70-73.

6. Espoy, H.M. 1972. US Patent 3,689,275.

7. Goldblatt, L.A. 1967. Journal of the American Oil Chemists' Society 15:679.

International Survey on Natural Aflatoxin Occurrence in Maize

M.S. Zuber, Department of Agronomy, L.L. Darrah, Agricultural Research Service, US Department of Agriculture, both of the University of Missouri, Columbia, Missouri, and E.B. Lillehoj, Agricultural Research Service, US Department of Agriculture, New Orleans, Louisiana, USA

Abstract
In 1979 and 1980, maize grain samples were obtained for an aflatoxin survey from six and ten countries, respectively, located in the tropical and subtropical areas of the world. These samples came from the international maize testing program conducted by CIMMYT. Incidence of aflatoxin ranged from 42 to 100% in samples from Mexico, Thailand, Colombia, India and Costa Rica. No aflatoxin was detected in samples from Burma, Republic of South Africa, Ivory Coast, Ecuador and the Philippines. Aflatoxin levels associated with genotypes over locations were not significant.

Resumen
En 1979 y 1980 se obtuvieron muestras de granos de maiz de 6 y 10 paises, respectivamente, para efectuar una encuesta sobre la aflatoxina; dichos paises se localizan en las zonas tropicales y subtropicales del mundo. Las muestras procedian del Programa de Ensayos Internacionales de Maiz realizado por el CIMMYT. La incidencia de la aflatoxina varió del 42 al 100% en las muestras procedentes de México, Tailandia, Colombia, India y Costa Rica. No se detectaron aflatoxinas en las muestras procedentes de Birmania, República de Sudáfrica, Costa de Marfil, Ecuador y Filipinas. No fueron significativos los niveles de aflatoxinas relacionados con los genotipos en todas las localidades.

Although aflatoxin contamination in maize (*Zea mays* L.) has been considered a worldwide problem, only fragmentary data support this assumption. To acquire definitive data on an international basis, a survey was conducted through the worldwide international maize testing program administered by the International Maize and Wheat Improvement Center (CIMMYT). Grain samples were obtained from the same maize genotypes grown in replicated trials in a number of countries. The primary objective of the study was to utilize grain from the CIMMYT trials to obtain information on the occurrence of natural aflatoxin contamination from diverse regions of the world.

Previous investigations of material grown in the USA and artificially inoculated with *Aspergillus flavus* Link ex Fries showed that aflatoxin levels in maize kernels appear to be genetically controlled (1,2). These results led to the second objective of the study, i.e., to determine if maize varieties grown in the tropics and subtropics might have inherent tolerance of or resistance to kernel infection by toxin-producing *Aspergillus* spp.

CIMMYT has had long-term, ongoing, replicated selection performance trials throughout the world that involve several indigenous varieties. Through selection and recombination, several pools and populations have been developed. The 1979 study involved six entries in EVT 13A (the late, tropical, yellow, normal endosperm group). Grain samples were obtained from three countries: Colombia, Costa Rica and the Philippines. In 1980, six entries were sampled in EVT 13A and EVT 15A (tropical, high-lysine, modified opaque-2). Grain samples were obtained from ten countries: Bolivia, Brazil, Burma, India, Ivory Coast, Mexico, Ecuador, Ghana, Thailand and the Republic of South Africa.

The sampling procedure at each location involved obtaining 2.5-kg samples by combining replications one and two and another 2.5 kg from replications three and four for each of the six entries. Shelled grain was dried to 12% moisture as rapidly as possible after harvest.

Grain samples were returned to the USA through the cooperative efforts of US agricultural attachés and US Agency for International Development missions in participating countries. Local cooperators delivered the grain, and it was sent by diplomatic air pouch to the USDA Foreign Agricultural Service in Washington, D.C. Samples were transferred to the USDA Animal and Plant Health Inspection Service for inspection and clearance, and finally sent to the Southern Regional Research Center, ARS, USDA, New Orleans, Louisiana, for aflatoxin analyses.

Results from the 1979 study (EVT 13A) showed all 12 samples from the Philippines to be aflatoxin negative and all samples from Costa Rica positive; 83% of the samples from Colombia were positive (Table 1).

Only three of the ten countries (Brazil, India and Mexico) participating in the 1980 study had samples that were positive for aflatoxin, and the incidence was quite low, except in India (Table 2). Since most of these countries are located in the tropics and subtropics, a higher incidence of aflatoxin contamination was anticipated. It is obvious that differences would not be statistically significant among the six entries with such low incidence and levels of aflatoxin contamination. However, Ferke 7928 and Poza Rica 7928 showed no aflatoxin contamination in any of the ten countries.

For the EVT 15A tropical, high-lysine, opaque-2 evaluation, a higher incidence of aflatoxin contamination was detected than for the EVT 13A

Table 1. Presence, range and means of aflatoxin B_1 levels in EVT 13A maize samples [a/] from three countries, Southern Regional Research Center, New Orleans, Louisiana, 1979

Country	No. of samples	Aflatoxin presence (%)	Aflatoxin B_1 (ng/g^{-1}) Range	Mean
Colombia	12	83	2-26	12
Costa Rica	12	100	11-209	71
Philippines	12	0	—	ND[b/]

a/ Tropical, yellow, normal endosperm
b/ ND = no data

late, tropical, yellow, normal endosperm. Six of the 10 countries showed aflatoxin contamination for the high lysine maize, compared with only three of 10 for the yellow, normal endosperm types (Table 3).

Reasons for the limited aflatoxin contamination in these studies might be related to the better-than-average agronomic practices, for example, insect control, associated with the CIMMYT maize performance trials. It is

Table 2. Presence, range and means of aflatoxin B_1 levels in EVT 13A maize samples [a] from ten countries, Southern Regional Research Center, New Orleans, Louisiana, 1980

Country	No. of samples	Aflatoxin presence (%)	Aflatoxin B_1 (ng/g^{-1}) Range	Mean
Bolivia	12	0	—	—
Brazil	12	8	Trace	Trace
Burma	12	0	—	—
Ecuador	12	0	—	—
India	11	27	4-121	13
Ivory Coast	12	0	—	—
Mexico	12	8	0-16	1
Philippines	6	0	—	—
Rep. of South Africa	12	0	—	—
Thailand	12	0	—	—

[a] Tropical, yellow, normal endosperm

Table 3. Presence, range and means of aflatoxin B_1 levels in EVT 15A maize samples[a] from ten countries, Southern Regional Research Center, New Orleans, Louisiana, 1980

Country	No. of samples	Aflatoxin presence (%)	Aflatoxin B_1 (ng/g^{-1}) Range	Mean
Bolivia	12	8	0-3	Trace
Brazil	12	0	—	—
Burma	12	0	—	—
Ecuador	12	0	—	—
Ghana	12	8	0-24	2
India-1	12	100	5-293	98
India-2	12	17	0-17	3
Ivory Coast	12	0	—	—
Mexico	12	58	3-48	9
Rep. of South Africa	12	0	—	—
Thailand	12	42	Trace	5

[a] Tropical, high-lysine, opaque-2

also possible that the genotypes developed by the CIMMYT maize breeding program and chosen for this study may have better than average natural resistance to fungal infection by *A. flavus*.

Conclusions

Problems of moving whole maize grain samples across national borders were encountered. For future studies, grinding samples at each location and taking small random aggregate samples for aflatoxin determination would be less costly, and fewer problems with quarantine regulations in the various countries would likely be encountered.

References

1. Widstrom, N.W., and M.S. Zuber. 1983. Prevention and control of aflatoxin in corn: Sources and mechanisms of genetic control in the plant. *In* Aflatoxin and *Aspergillus flavus* in Corn, U.L. Diener, R.L. Asquith and J.W. Dickens, eds. Southern Cooperative Series Bulletin 279. Alabama Agricultural Experiment Station, Auburn, Alabama, USA. Pp. 72-76.

2. Zuber, M.S., O.H. Calvert, W.F. Kwolek, E.B. Lillehoj and M.S. Kang. 1978. Aflatoxin production in an eight line diallel of *Zea mays* L. infected with *Aspergillus flavus*. Phytopathology 68:1346-1349.

Aflatoxin in Costa Rica

M.A. Mora, Grain and Seed Research Center, University of Costa Rica, San José, Costa Rica

Costa Rica, with an area of 51,100 square kilometers and a population of 2.6 million inhabitants, is the second smallest country in Central America. As is the case of many developing countries, it has a large agricultural economy. The agricultural sector accounts for 20.4% of the gross national product (GNP), and basic grains represent 10% of that amount. Grains are important not only economically, but also for their contribution to nutrition; they are the source of nearly 32% of the proteins and 35% of the calories consumed by Costa Ricans (not counting indirect consumption in the form of livestock products). Basic grains are also an important source of income for small-scale farmers, generally a very low-income group.

Of the total volume of basic grains consumed in Costa Rica in 1983-84, all of the wheat and nearly 20% of the beans were imported; the remainder was grown domestically (Figure 1). Twenty-four per cent of the rice produced was exported. This situation may vary, depending on the growing season, as can be seen in Table 1.

Grains are grown in the four geographical regions of the country, with either one or two crops produced per year. Certain areas produce larger amounts of a specific grain than others. The Chorotega region (northern Pacific) is an important producer of rice and sorghum, and the Atlantic and southern regions are the major maize-producing areas.

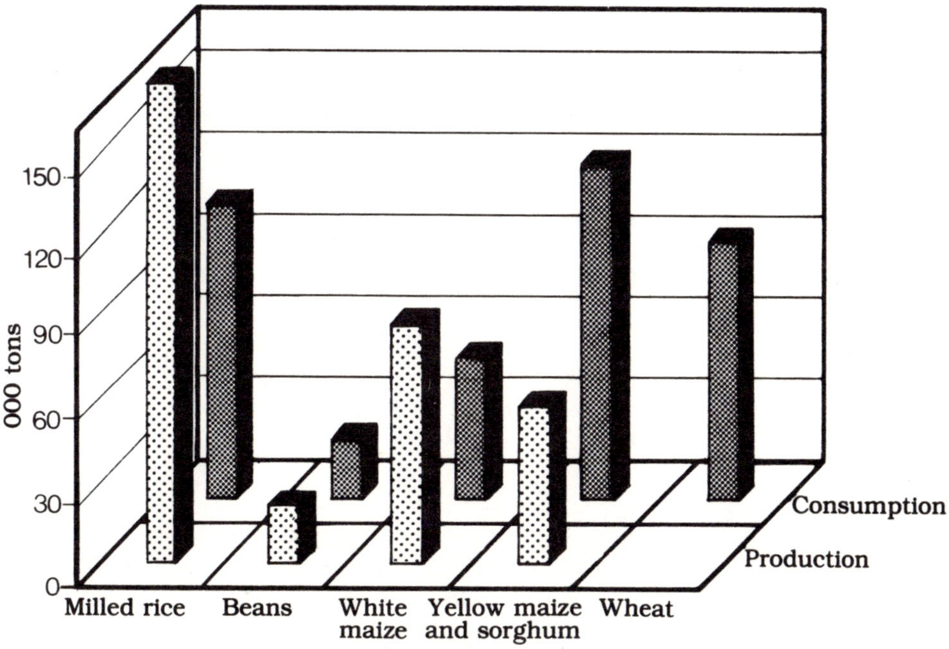

Figure 1. Grain production and consumption in Costa Rica, 1983-84

The relationship between grain-production cycle and weather in the various regions often results in the harvest season coinciding with periods of high precipitation and accentuated fungal growth (Figure 2). Temperature is always favorable for growth of the fungus (Figure 2).

Costa Rican agriculturists became aware of the aflatoxin problem in 1974, when a shipment of 20,000 tons of maize was received that contained hundreds of μg/kg of aflatoxins. The entire shipment was destroyed. As a result, the Grain and Seed Research Center (CIGRAS) began exploratory testing for aflatoxin, periodically analyzing samples of imported maize in which the toxin was suspected, and cooperating with other laboratories to develop techniques for aflatoxin analysis.

Research on aflatoxins during the past few years has led to certain basic conclusions, one of which is that post-harvest systems in Costa Rica unquestionably lead to aflatoxin contamination. Also, white-grain maize has been found to have higher contamination than yellow grain. The aflatoxin project described here is based on these findings.

The Aflatoxin Project

The overall objective of the aflatoxin project is to evaluate levels of aflatoxin contamination in white maize in the various stages of the post-harvest

Table 1. Grain production by region and crop, Costa Rica, 1985-86

	Crop (tons)				
Region	**Rice**	**Beans**	**Yellow maize**	**White maize**	**Sorghum**
Central					
First crop	49,833	414	2,746	11,588	874
Second crop	1,141	5,888	764	3,662	2,377
Total	50,974	6,302	3,510	15,250	3,251
Chorotega (north Pacific)					
First crop	108,786	-	2,349	11,568	39,769
Second crop	16,235	7,533	2,123	7,258	25,522
Total	125,021	7,533	4,472	18,826	65,291
Brunca (south)					
First crop	47,305	2,789	4,645	22,195	-
Second crop	1,059	8,001	2,869	10,114	2,503
Total	48,364	10,790	7,514	32,309	2,503
Huetar (Atlantic)					
First crop	18,341	-	832	11,046	608
Second crop	10,770	239	5,211	23,173	832
Total	29,111	239	6,043	34,219	1,440
Total	253,470	24,864	21,539	105,604	72,485

system and to discover the factors that foster this contamination. The project is conducted by CIGRAS at the University of Costa Rica. It began in August 1985, and will continue for 30 months. Nearly 3,000 samples will be tested for aflatoxin during that period. The overall cost of the project is estimated at US$ 69,500, of which $37,300 are being paid by the International Development Research Centre of Canada (IDRC), $29,200 by the University of Costa Rica, principally through the use of existing staff and equipment, and $3,000 by the National Production Board, a government marketing agency.

The methodology employed includes:

- Sampling. The minimum sampling weight is 5 kg; samples are taken both within and ouside of farm boundaries and both before and after maize is dried. Samples are also obtained from the government agency, from private farmers and from commercial storage facilities.

Handling samples. The grain is taken to the laboratory as soon as possible after harvest (a maximum of 24 hours), and dried when necessary. The sample to be used for establishing moisture content is transported in an air-tight container, and the rest in ventilated containers. If the maize grain has been received on the ear, it is shelled and then dried. From each sample, portions are separated out for moisture testing and for analysis. The grain is ground until it can pass through a screen with 0.8 mm openings, and is blended mechanically for at least five minutes. Whenever a sample has to be stored at any stage of the process, it is kept in air-tight chambers at a temperature of 5°C.

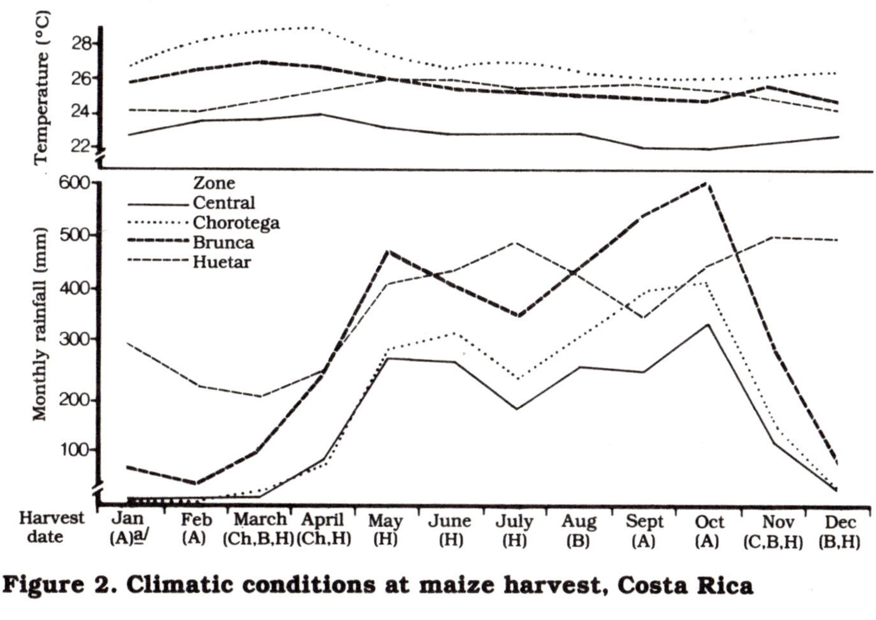

Figure 2. Climatic conditions at maize harvest, Costa Rica

a/ A = all locations, CH = Chorotega, B = Brunca, H = Huetar, C = Central

Source: Costa Rican meteorological stations

- Analysis. The method of analysis used is shown in Figure 3.

Conclusions

It is still too early in the project to be able to report much progress. However, nearly half of the 400 samples tested to date have been found to be contaminated with aflatoxin. Most of these samples have shown levels of from 100 to 200 μg/kg, and some have had levels as high as 800 μg/kg.

The amount of aflatoxin contamination found in Costa Rica's 1985 maize harvest was probably not typical, due to a number of problems, particularly the lack of drying capacity caused by the country's increased production.

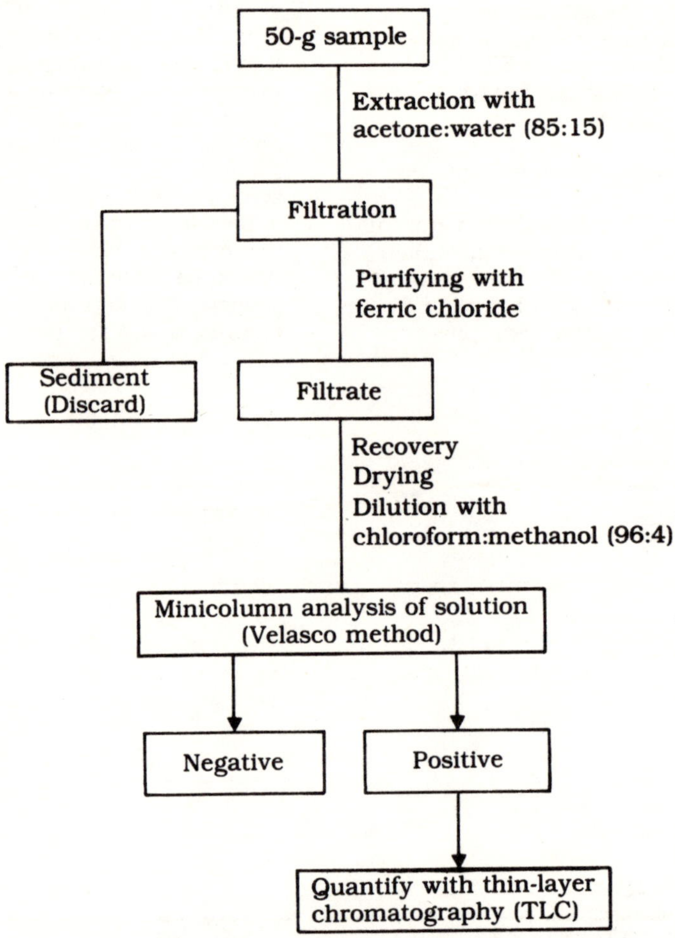

Figure 3. Flow chart, analysis of maize for aflatoxin contamination, Grain and Seed Research Center, University of Costa Rica, San José, Costa Rica

Aflatoxin in El Salvador

G.R. Calderon, Laboratory for Toxic Residues, Agricultural Technology Center, San Salvador, El Salvador

El Salvador, a small country in Central America, covers approximately 36,260 square kilometers and has a population of five million. At present, problems and conflicts of a social nature have had serious consequences that influence the country's food supply and sources of food.

Mycotoxin Research in El Salvador

Exports

Mycotoxin research, especially on aflatoxins, began in the country in 1971, principally in the analysis of export materials (grated coconut, peanuts and sesame seed). These products were not found to have high levels of contamination since they had been preselected. However, in 1982 and 1983, some exports, such as the cashew nuts produced by the agrarian reform sector, did present serious problems; aflatoxin contamination in the range of 150 to 200 ppb was found.

Government storage facilities

In view of the health hazards associated with mycotoxin-contaminated foods, testing was done at 98 government-owned storage facilities throughout the country in 1979. This represented 166,099 tons of foodstuffs with a value of US$ 1,910,133 (Table 1). The results showed that samples from 72 of the 98 facilities (73.4%) were contaminated

Table 1. Aflatoxin content in cereal samples from government storage facilities, El Salvador, 1979

Cereal	No. of contaminated samples per warehouse	Aflatoxin (ppb)[a]
Beans		
Red	16/23	<0.01 to 63
Black	33/36	<0.01 to 79
Maize		
White	16/32	<0.01 to 22
Yellow	2/2	<0.01
White sorghum	4/4	<0.01 to 28
Golden rice	1/1	<0.01
Total of contaminated samples	72/98 (73.4%) [b]	

Number of samples per warehouse exceeding the tolerance level of 30 ppb: 11

[a] Maximum values found
[b] Equivalent of 166,099 tons (US$ 1,910,133)

with aflatoxin B_1, with values reaching 79 ppb for black beans and 63 ppb for red beans. The highest quantities were in 11 warehouses storing imported materials; there the tolerance level of 30 ppb was surpassed.

The farm level

In 1983 and 1984, sampling was done nationwide to obtain more information about the state of mycotoxin contamination at the farm level. Sampling was conducted on agricultural products classified as celluloses, proteins and carbohydrates (154 samples from 25 different locations) (Table 2). Of the 98 samples found to be contaminated with aflatoxin, more than half contained both aflatoxins B and G; the highest percentage of contaminated samples was among the carbohydrates (Table 3). The amount of aflatoxin found varied from nondetectable (ND) to 332.7 ppb for aflatoxin B and 90 ppb for G. Carbohydrate materials showed the greatest concentration with aflatoxins B and G (Table 4). Of the contaminated samples, 32 surpassed the established tolerance limit of 30

Table 2. On-farm testing of agricultural products for aflatoxin contamination, El Salvador, 1983 and 1984

Cellulose material	No. of samples	Protein material	No. of samples	Carbohydrate material	No. of samples
Ground whole maize plant	1	Cottonseed meal	10	Whole white maize kernels a/	17
Ground bean stems leaves and pools	2	Feed for dairy cows, a/	22	Cornmeal a/	6
Ground sugar cane whorl a/	2	laying hens, a/, rabbits,		Ground maize ears	1
Sugar cane bagasse	6	beef cattle, a/ chickens,		Sorghum grain a/	6
Pangola hay a/		calves and swine		Sorghum flour a/	6
Pangola silage	1	Red, a/ black and mung beans	15	Rice a/	12
Pangola grass	1	Meat, a/ fish, soy, coconut	7	Rice dust a/	2
Cotton seed husk a/	6	and peanut meal a/		Fine broken rice	3
Maize stalks a/	2	Whole milk	1	Bran a/	5
Broken corncobs a/	1	Brewers' yeast	1	Wheat flour a/	4
Dehydrated coffee pulp a/	1	Barley chaff	2	Fine broken wheat grain	1
Estrella hay	2	Pumpkin seed	1		
Rice chaff a/	1	Chicken manure	1		
Peanut skin	1	Cow manure	1		
Total	30	Total	61	Total	63

a/ Samples that surpassed the permissible aflatoxin level (30 ppb)

Table 3. Results of on-farm testing for aflatoxin in samples of agricultural products, El Salvador, 1983 and 1984

Product	No. of samples	Aflatoxin-contaminated samples			
		B	G	B + G	Total
Cellulose materials	30	3	5	10	18
Protein materials	61	6	12	15	33
Carbohydrate materials	63	8	13	26	47
Total	154	17	30	51	98

ppb; the greatest number of these were from the carbohydrate group. Of the samples exceeding the limit, aflatoxin B was most prevalent (Table 5).

Some of the variables for the interpretation of these results were such factors as temperature and humidity. For example, temperatures of various stored products, according to storage method and location, were shown to be favorable for the development of toxin: average temperatures ranged from 29° to 31° C for the products, 29.6° to 29.7°C for the storage areas and 25° to 26°C for different locations. The relative humidity of the locations where the crops were grown reached 87%, which favors fungal growth and toxin formation (Tables 6 and 7).

Food aid

In recent years, especially 1985 and 1986, social conflicts have displaced approximately 650,000 rural people who now need to be fed. Many international organizations have offered to help feed these people, but unfortunately, of the shipments of food aid, between 1000 and 1400 tons received over a few months had to be destroyed. This resulted in economic losses of US $215,600 per shipment, according to local market prices. The principal problem was the high occurrence of the fungi *Aspergillus*, *Penicillium* and *Fusarium* spp. Aflatoxin in quantities greater than 100 ppb was detected, especially in samples of maize and beans.

Table 4. Results of on-farm testing for maximum and average aflatoxin contamination in agricultural products, El Salvador, 1983 and 1984

Product	Aflatoxin B (ppb)		Aflatoxin G (ppb)	
	Maximum	Average	Maximum	Average
Cellulose materials	100.6	19.75	61.0	9.14
Protein materials	209.7	13.55	71.7	5.16
Carbohydrate materials	332.7	24.56	89.9	12.45

Table 5. Results of on-farm testing of agricultural products for levels of aflatoxin exceeding the tolerance level of 30 ppb, El Salvador, 1983 and 1984

Product	No. of samples	No. of aflatoxin samples with over 30 ppb			Total
		B	G	B + G	
Cellulose materials	18	8	2	1	11
Protein materials	33	7	1	0	8
Carbohydrate materials	47	8	0	5	13
Total	98	23	3	6	32

Part of this loss was due to improper handling and storage between the ports of embarcation and the storage facilities in El Salvador, combined with inadequate distribution procedures that caused foodstuffs to be held too long in storage. This situation has had severe effects on human health. Although there are no official data on deaths from aflatoxin, hepatic and gastrointestinal diseases have occurred in those who have eaten these foods.

The livestock industry

Analyses of feed for cattle, turkeys and geese have shown high levels of fungus and amounts of aflatoxin above 300 ppb. This has caused the death of livestock, with great economic losses.

Conclusions

It may be concluded from the foregoing discussion that there is a lack of knowledge about mycotoxins and storage of foodstuffs in El Salvador.

Table 6. Average temperatures of on-farm stored agricultural products and storage areas compared to atmospheric temperatures, El Salvador, 1983 and 1984

	Temperature (°C)a/		Temperature of locations (°C)b/		
Product	**Product**	**Storage area**	**Minimum**	**Maximum**	**Mean**
Cellulose materials	31.0	29.7	21.0	31.4	26.1
Protein materials	29.4	29.6	17.0	32.1	25.0
Carbohydrate materials	29.1	29.1	18.0	33.0	25.5

a/ Determined at time of sampling
b/ Meteorological service information

Note: 25° to 32°C optimum for fungal development

Table 7. Relative humidity (RH) of on-farm storage of agricultural products according to storage area and location, El Salvador, 1983 and 1984

	RH of product a/ (%)		RH of area a/ (%)		RH of location b/ (%)	
Product	**Minimum**	**Maximum**	**Minimum**	**Maximum**	**Minimum**	**Maximum**
Cellulose material	7.0	14.0	52.0	82.0	56.0	86.0
Protein materials	5.9	19.5	48.5	81.0	58.0	87.0
Carbohydrate materials	9.4	14.5	48.5	82.0	58.0	86.0

a/ Determined at time of sampling
b/ Meteorological service information

Note: Relative humidity >85% optimum for aflatoxin formation

The problem is complicated by temperature and humidity of the product in the storage areas. Temperature and humidity are both critical factors for fungal growth and toxin formation, especially when products have been dried in the open air under unfavorable climatic conditions.

Training and technical assistance for aflatoxin control in El Salvador could be conducted by establishing agreements among interested countries and their respective government agencies. Offers of aid could be used to acquire the necessary reagents and equipment for mycotoxin assay, or perhaps for paying to have the analysis done, since there are problems in finding the reagents in El Salvador.

It would be useful to form an international cooperative network among mycotoxin workers to standardize regulations and methods of analysis. Also, a quality control program could be started among the various laboratories in the country to evaluate the precision and effectiveness of their analyses, as well as to correct mistakes that might occur. A newsletter could be established to include all reports related to mycotoxin research; the Agricultural Technology Center could contribute by publishing these articles in *El Investigador Informa*, its monthly publication.

References

1. Calderon, R. 1985. Evaluación preliminar de la presencia de aflatoxinas en materiales celulolíticos, proteínicos y energéticos almacenados para consumo humano y animal. XXXI Annual Meeting of the PCCMCA, San Pedro Sula, Honduras.

2. Centro de Tecnología Agrícola. Reportes de asistencia técnica, 1971 a 1986. Ministerio de Agricultura y Ganadería, San Salvador, El Salvador.

Aflatoxin and *Tortilla* Preparation in Guatemala

M.C. de Arriola, E. de Porres, S. de Cabrera, M. de Zepeda and C. Rolz, Applied Research Division, Central American Industrial Research Institute, Guatemala, Guatemala

Maize is the cereal that has both the highest production and the highest consumption in Central America. Present production in the region is above two million tons, most of which is used for human consumption; from 5 to 20% of the crop, depending on the area, is used for livestock feed, especially for poultry. About one-third of the crop is harvested during the rainy months. In the field, major maize damage is caused by insects, birds and molds, the latter often aggravated by bad weather before and during harvest. During storage, especially on small farms, the grain is subject to insect infestations and storage problems because grain is poorly dried. The storage problem is especially severe with early harvested maize (May to November) and in areas where rainfall is scattered throughout the year, as is the case in certain parts of the Atlantic coast.

In Guatemala, where about half of the maize in Central America is produced, maize is mostly consumed as *tortillas* (flat, unleavened maize cakes), which are made by the process called nixtamalization (Figure 1). With this procedure, maize grain is boiled in lime water (CaO) and then left to soak overnight. After draining and washing, the grain is lightly pressed to remove the seed coats (pericarp) and excess lime. The resulting *nixtamal* is then ground to prepare the *masa* (dough) and small amounts are shaped and baked on a *comal* (flat clay or metal plate) on top of the stove for a few minutes. Most of the processing of *tortillas* is done on a small scale, although there is an industrial operation that follows the same procedure for producing instant *tortilla* flour.

Fungal Contamination of Guatemalan Maize

Studies have shown that maize harvested in Guatemala during the early harvest period (May to November) has considerable fungal contamination. Among the fungi are species of the genus *Aspergillus* (8,9,36), which produce some of the most harmful mycotoxins, the hepatoxic aflatoxins (7,20,30,37). Maximum contamination with aflatoxin occurs during the rainy season and in storage. Samples analyzed 20 days after harvest had aflatoxin levels of 130 ppb, and levels up to 1680 ppb when analyzed 60 days later (8).

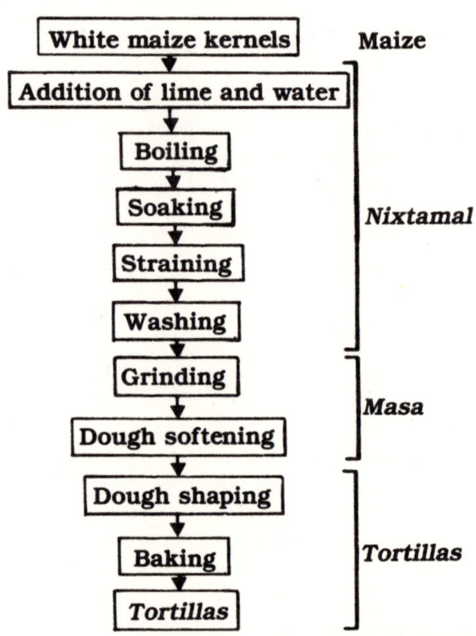

Figure 1. Flow chart, *tortilla* making, Guatemala

Aflatoxin Deactivation

Much research has been carried out on the deactivation of aflatoxin by various means, including insecticides (26,43), chemicals (4,12,14,18,19,24,25,31,34, 41,45,46,49), radiation (19,49), biological inactivation by acid-producing fungi (10) and by physical methods (4,13,14,21,33,38,42,46, 47,54). Chemical agents that have been tested include the bases calcium hydroxide (12) and sodium hydroxide (18,24,25,34).

In addition to reducing aflatoxin contamination (35,51,52), treatment with lime has been found to improve the nutritional value of maize by increasing calcium values and lysine availability in the glutelin fraction of the protein (50). It also promotes some favorable changes in amino acid content (5,22,29).

The experiments described in this paper were designed to determine the effect of the nixtamalization process (alkaline treatment) on aflatoxin-contaminated maize, quantify aflatoxin reduction, and determine whether treated material was safe for human consumption. Two different assays were made, differing in spore concentration of the inoculum, incubation time and temperature.

Materials and Methods

Intact white maize grain of the improved variety Nutricta was used in the two studies discussed here; it was not treated with fungicides. Nutricta has a high nutritional value, with lysine and tryptophane levels twice that of normal maize. The grain was first dried in a pilot plant-tray dryer using warm air for one hour at 80° to 85°C. This was followed by a period of live steam humidification for two to three hours, until the grain reached a final moisture level of 25%, suitable for fungal growth and aflatoxin production.

Inoculation

Spore preparation for inoculation was carried out according to the procedures of Hayne *et al.* (26). Petri dishes containing potato dextrose agar (PDA) were inoculated with 0.1 ml spore suspension of *Aspergillus parasiticus* strain NRRL 2999 and incubated for 10 to 30 days at 30°C for spore production. The petri dishes were isolated in metal containers and then in plastic bags to avoid contamination. To harvest the spores, the agar surface of each petri dish was washed with 5 ml of 0.1 M phosphate buffer solution at pH 7.0 with 0.1% Tween 80.

To avoid germination, the maize grain was treated to inactivate the germ and inoculated with the spore suspension. The inoculum was distributed in petri dishes containing 25 g of grain and incubated at 21°, 28° and 35°C. The grain was sampled after 10, 17 and 24 days and analyzed to determine aflatoxin levels at different temperatures.

The germ-inactivated maize grain (240 g) was placed in wide-mouthed glass jars and inoculated with a spore suspension of *A. parasiticus*. Concentrations of 2.2 x 10^6 spores per gram of dry maize were used for the first assay, and 4.86 x 10^6 spores for the second. Sterile water was added to obtain 25% grain moisture. The jars were shaken to distribute the spores and incubated at 35°C. Controls were prepared without spore suspension. Samples were taken at three, seven and ten days (first assay) and four, seven and ten days (second assay) to be submitted to the nixtamalization process.

Nixtamalization

In the first trial, the amount of lime commonly used in the nixtamalization process in home *tortilla* making in Guatemala (1.87% W/V, 3.00% W/W) was tested as well as several higher and lower lime concentrations (0.03 to 10.0% W/W). The objective was to find a concentration level that would permit easy hand removal of the pericarp, without alteration of the organoleptic characteristics of the dough and the *tortilla*. Lime of industrial quality was pulverized and sifted through a no. 14 mesh screen. Cooking times from 20 to 40 minutes were studied with an open kettle (domestic) process and an autoclave (industrial) process. Data were recorded on pH changes, the amount of water added to the cooked grain to soften the dough, and the time and temperature necessary to bake the *tortillas*.

In the second study, four 250-g samples of contaminated grain were mixed with lime, two with 0.6% W/V (1.0% W/W), the optimum found in the first trial, and the other two with the home level of 1.87% W/V (3.0% W/W). One sample of each concentration was cooked by one of the two methods, in an open kettle at 94°C for 40 minutes and in an autoclave at 121°C (15 psig) for 30 minutes. All of the cooked samples were left at room temperature overnight and then washed several times with tap water and ground. Water was added to the dough to soften it, and the *tortillas* were made.

Chemical analysis

Moisture was determined for duplicate samples of contaminated grain, dough and *tortillas*, by drying at 100°C to constant weight (1). Aflatoxin levels were determined for the samples for lime level and cooking method; aflatoxin content was determined by the AOAC method (2). Grain contaminated with a known amount of pure aflatoxin was extracted (92%) and analyzed. Quantitative thin-layer chromatography was carried out using a densitometer (Kontes, model 800), and measurements made following the procedure cited by Stubblefield *et al.* (48). Filter excitation and emissions of 365 nm and 436 nm were used.

Results and Discussion

Aflatoxin levels according to temperature and incubation time

Production of aflatoxins B_1, B_2, G_1 and G_2 was determined for three temperatures sampled at 10, 17 and 24 days (Figure 2). Diener and Davis have reported that the optimum temperature range for aflatoxin production of *A. parasiticus* on natural and semisynthetic media is 25° to 30°C for 7 to 21 days of incubation (16,17); maximum production of aflatoxin B_1 is obtained at 30° to 35°C, and G_1, 25° to 30°C (15). The ratios between G_1 and B_1 and between G_2 and B_1 have been reported to vary with respect to temperature (17) and incubation time (44) (Figure 2).

Results of the first of the studies demonstrated similar patterns. Most aflatoxins increased with incubation time, although some leveled off after 17 days. In only one case, G_2 at 35°C, a sharp and unexplained decrease occurred. In general, more B_1 and B_2 were produced at higher temperatures, and more G_1 and G_2 at lower temperatures. Routinely, a higher production of the four aflatoxins was observed at 35°C (Figure 3). For this reason, 35°C was used in the second experiment.

Lime levels in the nixtamalization process

The pH of the original maize-water mixture was found to be 6.0 to 6.5, which increased to 12.0 when lime was added. Due to the buffering capacity of maize grain, the pH did not change significantly as more lime was added. The pH decreased to 11.2 after the maize was cooked and to 9.2 after three washings.

Organoleptic characteristics (color, texture and flavor) of the *nixtamal*, dough and *tortillas* were normally good with lime levels up to 1.87% W/V (3.0 W/W). When higher amounts, 2 to

10%, were used, the *nixtamal*, dough and *tortillas* were yellow, becoming darker as lime increased. The flavor of lime also become objectionable. With the optimum lime level (0.6% W/V, 1.0% W/W), kernels were soft and easy to peel and had a normal color, the dough and *tortillas* were fine textured, and the *tortillas* were elastic and did not break easily.

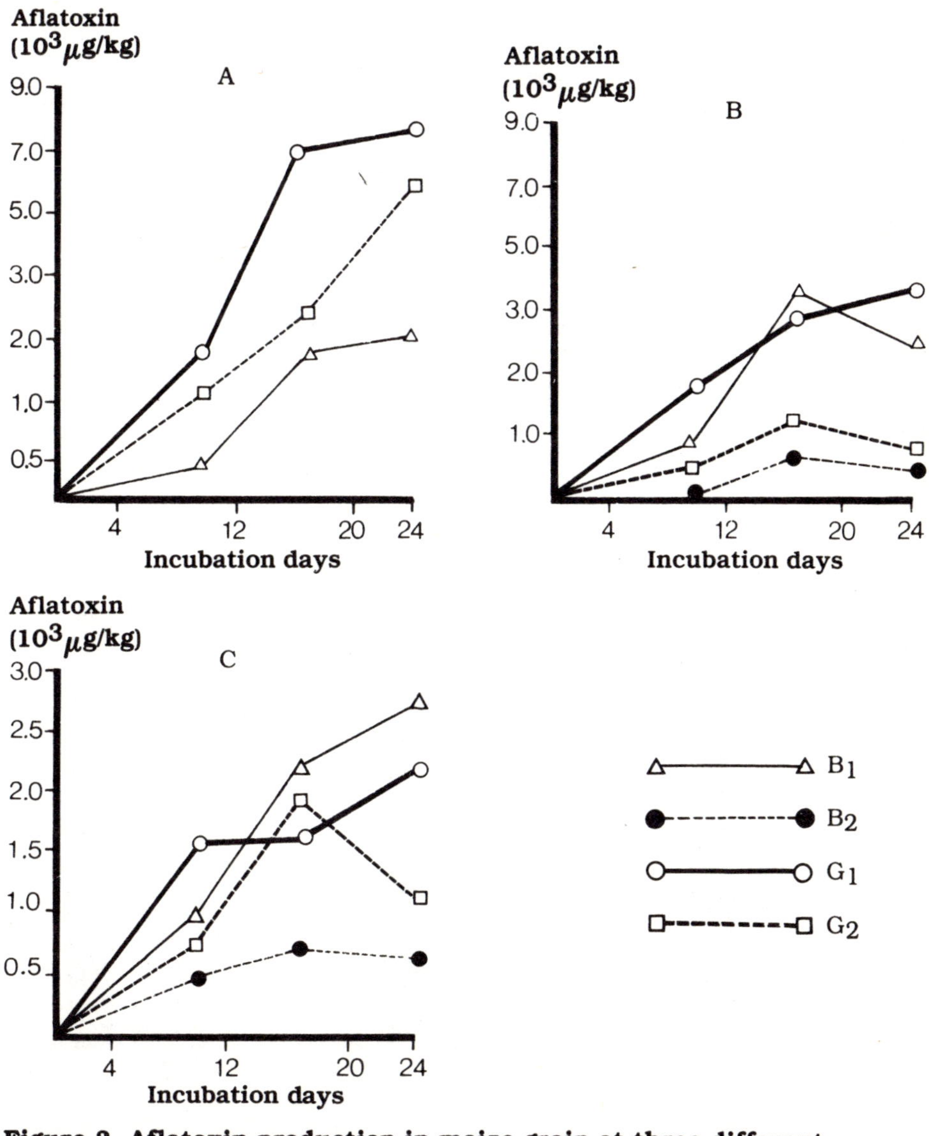

Figure 2. Aflatoxin production in maize grain at three different temperatures, study no. 1, Central American Industrial Research Institute, Guatemala[a]

[a] Aflatoxin production at 21° (A), 28° (B) and 35°C (C)

Cooking and baking

Optimum autoclave cooking time and temperature were found to be 30 minutes at 121°C (15 psig); a longer time was necessary when the maize was cooked under atmospheric conditions, the optimum being 40 minutes at 95°C. The baking time for *tortillas* ranged from 1.5 to 3 minutes on a *comal* at temperatures of 180° to 250°C. During baking, the internal temperature of the *tortillas* reached 94°C.

Aflatoxin Levels in the Two Studies

Aflatoxin levels were determined at 35°C for three, seven and ten days (Figure 4). The levels of toxin produced were higher in the second experiment, when a higher spore concentration was used in the inoculum. Comparison of these data with those of the first study (Figure 3) shows striking differences. In the second study, aflatoxin B_1 was predominant, while G_1 was highest in the first. Practically all aflatoxins were produced in much greater amounts in the second study, undoubtedly because of better fungal growth caused by the higher spore concentration. Aflatoxin B_1 had a seven-fold increase, B_2 had a three-fold increase, G_1 increased about 1.4 times, and G_2 remained almost the same. Hesseltine *et al.* (29) have pointed out that G_1 is always biosynthesized along with B_1; if neither is present, B_2 and G_2 are not found.

Aflatoxin Destruction During Nixtamalization

Aflatoxin destruction in contaminated maize treated with the two lime levels was determined (Figure 5). A decrease in aflatoxin levels in dough and

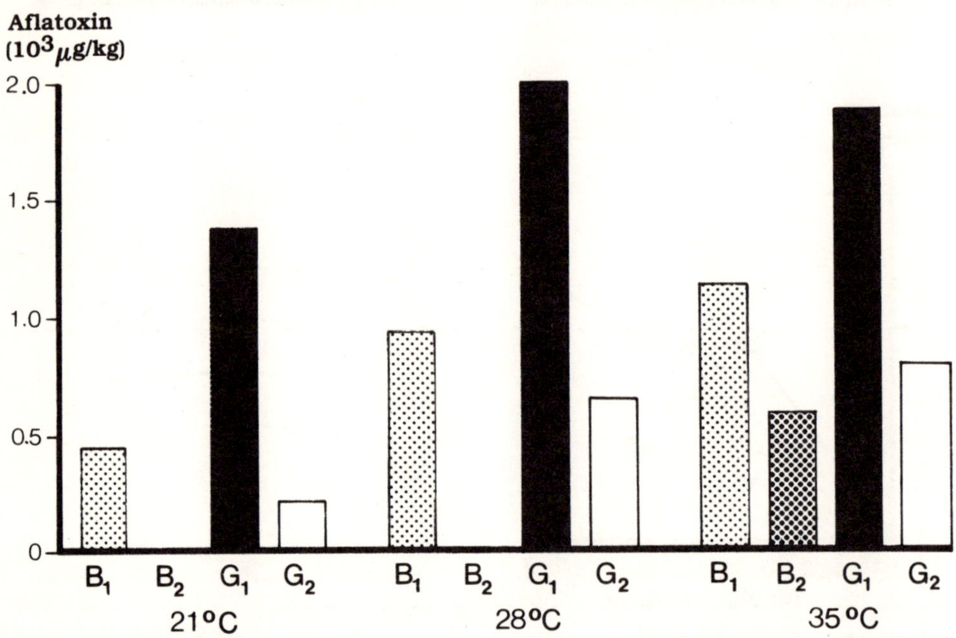

Figure 3. Aflatoxin production in maize grain by *Aspergillus parasiticus* at three different temperatures, study no. 1, Central American Industrial Research Institute, Guatemala

tortillas was found for all levels of contamination, the decrease being greater from maize to dough than from dough to *tortillas*. The reduction was more pronounced with the higher level of lime. The effect of alkalinity was more important than that of temperature, although temperature has also been reported as important (22,34,48,54).

Results of the current study confirm reports by Ulloa and Herrera (52), Ulloa-Sosa and Schroeder (53) and Price *et al.* (41) regarding aflatoxin reduction with alkalinity. This is mainly caused by the opening of the lactone ring and its irreversible binding with the protein (3,33).

Data analysis

The data were submitted to least square analysis employing a trifactorial design (11,39). The parameters and levels used included incubation times of three, seven and ten days for the first assay and four, seven and ten days for the second, two lime (CaO) concentrations, 0.6% W/V (1.0% W/W) and 1.87 W/V (3.0 W/W), and two cooking methods, open kettle and autoclave.

At significance levels of 0.05 and 0.01, no significant difference in amount of aflatoxin reduction was found between the two cooking methods or the two lime levels; however, the difference was highly significant for the three incubation periods. There were no significant differences among the various interactions of incubation time and lime level, incubation time and cooking method, lime level and cooking method, and time, lime level and cooking method. Although there was a decrease in aflatoxin content during the nixtamalization process, none of the treatments lowered it sufficiently to meet the 20 μg/kg standard suggested as safe by FAO/WHO/PAG/UNICEF (22) and the US Food and Drug Administration (23).

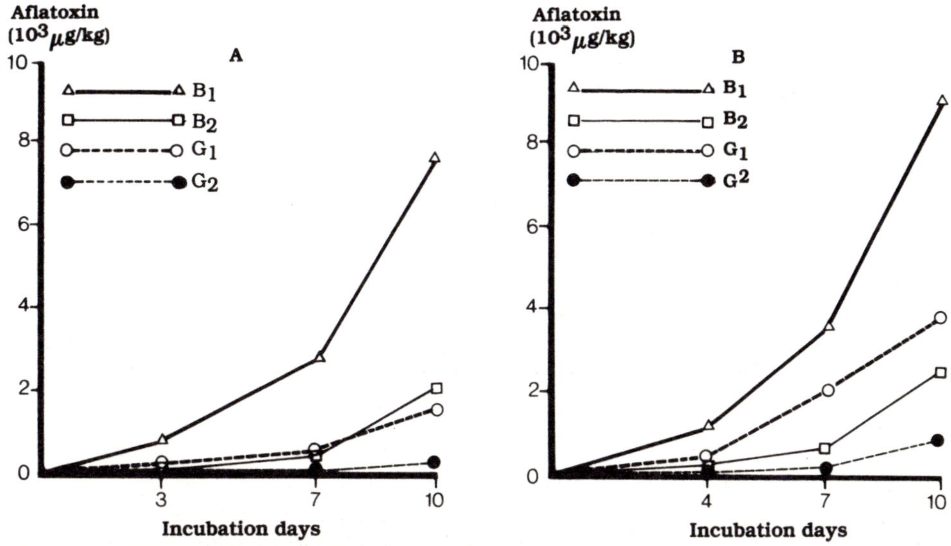

Figure 4. Aflatoxin production in maize grain at 35°C, Central American Industrial Research Institute, Guatemala[a/]

a/ A = study no. 1, B = study no. 2

Aflatoxins G_1 and G_2 were lowered more by alkaline hydrolysis than were B_1 and B_2; in most cases, they were reduced by as much as 100%. These results confirm those of Codifer *et al.* (12). Through the nixtamalization process, aflatoxin reduction from contaminated grain to dough stage was 82.5% on a wet-weight basis (three days of incubation), 89.2% (seven days) and 89.6% (ten days of incubation). On a dry-weight basis, it was 86.1% (three days of incubation), 95.6% (seven days) and 97.3% (ten days). The equivalent values for the reduction from contaminated grain to *tortillas* were 89.6% (three of days incubation), 89.2% (seven days) and 93.5% (ten days) on a wet basis, and 91.8% (three days), 95.6% (seven days) and 97.1% (ten days) on a dry-weight basis. These results are an average of the two assays at the two lime levels, two cooking methods and different numbers of days of incubation.

Conclusions

The levels of lime normally used to prepare *nixtamal* in rural and urban areas of Guatemala do not reduce aflatoxin levels in contaminated grain sufficiently to make it safe for human consumption. When lime levels above 2% are used, high aflatoxin reduction

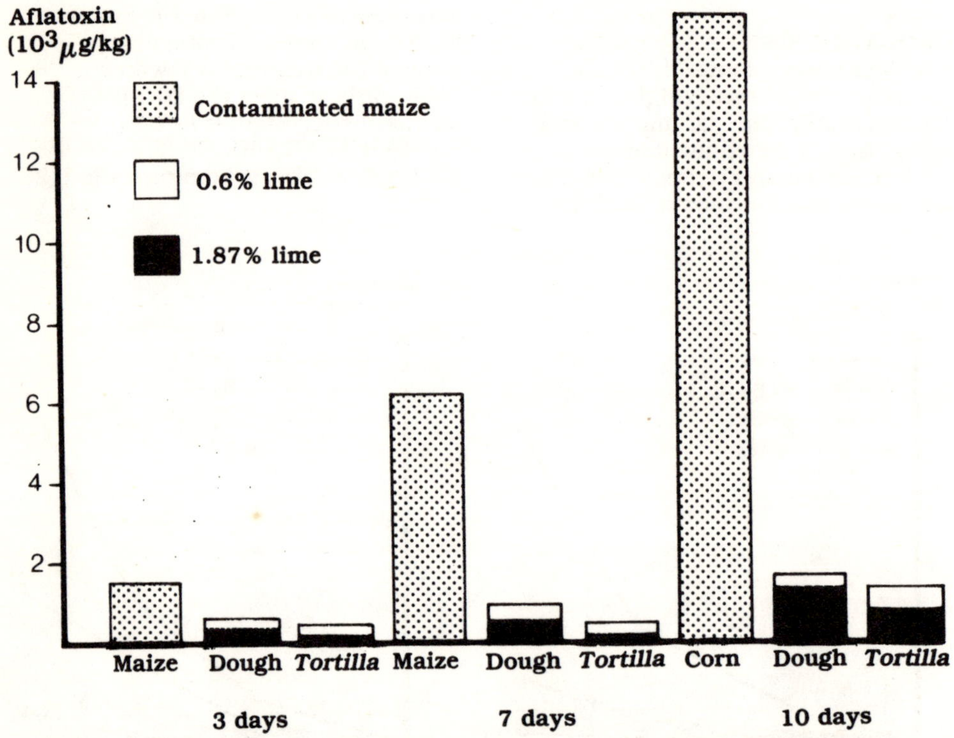

Figure 5. Aflatoxin content of dough and *tortillas* made with contaminated maize after nixtamalization with two different lime levels, studies no. 1 and 2, Central American Industrial Research Institute, Guatemala[a]

[a] Determined at 35°C

is achieved; however, the resulting *tortillas* are undesirable. During alkaline cooking, aflatoxins G_1 and G_2 are reduced more than B_1 and B_2; B_1 is reduced the least. In all cases, even at the lowest lime levels, a decrease is shown for total aflatoxins during the nixtamalization process.

Acknowledgements

This project was supported through contract 936-5542-G-00-3066-00, USAID Program in Science and Technology Cooperation (PSTC). Their encouragement and support is sincerely appreciated.

References

1. AOAC. 1975. Manual, 12th ed. Chapter 14, Cereal foods. Association of Official Analytical Chemists.

2. AOAC. 1975. Manual, 12th ed. Chapter 26, Natural poison. Association of Official Analytical Chemists.

3. Beckwith, A.C., R.F. Vesontes and A.C. Ciegler. 1975. Journal of Agricultural and Food Chemistry 23:582.

4. Brekke, O.L., A.J. Pepelinski and E. Lancaster. 1975. Paper 74-30705 presented at the meeting of the American Society of Agricultural Engineers.

5. Bressani, R., R. Paz y Paz and N.S. Scrimshaw. 1958. Journal of Agricultural and Food Chemistry 86:770-774.

6. Bressani, R., and N.S. Scrimshaw. 1958. Journal of Agricultural and Food Chemistry 86:774-777.

7. Butler, W.H. 1965. Mycotoxins in Foodstuffs. Massachussetts Institute of Technology Press, Cambridge, Massachussetts, USA. Pp. 175-186.

8. de Campos, M., J.S. Crespo and E. Olszyna-Marzys. 1980. Bulletin of Environmental Toxicology 24:789-795.

9. de Campos, M., and E. Olszyna-Marzys. 1979. Bulletin of Environmental Toxicology 22:350.

10. Ciegler, A., and R. Peterson. 1968. Applied Microbiology 16:665-666.

11. Cochran, W.G., and G.M. Cox. 1966. Experimental Designs, 2nd ed. John Wiley and Sons, New York, N.Y., USA. Pp. 148-182.

12. Codifer, L.P., G.E. Mann and F.G. Dollear. 1976. Aflatoxin inactivation: Treatment of peanut meal with formaldehyde and calcium hydroxide. Journal of the Association of Official Analytical Chemists 53:204-206.

13. Coomes, T.J., P.C. Crowther, A. Ferrel and B.J. Francis. 1966. Nature 209:406-407.

14. Diener, U.L., R.L. Asquith and J.W. Dickens. 1983. Aflatoxins and *A. flavus* in Corn. Craftmaster Printers, Opelika, Alabama, USA. Pp. iv, 112.

15. Diener, U.L., and N.D. Davis. 1969. Production of aflatoxin on peanuts under controlled environments. Journal of Stored Products Research 5:251-258.

16. Diener, U.L., and N.D. Davis. 1970. Journal of the American Oil Chemists' Society 44:259-263.

17. Diener, U.L., and N.D. Davis. 1966. Aflatoxin production by isolates of *Aspergillus flavus*. Phytopathology 56:1390-1393.

18. Dollear, F.G., G.E. Mann., L.P. Codifer, H.K. Gardner, Jr., S.P. Koltun and H.L. Vix. 1968. Elimination of aflatoxins from peanut meal. Journal of the Association of Official Analytical Chemists 45:862-865.

19. Donwie, T., and F. Suchu. 1973. Journal of the Association of Official Analytical Chemists 56:1425-1430.

20. Enomoto, M., and M. Saito. 1972. Annual Review of Microbiology 26:279-312.

21. Escher, F.E., P.E. Koehler and J. Caynes. 1973. Journal of Food Science 38:889-892.

22. FAO/WHO/UNICEF/PAG. 1969. PAG recommendation on aflatoxin. *In* PAG Statement 2. United Nations, New York, N.Y., USA.

23. FDA. Aflatoxin in milk: Establishment of action level. Federal Registry 42.61630. Food and Drug Administration, Washington, D.C., USA.

24. Goldblatt, L.A. 1966. Advances in Chemistry Series 57:216-227.

25. Goldblatt, L.A. 1968. Botany 22:51-61.

26. Gunda, H., and P. Harein. 1973. Environmental Cont. Tox. 10:112-115.

27. Hayne, W.C., L.J. Wickerham and C.W. Hesseltine. 1975. Applied Microbiology 3:361-368.

28. Hesseltine, C.W., O.L. Shotwell, J.J. Ellis and R.D. Stubblefield. 1966. Aflatoxin formation by *Aspergillus flavus*. Bacteriological Review 30:795-805.

29. Katz, S.H., M.L. Hediger and L.A. Vallero. 1974. Science 184:765-773.

30. Le Breton, E., C. Frayssinet and J. Boy. 1962. Humaine et Cancerologia Experimentales CR. Academy of Science 25:784-786.

31. Lee, L.S., J.B. Stanley, A.F. Cucullu, W.A. Pons and L.A. Goldblatt. 1974. Journal of the Association of Official Analytical Chemists 57:626-631.

32. Lee, L.S., L.Y. Yatsu and L.A. Goldblatt. 1967. Journal of the American Oil Chemists' Society 44:331-332.

33. Mann, G.E., L.P. Codifer and F.G. Dollear. 1967. Journal of Agricultural and Food Chemistry 15:1090-1092.

34. Mann, G.E., L.P. Codifer, H.K. Gardner, S.P. Koltun and F.G. Dollear. 1970. Journal of the Association of Official Analytical Chemists:173-176.

35. Martínez, M.L. 1968. Efecto de ciertos hongos sobre el valor nutritivo, calidad y conservación del maíz en Guatemala. Thesis. Facultad de Agronomía, USAC, Guatemala.

36. Martínez, M.L., E. Schieber, R. Gómez, R. Bressani and R. Turrialba. 1970. 20(3):311-319.

37. Newberne, P.M., W.W. Carlton and G.M. Wogan. 1964. Pathol. Vet. 103.

38. Newberne, P.M., A.N. Wo and G.N. Hall. 1966. Journal of Nutrition 90:123-130.

39. Ostle, B., and R.W. Mensing. 1975. Statistics in Research, 3rd ed. Iowa State Press, Ames, Iowa, USA. Pp. 382-384.

40. Price, R.L., and K.V. Jorgensen. 1985. Journal of Food Science 50:347.

41. Rayne, E., and F.G. Dollear. 1968. Journal of the Association of Official Analytical Chemists 45:622-624.

42. Schroeder, H.W., and H. Boller. 1968. Cereal Chemistry 45:574-580.

43. Schroeder, H.W., R. Cole, R. Grigslay and H. Hein. 1974. Applied Microbiology 27:394-399.

44. Sorenson, W.G., C.W. Hesseltine and O.L. Shotwell. 1967. Effect of temperature on production of aflatoxin on rice by *Aspergillus flavus*. Mycopathologia et Mycologia Applicata 33:49-55.

45. Southern, L.L., and A.J. Clawson. 1980. Ammoniation of corn contaminated with aflatoxin and its effects on growing rats. Journal of Animal Science 50:459-466.

46. Stoloff, L., and W. Trager. 1963. Journal of the Association of Official Analytical Chemists 48(3):681-683.

47. Stoloff, L., and M.W. Trucksess. 1981. Journal of the Association of Official Analytical Chemists 64(3):678-680.

48. Stubblefield, R.D., O.L. Shotwell, C.W. Hesseltine, M.L. Smith and H.H. Hall. 1976. Applied Microbiology 15(1):186-190.

49. Tutherman, H., H. Banhart and J. Hatun. 1973. Journal of Food Science 38:1238-1240.

50. Trejo-Gonzalez, A., A. Feria-Morales and C. Wild-Altamirano. 1982. Advances in Chemistry Series 198:245-263.

51. Ulloa, M., and T. Herrera. 1970. Revista de Lat Microbiologia 12:19-25.

52. Ulloa-Sosa, M., and H.W. Schroeder. 1969. Cereal Chemistry 46:397-460.

53. Whitaker, T.B., J.W. Dickens and R.J. Monroe. 1979. Variability associated with testing corn for aflatoxin. Journal of the Association of Official Analytical Chemists 56:789-794.

54. Yahl, K.R., S.A. Watson, R.J. Smith, and R. Barabalak. 1971. Laboratory wet-milling of corn containing high levels of aflatoxin and a survey of commercial wet-milling products. Cereal Chemistry 48:385-391.

Aflatoxin in Haiti

F. Justafort, Research in Mycotoxins and Foodstuff Toxins, Flour Milling of Haiti, S.A., Port-au-Prince, Haiti

Haiti depends heavily on agriculture, and it is important that special attention be given to promoting production of domestic crops. The country is tropical, with the main crops being rice, peas, millet and maize. Among the cereals, maize is the least expensive. However, samples of maize from across the country have demonstrated a significant percentage of aflatoxin contamination. The causes of the contamination and appropriate solutions have been considered. In order to protect the health of the Haitian people and livestock, the consumption of contaminated maize must be stopped. The impact of aflatoxin contamination of maize on the national economy must also be assessed.

Maize is produced throughout Haiti. Although farmers do not use improved agronomic practices to raise the crop, they nevertheless manage to meet the daily maize needs of almost half of the population. The methods they use are not very different from those used 100 years ago, and the systems for drying and storing maize are very simple.

The problem of aflatoxin contamination in maize is not new in Haiti. Investigations between 1982 and 1984 showed that the high levels of aflatoxin in maize are principally due to poor storage conditions. For example, for drying, farmers hang maize ears under trees, inside their homes, or place them on their roofs. In hot, humid weather, maize usually molds during drying.

The maize is taken to the market for sale and is left in the open, exposed to the sun. If not sold, it is taken back and left on the floors of the houses until the next day. Under these conditions, mold growth and aflatoxin formation are quite likely to occur. The climate of Haiti, with high temperatures and high humidity, favors the growth and proliferation of *Aspergillus flavus* and *A. parasiticus*.

Plans are being made to build proper storage facilities throughout the country for locally grown maize. Methods of decontaminating maize that contains aflatoxin, so that the maize can be used for human consumption, are also under consideration. Merely destroying contaminated maize would impose unacceptable economic hardships. In addition, a factory to produce precooked foods made of maize and peas is being built in the southern part of the country, and it should help meet the needs of the population for a safe and balanced diet.

Currently, the effects of aflatoxin on human and animal health in Haiti are being determined with the help of medical doctors, but the task is not easy. According to the Catholic Church, 85% of the population is illiterate, which presents a major barrier to any program of scientific education and information. The people believe that if Haitian maize were really contaminated, many people would already have died. Therefore, regardless of the amount of contamination in their maize, farmers always find buyers. Another problem is superstition and the belief in magic. Sickness and disease are attributed to evil spirits, and often scientific medicine is not called upon. Sometimes, when a disease is not cured by scientific medicine, the witch doctors with their African gods (loas) do have success because of the strength of the people's belief.

The problems associated with aflatoxin contamination in Haiti are numerous, and the Haitians invite suggestions, advice and collaboration to help solve them.

Aflatoxin in Maize and *Tortillas* in Mexico

A. Torreblanca R., Department of Nutrition and Food Science, Ibero-American University, and H. Bourges R., Department of Experimental Nutrition, and J. Morales, Department of Food Technology, both of the National Institute of Nutrition, Mexico, D.F., Mexico

Maize accounts for nearly half of the total volume of food consumed annually in Mexico (2), which ranks high among the countries with the largest per capita maize consumption. There is an inverse relationship between consumption of maize and income in the country; as a result of the low cost and the many dishes that can be made from it, maize is the principal source of food among the lower socioeconomic groups (11). Although maize is used in many ways, the most important is as *tortillas*, flat, unleavened maize cakes. It is estimated that 700 million *tortillas* are produced each day, with an average daily per capita consumption of ten *tortillas* (4).

Survey of Mexico City *Tortillerias*

In a country such as Mexico, it is of utmost importance that maize quality be ensured, especially in maize used for making *tortillas*. However, polls of merchants and consumers (7,8) have shown that the quality of *tortillas* available in Mexico City leaves much to be desired. In 1983, the National Institute of Nutrition and the Ibero-American University collaborated in a preliminary survey to evaluate the maize storage conditions at *molinos*, mills where *nixtamal* (maize kernels boiled with lime) is ground to make *masa* (maize dough), and *tortillerias*, the small neighborhood factories where *tortillas* are made and sold (12).

According to the 1980 census, there were 24,628 *molinos* and 23,216 *tortillerias* throughout the Republic of Mexico. Of these, 1100 were located in Mexico City. The study included 50 establishments in two sections of the city (Coyoacan and Ixtapalapa); it was felt that these would be representative of the city as a whole. A questionnaire was developed consisting of two parts, the first to be filled in by the investigators as to conditions found in the establishments (Figure 1a), and the second consisting of questions to be answered by the managers (Figure 1b).

Results of the Survey

Only one person of the 50 interviewed reported having received training in grain handling, preparing dough or making *tortillas*. Although the employees stated that they seldom found insects in the grain, the interviewers saw both live or dead insects in the grain at most of the sites. Forty percent of those interviewed reported that they frequently encountered rodents.

Thirty-six percent of the *molinos* and *tortillerias* lacked a special area for storing grain; among the others, storage conditions were inadequate. In 43 (86% of the total), bags of maize grain were piled against a wall on the floor; only 14% had wooden platforms for storing the bags above the floor. Ceilings, walls and floors were cracked in 86% of the sites, permitting the entrance of rodents and insects. Also, the presence of cobwebs and garbage, as well as high humidity, were factors that contributed to poor grain quality. The personal hygiene of the employees (hands, hair and clothing) was found to be satisfactory in only 24% of the businesses. Since this preliminary survey showed many inadequate grain storage and handling practices, further collaborative studies were planned for the evaluation of the maize used for making *tortillas*.

Maize Evaluation Studies

The objectives of the studies on maize used for making *tortillas* in Mexico City were to evaluate the physical condition of the maize grain, quantify the amount of aflatoxin contamination present in grain and the *tortillas*, and identify which of the mycotoxins found in maize carried over to *tortillas*. Since wheat is used for *tortillas* instead of maize in some parts of Mexico, a study was also conducted on *tortillas de harina* (wheat flour).

1. The maize is stored in:
   - a) milling area ____
   - b) special area ____

2. Maize is stored:
   - a) in bulk ____
   - b) in bags ____

2.1 In bulk:
   - a) in silos ____
   - b) in the milling area ____

2.2 In bags:
   - a) on raised platforms ____
   - b) on the floor ____
   - c) against a wall ____

3. Storage area construction:

3.1 Roof:
   - a) sheet asbestos ____
   - b) cement ____
   - c) wood ____
   - d) plaster ____

3.2 Walls:
   - a) cement and tile ____
   - b) brick ____
   - c) cement ____
   - d) tile ____
   - e) plaster ____

3.3 Floor:
   - a) cement ____
   - b) tile ____

4. Condition of the milling and storage areas:

4.1 Ceilings and walls:
   - a) cracks ____
   - b) cobwebs ____
   - c) mildew ____
   - d) dampness ____

4.2 Floor:
   - a) swept ____
   - b) spilled kernels ____
   - c) spilled water ____
   - d) cracks ____

4.3 Housekeeping:
   - a) litter ____
   - b) garbage containers ____
   - c) pesticides in storage areas ____
   - d) pesticides in milling areas ____

5. Restrooms:
   - a) toilet and lavatory ____
   - b) toilet only ____
   - c) none ____

5.1 Condition of restrooms:
   - a) good ____
   - b) fair ____
   - c) poor ____
   - d) very poor ____

6. Personal hygiene of employees:
   - a) good ____
   - b) fair ____
   - c) poor ____

6.1 Condition of:
   - a) clothing ____
   - b) hands ____
   - c) hair ____

Figure 1a. Form used for the survey of *molinos* and *tortillerías*, National Institute of Nutrition and Ibero-American University, Mexico, D.F., 1983

The physical and microbiological quality of maize used in the *tortillerias* located in Mexico City was evaluated (Figure 2) using a procedure developed by the national storage agency, Almacenes Nacionales de Deposito (ANDSA) (9). Grain temperature was measured with the minitherm, and humidity with the ceratester. Internal fungal contamination was evaluated

---

1. Are you in charge?
   a) yes______ b) no______

2. Where do you buy maize? ____________________

3. How often is maize delivered?
   a) once a week ______ b) every two weeks ______
   c) every three weeks ______ d) once a month______

4. When was your last delivery? ____________________

5. What is the longest time that a bag of maize lasts?
   ____

6. Have you or your employees received training in:
   a) handling grain______ b) preparing dough______ c) making tortillas______

7a. What did this training include?
   ____

7b. Who gave the training?____________________

8. How often are insects found in the maize?
   a) often ______ b) sometimes______ c) never______

9. How are insects controlled?
   a) pesticides______ b) other______

10. How often are rodents found in the maize storage area?
   a) often______ b) sometimes______ c) never______

11. How are rodents controlled?
   a) traps______ b) poison ______ c) other______

12. Is the maize treated in any way before the nixtamalization process?
   ____

13. How often is cleaning done?
   a) storage area ______ b) milling area ______

14. How often is maintenance work done?
   a) ceiling______ b) walls______ c) floor ______

Figure 1b. Questionnaire for managers of *molinos* and *tortillerías*, National Institute of Nutrition and Ibero-American University, Mexico, D.F., 1983

using the techniques of Christensen and Lopez (3). The homogenization and separation of the sample was done with a Boerner-type mixer-separator (12). Tests for impurities, specific weight, infestation, etc., were carried out using ANDSA techniques, and the tests for aflatoxin with the procedure outlined by Shotwell (10). For the detection of aflatoxins in the grain and the resulting *tortillas*, the mycotoxins were extracted by the Heyde technique (6) and quantified in duplicate by thin-layer chromatography (TLC).

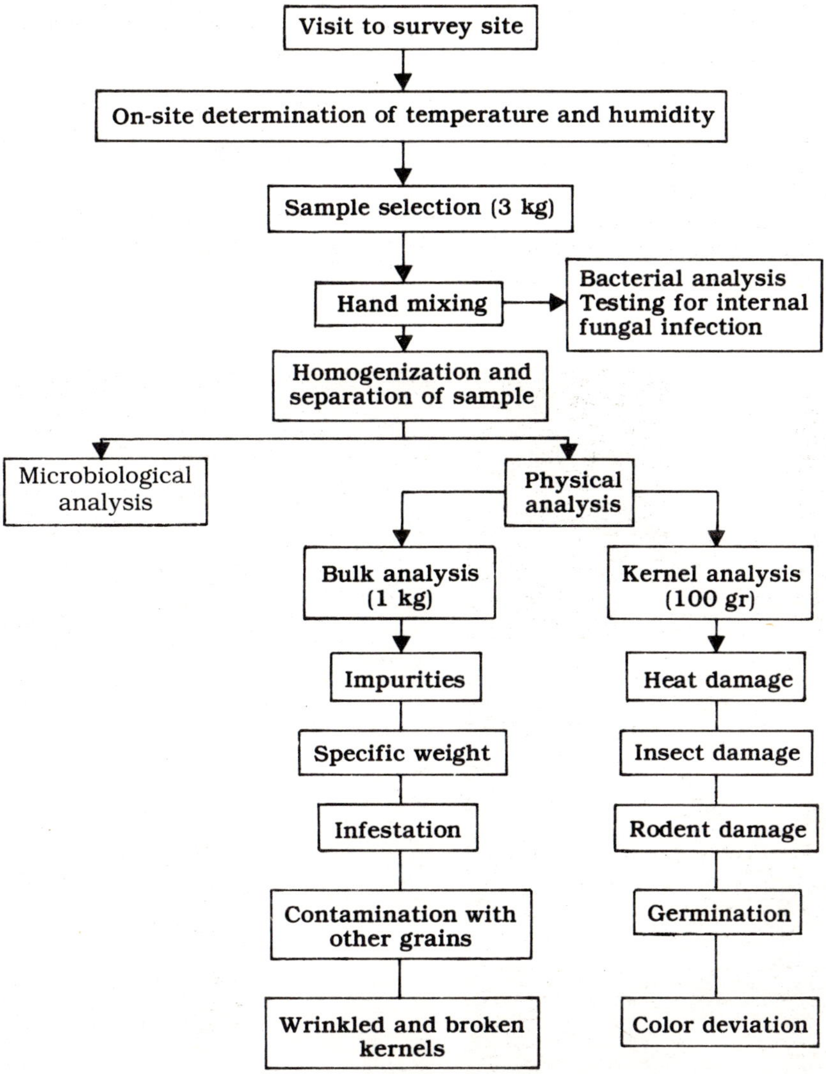

Figure 2. Flow chart, testing for physical and microbiological quality of maize in *tortillerías*, National Institute of Nutrition and Ibero-American University, Mexico, D.F.

Results

Physical condition of the maize grain

The study of the physical condition of maize grain used by the Mexico City *tortillerias* provided the following observations:

- Average grain moisture content was 12.3%, within the limits of the Mexican quality standard which accepts levels up to 13.5% for maize grain;

- Temperatures fluctuated greatly, with the average temperature of the grain being 22.5°C (the variation in air temperature around the grain was not greater than 5°C);

- Fifty-five percent of the samples exceeded the permissible limit of 1.5% of total weight for impurities and foreign matter;

- Eighty percent of the samples were infested with dead insects, and 20% contained live insects as well;

- The specific weight of the grain was found to be within accepted limits; and

- The number of kernels of other grains found in the samples was negligible (0.046%).

Grain damage was found in the following amounts (median ± standard deviation):

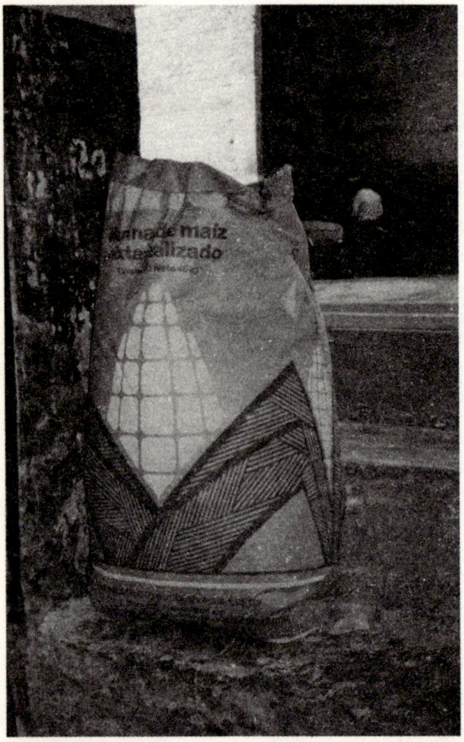

Grain storage and handling practices were evaluated at *molinos* and *tortillerías* in Mexico City

Heat damage	11.8% ± 4.10%
Insect damage	10.0% ± 2.69%
Rodent damage	2.2% ± 1.03%
Germinated kernels	0.5% ± 0.50%
Total kernel damage	24.5% ± 5.2%

The amount of grain damaged by heat and insects and total damaged grain surpass the maximum limits permitted by Mexican standards, which are 5.5%, 4.0% and 10.0%, respectively. The ruling on grain quality states: "Maize grain which exceeds any of the limits established for the standard of Mexican quality 3, that smells of mold, fermentation, putrefaction or any other commercially objectionable odor, or that contains stones, glass, rodent excrement or any any other foreign matter, is considered as substandard" (5).

Aflatoxin presence in the maize grain

To test for aflatoxin presence, whole grains were observed under long-wave ultraviolet radiation; 77% of the samples had levels of from one to six fluorescent particles per 100 grams of sample. Shotwell *et al.* (10) have reported that when over 20 fluorescent particles are found in 4.54 kg of grain, 65% of the samples will have more than 20 mg of aflatoxin per ton of maize. The amounts found in the present study were much higher than this; the number of fluorescent particles per 4.54-kg sample ranged

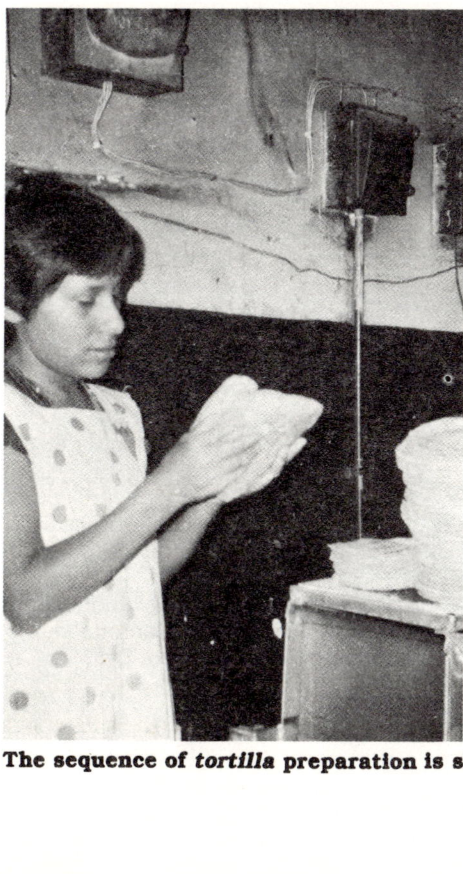

The sequence of *tortilla* preparation is shown above

from 45.4 to 272.4. Aflatoxin B_1 was found in 72% of the maize or *tortilla* samples tested; aflatoxins B_2, G_1 and G_2 were not found in any of the samples. The amount of aflatoxin B_1 (μg/kg) by sample groups are shown in Table 1.

The samples were also analyzed for zearalenone. Twenty-four percent had a positive reaction, and 14% had other unidentifed metabolites with a higher mobility on TLC plates.

Testing is currently in progress to obtain more information on aflatoxin contamination in maize and *tortillas* (see Carvajal *et al.*, this proceedings). However, a permanent detection program is needed in Mexico to fully understand the scope of the aflatoxin-contamination problem in foodstuffs, especially maize.

Table 1. Aflatoxin B_1 content in samples of maize and *tortillas* from 50 *molinos* and *tortillerías*, Mexico, D.F.

Aflatoxin B_1 content (μg/kg)	Number of locations
0	14
5	1
10	3
20	5
40	12
80	9
100	1
160	4
500	1

References

1. Almacenes Nacionales de Depósito, S.A. 1977. Certificación de maíz: Folleto de divulgación. México, D.F.

2. Bonfil, G. 1962. El Maíz, Fundamento de la Cultura Popular. Museo de las Culturas Populares. Secretaría de Educación Pública, Mexico, D.F.

3. Christensen, C., and I. López. 1962. Daños que causan en México los hongos de granos almacenados. Instituto Nacional de Investigaciones Agrícolas. Folleto 44. Mexico, D.F. Pp. 12-13.

4. Comisión Nacional de la Industrialización del Maíz para Consumo Humano. 1977. La industria del maíz. Mexico, D.F.

5. Diario Oficial de la Federación. 1982. Norma de calidad mexicana para la comercialización del maíz en grano. NCM-AG/3-1981. Mexico, D.F.

6. Heyde, W., *et al.* 1977. Analytical Toxicology Methods Manual, H.M. Starh, ed. Iowa State University Press, Ames, Iowa, USA. Pp. 164-173.

7. Lomelí, A. 1979. Las tortillas de maíz y trigo. Guía del Consumidor 1(3). Mexico, D.F. Pp. 2-9.

8. Reyes, J.G. Excelsior newspaper. Mexico, D.F. Pp. 5 and 21.

9. Salazar, J. Muestreo y análisis de granos. Almacenes Nacionales de Depósito, S.A. Mexico, D.F.

10. Shotwell, O.L., and C.W. Hesseltine. 1981. Use of bright greenish-yellow fluorescence as a presuntive test for aflatoxin in corn. Cereal Chemistry 58 (2):124-127.

11. STPS. 1977. Encuesta de ingresos y gastos familiares de 1975. CENIET, Mexico, D.F.

12. Torreblanca, R.A., and M.E. Rivera-Breehm. 1984. Evaluación de las condiciones de almacenamiento del maíz en los molinos de nixtamal del Distrito Federal. *In* Tecnología Alimentaria. Mexico, D.F. Pp. 15-20.

Mycotoxin Carryover from Grain to *Tortillas* in Mexico

M. Carvajal, Institute of Biology, and R. Rosiles M., Department of Veterinary Medicine and Animal Science, both of the National Autonomous University of Mexico, Mexico, D.F., Mexico, and H.K. Abbas and C.J. Mirocha, Department of Plant Pathology, University of Minnesota, St. Paul, Minnesota, USA

Maize is probably used for human consumption in Mexico more than in any other country. Therefore, the study of mycotoxins in maize is of great importance to the health of 80 million Mexicans, as well as a large number of Mexican-Americans in the USA. Maize is principally consumed as *tortillas*, flat, unleavened maize cakes. They are rolled and eaten as *tacos* or *enchiladas*, cut in pieces and eaten in soups and casseroles, or eaten plain as bread with meals. Maize is also eaten in *tamales*, maize dough wrapped around a filling and steamed in husks or banana leaves. Many other regional dishes are made with maize, as well as the beverage *atole* and fermented beverages used by several indigenous groups.

The most important mycotoxins in maize are the aflatoxins produced by *Aspergillus flavus* and *A. parasiticus*. The aflatoxin B_1 is a potent mutagen and carcinogen in humans and animals (4). The mycotoxins produced in maize by *Fusarium* spp. include zearalenone, as well as the trichothecenes, deoxynivalenol (DON), T-2 toxin, diacetoxyscirpenol (DAS) and nivalenol (NVL).

Tortilla Study

A study of mycotoxin carryover from maize and wheat grain to *tortillas* was conducted by the National Autonomous University of Mexico as part of a larger study of mycotoxins in *tortillas* and their relation to human cancer; this was a project of the National Commission of Science and Technology (CONACYT) (see Torreblanca, this proceedings). This part of the study involved cooperation with the University of Minnesota, St. Paul, Minnesota, USA, and was carried out in their laboratories.

The objective of the current study was to determine whether the aflatoxins B_1, B_2, G_1 and G_2, zearalenone and DON carry over from contaminated maize and wheat grain to *tortillas*, or whether the lime and heat treatments used in their preparation detoxify the mycotoxins.

Maize and wheat grain naturally contaminated with mycotoxins from the USA were utilized in the study. For aflatoxins B_1 and B_2, the maize variety Georgia 807 was analyzed by the De Vries and Chang method (3). For zearalenone and zearalenol, the maize varieties tested were Indiana 362 and 808 and Georgia 807, by the method of Bennett *et al.* (1). For DON, the wheat varieties Nebraska (N1 and N2) and Missouri (MS1 and MS2) were used following the procedure of Chang *et al.* (2).

To make the maize dough (*nixtamal*), 100 g of maize grain was boiled and soaked in 2% lime water for 12 hours, with a resulting pH of 13 to 13.5. Grain from each variety was ground separately, and the grinder disinfected after each grain sample. *Tortillas* were made from the dough and baked at 110°C for seven to nine minutes. They were then dried and ground for analysis for mycotoxins. The wheat grain was not soaked; the flour was mixed with lard and water, the traditional method, and after baking, the tortillas were dried, ground and analyzed for DON. Thin-layer chromatography was then used for quantification of the mycotoxins.

Results

Through thin-layer chromatography, the aflatoxins B_1 and B_2 were detected in about equal amounts in *tortillas* made from contaminated maize; apparently aflatoxin is not destroyed by either the lime treatment or the cooking process. Aflatoxin B_1 and B_2 levels averaged about 20% less in *tortillas* than in the contaminated maize.

In analysis for zearalenone, the lime water used to boil maize for the nixtamalization process was found to contain transzearalenone. The *tortillas* made from maize heavily contaminated with zearalenone (4.23 ppm) contained both cis- and transzearalenone; however, those made from maize with only 1 mg/kg did not contain either of the mycotoxins.

Tortillas made with DON-contaminated wheat also contained the mycotoxin. In the wheat varieties N1 and N2, 40% of the *tortillas* contained DON; 20% of the *tortillas* made with the varieties M1, M2 and K1 were DON contaminated.

Conclusions

It was concluded that since the mycotoxins aflatoxin, zearalenone and deoxynivalenol (DON) are not destroyed by treatment with lime or by temperatures of 110°C, the temperature at which *tortillas* are baked, these mycotoxins represent a potential threat to human health.

References

1. Bennett, G.A., O.L. Shotwell and W.F. Kwolek. 1984. Determination of zearalenol and zearalenone in corn by high performance liquid chromatography. (Working paper.)

2. Chang, H.L., J.W. DeVries, P.A. Larson and H.H. Patel. 1984. Rapid determination of deoxynivalenol (vomitoxin) by liquid chromatography using modified Romer column cleanup. Journal of the Association of Official Analytical Chemists 66:520-54.

3. DeVries, J.W., and H.L. Chang. 1982. Comparison of rapid high pressure liquid chromatographic and CB methods for determination of aflatoxins in corn and peanuts. Journal of the Association of Official Analytical Chemists 65(2):206-209.

4. Mirocha, C.J., and C.M. Christensen. 1983. Mycotoxins. *In* Storage of Cereal Grains and their Products. Pp. 241-280.

Aflatoxin in Argentina

E. Planes de Banchero, Microbiology Department, National Institute of Agricultural Technology, Castelar, Argentina

The purpose of this paper is a discussion of the legislation, control and research on mycotoxins in general, and aflatoxins in particular, in the Republic of Argentina.

Argentina is a grain-exporting country, and so the commodities under scrutiny for aflatoxin contamination are those destined for export, including maize, wheat, peanuts, soybeans and sorghum and their by-products. Aflatoxin control measures were introduced in the 1970s and have become more stringent in the 1980s. They are administered by the laboratories of the Ministry of Agriculture and Fisheries and in other authorized labs, which use the AOAC method of analysis. Argentina mainly exports grain to the USSR, but also to the Netherlands, Egypt, South Africa, Czechoslovakia, Cuba, Colombia, Chile, Paraguay and Iran. The export agreement with each country specifies toxins for which commodities must be tested and also indicates their permissible levels. In some cases, countries provide their own inspectors. Tests are conducted principally for aflatoxins, although tests are also made for zearalenone, ochratoxin and T-2 toxin.

Institutionalized control still does not exist for grains sold on the domestic market, and the only standards included in the food code are for aflatoxins in soybean flour (30 μg/kg) and aflatoxin M_1 in milk (0.5 μg/liter). Plans are underway to include standards for other foods, such as maize flour and peanuts (which are used in the candy industry). The Department of Agriculture also regulates aflatoxin levels in feed for livestock when requested by manufacturers and/or consumers. Some companies have their own laboratories that test products destined either for export trade or for livestock feed.

Mycotoxicoses in Livestock

Aflatoxin effects have been detected in tissues of domestic animals. Photosensitization of cattle, caused by esporidesmina (*Phytomices chartarum*) in pasture grasses is one effect that has been observed. Sheep can develop disease symptoms after eating grass contaminated with a fungus that produces tremorgens. Cattle become poisoned by consuming fescue forage (*Festuca arundinacea*) infected with isolates of *Fusarium tricinctum* and *Acremonium ceonophyalum* (endophyte). The presence of *Claviceps pasplali*, linked to a nervous syndrome in cattle, has also been detected. In some cases it has been possible to isolate the fungi and the toxins they produce and to reproduce the resulting diseases in the laboratory.

Aflatoxin Research

Research on aflatoxin began in 1972 under the auspices of the Department of Public Health with a study of aflatoxin contamination in peanuts. However, before that time, various groups in government agencies and in the universities conducted research on mycotoxins. Among these groups were the National Institute of Agricultural Technology, the National Pharmaceutical Institute and the Food Technology Research Institute. Research was also conducted in the departments of the National University located in Buenos Aires, La Plata, Rosario, Santa Fe, Rio Cuarto, Tucuman and Misiones.

Research is presently being conducted on mycotoxins in the following areas:

- Incidence and determination of toxigenic levels for aflatoxins and zearalenone in grains (maize and sorghum), oil seeds (peanuts and sunflower) and their by-products, as well as in livestock feed; the studies have also included some other foods, such as corn flour (*polenta*), milk and apple juices and concentrates (for *patulina*);

- Analytical methods for determining levels of aflatoxin, zearalenone, ochratoxin, T-2 toxin and deoxynivalenol; the objective is to develop methods that are relatively simple and inexpensive (thin-layer chromatography and enzyme-linked immunosorbent assay) and that will, when possible, detect more than one toxin at a time;

- Designs for sampling methods in the field and in storage for each grain;

- Toxicosis studies in livestock, especially aflatoxin in chickens and zearalenone in swine; and

- Methods for preventing contamination through improved procedures (treating grain with insecticides, proper drying and appropriate storage) and varietal resistance to *Aspergillus flavus* contamination and the formation of aflatoxins (in peanuts and sunflower).

The research organization

The studies on incidence, methodology and toxicology have been conducted mainly by multidisciplinary research teams from various government institutions, so as to make optimum use of available human and physical resources. The National Institute of Agricultural Technology (INTA) plays an important part in the research system. In 1983, INTA set up a meeting of specialists from both the public and private sectors, which was sponsored by the Ministry of Science and Technology through its National Program for Food Technology Research. The result of the meeting was that priorities were established, the duplication of research efforts was prevented, subsidies were arranged and research results were disseminated. A summary of aflatoxin research in Argentina was published in 1985.

In all of the conferences on microbiology, mycology, veterinary science and toxicology, research on mycotoxins has been covered in papers, round table discussions, training sessions, specialized courses and working groups.

At the 1985 Latin American Microbiology Congress, a document dealing with all aspects of mycotoxin research was drafted by the mycotoxins working group. The National Institute of Agricultural Technology has always played an active role in these activities because they are so important to Argentina's largely agricultural economy. This year, under the auspices of FAO, INTA will begin work on aflatoxins in maize flour (*polenta*) intended for human consumption. The Department of Public Health will also participate in the project.

Research procedures

The methodology for detecting aflatoxins and zearalenone in maize and wheat grain is quite advanced in Argentina, but that for determining T-2 toxin and deoxynivalenol (mainly produced by fungi of the *Fusarium* group) needs to be developed.

Studies were conducted in the provinces of Buenos Aires and Santa Fe to determine levels of aflatoxin contamination in maize. In 1976 and 1977, out of 50 and 267 maize samples tested in Santa Fe, aflatoxins were detected in 5 and 10%, respectively. In 1980 in Buenos Aires, zearalenone was detected in 33% of 85 samples tested, at levels of 200 to 1600 μg/kg.

In 1982, from 41 samples obtained from grain storage facilities, aflatoxins were detected in 34% at levels of 3 to 64 μg/kg. Zearalenone (912 μg/kg) was found in only one sample. In 1983, maize was tested at harvest, and aflatoxins were detected in 33% of 87 samples (2 to 50 μg/kg) and zearalenone in 7% with a maximum of 300 μg/kg).

In 1981, 25% of 53 samples of maize grain intended for export were found to contain aflatoxin B_1 at 2 to 15 μg/kg. Nine percent had between 30 and 305 μg/kg of zearalenone. In 1984, aflatoxins were detected in 24% of 110 samples of maize flour (*polenta*), but only 4% contained over 20 μg/kg. Tests for zearalenone were negative in 50 of the samples.

Future Research

At the meeting of the Ministry of Science and Technology (SECYT) in Buenos Aires in April 1986, the mycotoxins group of the National Food Program reported on the results of research in 1984 and 1985 on the incidence of aflatoxins and zearalenone in stored maize and maize grain. Plans were made to set up programs in two regions of Santa Fe Province and in one region in Cordoba Province to evaluate the incidence of these toxins in the field (in 1986) and in stored maize flour (in 1987). The same methods will be used in sampling and analysis as were used in previous testing.

Maize hybrids have been found to possess different levels of sensitivity to fungi. Therefore, a breeding program for the development of more resistant varieties will also be initiated. This program will involve collaborative research groups working under the auspices of the Ministry of Science and Technology.

The study of the incidence of aflatoxins in peanuts is one of only a few research projects that may be considered to be complete; there is now sufficient information available. Studies must be undertaken on the incidence of aflatoxin in wheat, soybeans and other foods. The sampling program for aflatoxins needs to be expanded, as well as research on contamination by other toxins.

Aflatoxin in Colombia

R.G. Cuero, I. Hernandez, H. Cardenas, E. Osorio and L.C. Onyiah, Biology Department, Universidad del Valle, Cali, Colombia

Colombia is the northernmost country of South America, lying between 12°N and 2°S latitudes. Although it is generally tropical, temperate conditions are also found in some areas, but without a marked seasonal weather pattern. Altitude influences the climatic conditions and associated agricultural systems (Table 1). The highlands are cooler and more humid than the lowlands, but maize is grown in both areas. Maize is a dietary staple, and is the second major cereal grain (after rice) in Colombia. There was an average annual maize production of 850,000 tons/year between 1974 and 1982 (Table 2). Both yellow and white maize are produced and consumed in the country.

Although the varied tropical conditions of Colombia favor both fungal populations and mycotoxin production in maize, only limited information is available on toxigenic fungi and toxin production in grain. However, limited exploratory surveys on aflatoxin in stored maize (33,35) and in harvested sorghum (14) have been carried out in particular areas.

Table 1. Environmental conditions of the maize-growing highlands and lowlands, Colombia[a/]

Region and altitude (m)	Mean annual rainfall (mm)	Mean daily temperature (°C)	Relative humidity (%)
Highlands (1500 - 2000)	3100	21.2	22
Lowlands (900 - 1200)	2000	26.2	25

a/ Climatic data supplied by the regional weather bureaus

Table 2. Mean area planted to cereals, production and yield, Colombia, 1974-1982

Cereal	Area (000 ha)	Production (000 tons)	Yield (kg/ha)
Maize	600	850	1390
Wheat	-	-	-
Sorghum	200	500	2300
Rice	420	1800	4300
$\bar{x}$	1350	3500	2450

Source: FAO (9)

Several cases of aflatoxicosis in fowl have been reported in Colombia, 14 in 1975 and 41 in 1976. Diagnoses were based on necropsies of the birds. Poultry feeds are prepared primarily from maize, sorghum, soybean meal and cottonseed meal (8,9). Cuero (7) reported the death of four pigs that were fed moldy maize.

Previous studies in other countries have provided evidence of the occurrence of both toxigenic fungi and mycotoxin contamination in freshly harvested maize (6,22,24,28,39). Factors influencing the growth of the toxigenic fungal species and/or mycotoxin production have been investigated (20,21,29,30). Previous investigations have also demonstrated a wide distribution of toxigenic fungi in tropical and temperate climates (10,24).

Aflatoxin produced by *Aspergillus flavus* and *A. parasiticus* strains has been the most carefully studied mycotoxin (2,20). However, other mycotoxins have recently received attention, e.g., ochratoxin-A from *A. ochraceus* and some *Penicillium* spp. (17), citrinin and penicillic acid from *Penicillium* spp. (19) and other toxic metabolites produced by *Fusarium* spp. (5,12,27). It has also been shown that numerous fungal metabolites cause diseases in animals (8,11,12,16,17,19,20).

Environmental conditions in Colombia and other subtropical areas of Latin America are favorable for the occurrence of toxigenic molds and mycotoxin production. Therefore, the present investigation was carried out to determine the presence of common toxigenic fungi and their ability to produce toxins. Emphasis was placed on interactions of climatic conditions, insect and/or bird attack, mechanical damage and soil with toxigenic fungi and toxin production.

Materials and Methods

A total of 72 samples of harvested maize were collected from five different sites in two areas of different altitudes and geographic features. The two areas were representative of the variation in climatic conditions in Colombia, ranging from warm and dry to semidry in the lowlands (900 to 1200 meters altitude) to cool and humid in the highlands (1500 to 2200 meters altitude). Environmental temperature and relative humidity were recorded at the time of sampling. The moisture content of the maize samples was determined by using a Motomco moisture meter; grain temperature was also recorded. Random samples (approximately 10 kg) from each of the five test sites in the two areas were collected after the maize had been kept three to five days in stacked jute bags. The samples from different sites in each area were mixed, and a representative portion (about 200 g) was exposed to ultraviolet light ($\lambda = 365$ nm) to detect the characteristic bright greenish-yellow fluorescence (BGYF) associated with activity of *A. flavus* group fungi. Kernels were subsampled for mycological studies. Samples positive for BGY fluorescene were ground and blended for 15 to 30 minutes and assessed for aflatoxin B_1 by the Association of Official Analytical Chemists method (1).

Isolation and culture of fungi

Mixed samples of maize kernels from the same area (120 kernels per sampling area: 120 x 72 samples = 8640 kernels total for each sampled area) were separated into four portions of 30 kernels each. The four portions were selected according to kernel appearance: apparently sound kernels; kernels that had been attacked by molds; kernels that had been attacked by insects; and mechanically damaged kernels.

Each portion of 30 kernels was surface sterilized in a 2% aqueous solution of sodium hypochlorite and transferred to 3% malt extract agar. Six replicates of three agar plates per portion (30 kernels x 4 portions = 120 kernels/area/ sampling time) were incubated for seven days at 25°C. Sometimes maize samples were kept for one to three days at 40°C before mycological analysis. All colonies of the generas *Aspergillus, Penicillium* and *Fusarium* were identified according to the criteria of Raper and Fennell (34), Onions (32) and Booth (4). Colonies of *Aspergillus* and *Penicillium* were counted after isolation on fresh malt extract agar, and *Fusarium* spp. on potato dextrose agar (PDA). Pure cultures were used to inoculate (10^3 spore dilution) autoclaved rice. The rice culture was incubated with continuous shaking for seven days at 25°C and then used for mycotoxin analysis.

Culture of toxigenic molds on soil-extract agar

Soil extract was prepared from soil samples from the five different sites in the two geographic areas. Soil-extract agar plates were inoculated with spores of a toxigenic fungus. Triplicate plates were used for each toxigenic fungus selected, and the experiment was repeated. Inoculation of the soil-extract agar was carried out according to the ability of the fungus to produce toxin. For example, positive toxigenic isolates were inoculated on soil-extract agar made with soil from which negative toxigenic mold was isolated, and vice versa. Cultures were incubated at 25°C for six to seven days and then transferred to malt agar and incubated at 25°C and for another six to seven days. Spore dilution (10^3) was prepared and used to inoculate autoclaved rice in order to assess toxin production by each mold.

Soil Sampling and Analysis

All soil samples (3x25 g/site, 225 g/area, 450 g/two areas) were collected from the 0 to 20 cm layer within an area of 930 square meters, and composited according to the area of origin (highlands or lowlands). The subsamples were sieved through a 1/4-inch screen to remove debris and stored in polyethylene bags. Soil pH was determined on a 1:1 soil:water suspension using a Beckman pH meter. The soil mineral analysis was carried out following Jackson (13).

Mycotoxin Analysis

Isolates of *A. flavus, A. ochraceus, P. viriducatum* and *F. graminearum* grown on rice were assayed for aflatoxin B_1, ochratoxin-A, citrinin and zearalenone, respectively, by the AOAC (1), the Neishelm (31), the Marti *et al.* (26) and the Ware Thorpe (38) methods. Mycotoxin estimates were made by thin-layer chromatography (TLC) on silica gel G plates (Merck Silica Gel 60F-254). Absorption spectra were recorded with a spectrophotometer (Spekol-10- Carl Zeiss Jena).

Statistical Analysis

The normal approximation to the binomial was applied for comparison of proportions of toxigenic mold colonies isolated from maize kernels from highland and lowland climates at different periods of the year.

Results

Occurrence and geographical distribution of fungal species isolated from maize kernels

Mycological analysis showed that 54 of the 72 samples (75%) were contaminated by fungal strains of *A. flavus, A. ochraceus, P. citrinum* and *F. graminearum*; 27% of the isolates were unidentified species (Table 3). Incidence varied between the highlands and lowlands and from different sampling periods. The highest incidence of toxigenic species was found in lowland maize. However, *Fusarium* species were routinely isolated in both areas (Table 3). In general, frequency of isolation of toxigenic mold colonies was higher in samples from the lowlands at any period of the year, except for *Fusarium* spp., which were more prevalent in the highlands from September to December. Mold genera also showed different frequencies between the two areas and periods of the year (Tables 3 and 4).

There appeared to be a higher incidence of putative toxigenic molds in maize kernels with insect infestation and/or mechanical damage, especially among samples from the lowlands, which showed more insect than mechanical damage. All identified species isolated from maize samples showed clear differences in their

Table 3. Mold count of 72 maize samples of different geographic areas at different periods of the year[a]

No. of samples	No. of mold colonies	No. of colonies toxigenic [b]			
		Aspergillus	*Penicillium*	*Fusarium*	Other[c]
Jan. to March					
Highlands					
11	601	1(0%)	143(24%)	190(32%)	267(44%)
Lowlands					
11	565	29(5%)	38(7%)	335(60%)	165(29%)
March to Sept.					
Highlands					
17	707	93(13%)	125(18%)	450(64%)	40(6%)
Lowlands					
11	589	298(51%)	210(36%)	42(36%)	36(6%)
Sept. to Dec.					
Highlands					
7	402	5(1%)	100(25%)	150(37%)	147(37%)
Lowlands					
15	1100	127(12%)	240(22%)	336(31%)	397(36%)
Total					
72	3964	553(14%)	856(22%)	1502(39%)	1052(27%)

a/ 120 kernels were sampled from 3 different sites of each geographic area (highland and lowland) each time in 3 replicates (5 kernels/agar plate)

b/ *A. flavus, A. ochraceus, P. citrinum* and *F. graminearum* were the most common toxigenic species isolated

c/ Molds other than *A. flavus, A. ochraceus, P. citrinum* and *F. graminearum* were not identified in this investigation; the most prevalent were: highlands: *Penicillium* spp. and *Fusarium* spp., lowlands: *Cladosporium* spp., *Trichoderma* spp., *Phycomyces* spp., *Aspergillus* spp., *Penicillium* spp., *Fusarium* spp. and yeasts occurrence in both areas. The combined results from the two areas showed that *F. graminearum* was the most prevalent species (51%), followed by *P. citrinum* (26%), *A. ochraceus* (17%) and *A. flavus* (4%). Most of the isolates originated from lowland samples. Although the lowlands were warmer than the highlands, humidity, moisture content of the grain and grain temperature did not show striking differences between the two areas (Table 4).

Mineral composition and pH of highland and lowland soils

The comparison of the mineral composition and pH of highland and lowland soils is shown in Table 5. Although there were some variations in mineral composition of soils at any location, there was a general difference in the composition of soil minerals between the highlands and lowlands. The content of Ca, Mg and Fe was greater in the highlands, and Zn was noticeably greater in the lowlands. The average pH (about 6) was similar in both areas.

Colony growth (diameter) and pigmentation of molds changed markedly when isolates were grown on agar with soil extracts. Most of the molds showed faster colony growth, larger mycelial formation and brighter pigmentation after growth on soil-extract agar. The amount of colony

Table 4. Comparison of proportions of mold colonies isolated from maize kernels from highland and lowland locations at different periods of the year, Colombia[a/]

Period	*Aspergillus*	*Penicillium*	*Fusarium*	Other
Jan. - March	$P_L > P_H$[b/]	$P_L > P_H$	$P_L > P_H$	$P_L < P_H$
March - Sept.	$P_L > P_H$	$P_L > P_H$	$P_L > P_H$	$P_L = P_H$
Sept. - Dec.	$P_L > P_H$	$P_L > P_H$	$P_L < P_H$	$P_L < P_H$

a/ Data based on 72 maize samples
b/ The proportions P_L (lowland) and P_H (highland) were assumed to follow binomial distribution; the normal approximation of the binomial was applied

Note: Results significant at P = 0.01

Table 5. Comparison of pH and mineral composition of highland and lowland soils, Colombia[a/]

		Mineral composition			
Region	pH	Ca (meq/100g)	Mg (meq/100 g)	Zn(ppm)	Fe(ppm)
Highlands	6.6	22	11.2	2.9	7.7
Lowlands	6.5	12.4	5.6	3.6	6.8

a/ Data from mean of samples obtained from 3 different sites per area; 3 samples per site for each area were collected (3 x 3 = 9 per area) growth, mycelia formation and degree of pigmentation was the same on agar with highland or lowland soil.

Mycotoxin determination

Half of the 90 *A. flavus*, *A. ochraceus*, *P. citrinum* and *F. graminearum* isolates examined showed ability to produce toxin (Table 6). *P. citrinum* showed the highest percentage (65%) of citrinin-positive isolates, followed by *F. graminearum* (60%) and *A. ochraceus* (45%). Toxin production also varied between isolates from the highlands and lowlands, lowland toxin production being higher except for zearalenone (Table 7).

Toxin production also varied when the isolates were grown on agar with extracts of soil from a different area than its place of origin. For instance, some isolates from the highlands produced more toxin on agar with soil from the lowlands and some decreased toxin production; others showed no change (Table 8).

Production of aflatoxin $B_1 + B_2$ in maize was also greater in samples from the lowlands than from the highlands (Table 9). However, there was no correlation between BGY fluorescence in the maize kernels and aflatoxin in the seed. The few samples that showed positive BGY fluorescence were collected from lowland maize, which also showed a high *A. flavus* count.

Discussion

The results of the current investigation showed the incidence of toxigenic molds in freshly harvested maize that had been stored for three to five days in jute sacks before sampling. Samples were obtained from two environmentally unique areas of Colombia, the lowlands and the

Table 6. Isolates positive for toxin production isolated from maize samples from the highlands and lowlands, Colombia a/

Toxigenic fungi	Colonies isolated: Highlands	Colonies isolated: Lowlands	Molds analyzed: Total isolates	Molds analyzed: Toxic isolates
A. flavus	6/553(1%)	98/553(18%)	30(6 HL, 24 LL)b/	15(50%)
A. ochraceus	61/553(11%)	354/553(64%)	20(10 HL, 10 LL)	9(45%)
P. citrinum	237/856(28%)	278/856(32.5%)	20(10 HL, 10 LL)	13(65%)
F. graminearum	678/1502(45%)	350/1502(23%)	20(10 HL, 10 LL)	12(60%)

a/ Isolates selected from 72 samples analyzed for toxin production
b/ HL = highlands, LL = lowlands

Table 7. Mean toxin production of positive isolates from maize samples of the highlands and lowlands, Colombia a/

Region	Fungi and toxin production b/: *A. Flavus*	Aflatoxin* (ppm)	*A. Ochraceus*	Ochratoxin-A (ppm)	*P. Citrinum*	Citrinin** (ppm)	*F. Graminearum*	Zearalenone*** (ppm)
Highlands	2/15(13%)	1	0/9(0.0)	0	10/13(77%)	56	10/12(83%)	57
Lowlands	13/15(87%)	14	9/9(100%)	53	3/13(23%)	75	2/12(17%)	19

a/ Selected from 72 samples, number of toxin-positive isolates over total isolates analyzed
b/ Mean of isolates tested for the respective toxigenic mold

* Highly Significant at P = 0.01, ** significant at P = 0.01, *** significant at P = 0.05

LSD calculated by comparing absolute difference between highland and lowland means with the product of $t(n_1 + n_2 - 2, P = 0.01)$ and the pooled error highlands. The high prevalence of the genera *Fusarium, Penicillium* and *Aspergillus* isolated from maize in these areas suggests favorable conditions for potentially toxigenic fungi. The results suggest that aflatoxins and other mycotoxins may be important in maize directly used for food and feed in Colombia. The occurrence of the toxigenic molds in

Table 8. Toxin production by some isolates of *A. flavus, P. citrinum* and *F. graminearum* isolated from highland and lowland maize, before and after growing on soil extract agar, Colombia

Isolate and location[a]	Toxin production (ppb)[b] Before[c]	After[d]
A. flavus(HL)+ (aflatoxin)	840	850(HL)
A. flavus(LL)-	0	320(HL)
A. ochraceus(LL)+ (ochratoxin-A)	3910	1638(HL)
A. ochraceus(LL)-	0	230(LL)
P. citrinum(LL)+ (citrinin)	5880	6320(HL)
P. citrinum(HL)-	0	450(HL)
F. graminearum(LL)+ (zearalenone)	5770	5085(LL)
F. graminearum(HL)-	0	520(LL)

a/ HL = highlands, LL = lowlands
b/ Mean of 3 replicates per fungus
c/ Isolates grown on PDA and rice
d/ Isolates grown on soil extract agar and rice, sequentially; soil extracts made from soils of different areas of the highlands and lowlands

Table 9. Production of aflatoxin B_1 and B_2 in maize samples from the highlands and lowlands, Colombia

Region	No. of samples analyzed	BGYF-positive samples (%)	Mean aflatoxin(ppb)
Highlands	35	1(3%)	20
Lowlands	37	10(27%)	400[a]
Total	72	11(15%)	

a/ Only BGYF-positive maize samples for each area were analyzed for aflatoxin B_1 and B_2

harvested maize suggests a potential for mycotoxin formation in the field and at the time of harvest, as found in the USA (21,22,23).

Although the lowlands showed larger variations than the highlands in the type of fungal species, the total fungus population was not strikingly different between the two areas. *Fusarium* and *Aspergillus* spp. always varied between the areas. However, *Pencillium citrinum* appeared almost equally in the two areas. The proportion of toxigenic genera *Aspergillus, Penicillium* and *Fusarium* isolated from maize kernels was generally higher in samples from the lowlands than from the highlands in all periods of the year, except for September to December, when proportions of *Fusarium* spp. were greater in the highlands. This may be due to the higher rainfall and cooler temperature in the area during the latter part of the year. The varied incidence of other organisms seemed also to be affected by fluctuations in climatic conditions.

It appears that *Fusarium* and *Penicillium* strains are better adapted to the ecological conditions of the highlands, while *Aspergillus* strains are adapted to lowland conditions. Although *Fusarium graminearum* strains were prevalent in both areas, their incidence decreased in the lowlands, probably due to the higher temperature most of the year. The results show that *Penicillium* strains were best adapted to the continuous environmental changes of the two areas. Among the factors that might account for the occurrence of the toxigenic fungal species in maize harvested from these two areas are temperature and moisture content of the grain at harvest and during subsequent storage (30). Although there was slightly more rainfall in the highlands than in the lowlands, the high relative humidity was almost the same in both areas. Rainfall patterns may also account for variation of maize invasion and colonization by fungi.

Test results also show the role of insect and mechanical damage in the prevalance of toxigenic fungi in damaged as compared with intact grain, and are in accord with previous findings in the USA (3,18,21). The observations demonstrated a larger incidence of both insect and mechanical damage in lowland maize than in highland maize; this suggests more favorable conditions for insects and more frequent use of machines in the lowlands. *Aspergillus flavus* was more frequently isolated from whole damaged kernels. It appears that damage contributed to kernel invasion and colonization by *A. flavus*, and to subsequent aflatoxin production (21). The high prevalence of *Fusarium* and *Penicillium* spp. in whole undamaged kernels, especially in the highlands, suggests preharvest infection in maize by the two species.

White maize varieties showed higher mold counts than yellow. *Penicillium citrinum* isolates were most frequently isolated from white maize, followed by *A. flavus, A. ochraceus* and *F. graminearum*. It is possible that this variation between the two maize varieties is due to their different genetic makeup (7,40). Toxin production in the laboratory was significantly higher ($P = 0.01$) with molds isolated from the lowlands, except for zearalenone by *F. graminearum*, which was greater in the highland isolates ($P = 0.05$). The observation suggests that the trait for toxin production may be a characteristic of adaptation to ecological conditions in specific geographic areas (24).

The production of aflatoxin B_1 and B_2 (400 ppb) from lowland maize samples is perhaps due to the higher *A. flavus* population in damaged grain and more suitable conditions for toxin formation. The absence of a positive correlation between BGY fluorescence in undamaged kernels, *A. flavus* infection and aflatoxin content is in agreement with results from other workers (21).

Lee *et al.* (18) explained these variations as the ability of the fungus to produce the metabolite(s) in the kernel associated with BGY fluorescence, independently of aflatoxin synthesis. Lee *et al.* (18) and Lillehoj (21) found many examples of BGY fluorescence in maize samples that did not contain aflatoxin. The presence of BGY fluorescence in damaged kernels clearly shows the existence of some interaction between *A. flavus* metabolites and enzymes inside the maize kernel, as suggested by Marsh (25).

Although results on the effect of different soil extracts on the morphological characteristics and toxin production by toxigenic fungi were not consistent, they suggest some interrelationships between mineral composition of the soil, toxigenic fungi and soil microflora. Although soil may not be a major inoculum source of the toxigenic fungi, it perhaps provides some necessary elements that influence either the host crop and/or the fungus (36,37). Since toxigenic fungi spend a portion of their life cycle in the soil, it is probable that mineral composition influences metabolism and geographic distribution of fungi. Inconsistency of results may be due to the variation in mineral composition of soils, even between samples from the same site. Similar variations have been found by other workers (36,37).

Conclusions

In Colombia, no generalizations can be made about toxigenic fungi and aflatoxin contamination in harvested maize without considering both regional factors and time of year. Aflatoxigenic fungi and aflatoxin contamination are more common in the lowlands. Although aflatoxicosis in humans has not been reported in Colombia, it frequently occurs in livestock, mostly associated with the feeding of damp, moldy grain of poor quality.

References

1. Association of Official Analytical Chemists. 1980. Official Methods of Analysis. Washington, D.C., USA. P. 36.

2. Austwick, P.K.C. 1975. Mycotoxins. British Medical Bulletin 31(4):415-416.

3. Ashworth, L.J., Jr., R.E. Rice, J.L. McMeans and C.M. Brown. 1971. The relationship of insects to infection of cotton bolls by *Aspergillus flavus*. Phytopathology 61(5):488-493.

4. Booth, C. 1971. The genus *Fusarium*. CMI, Surrey, England.

5. Burroughs, R., and D.B. Sauer. 1971. Growth of fungi in sorghum grains stored at high moisture contents. Phytopathology 61(7):767-772.

6. Caldwell, R.W., and J. Tuite. 1974. Zearalenone in freshly harvested corn. Phytopathology 64:752-753.

7. Cuero, R.G., E.B. Lillehoj, W.F. Kwolek and M.S. Zuber. 1985. Mycoflora and aflatoxin in pre-harvest maize kernels of varied endosperm types. *In* Trichothecenes and Other Mycotoxins, J. Lacey, ed. John Wiley and Sons, New York, N.Y., USA. Pp.109-117.

8. FAO. 1979. La vigilancia de las micotoxinas. Directrices, Inspección de los Alimentos 4:76.

9. FAO. 1983. Monthly Bulletin of Statistics 6(6):12-19.

10. Hesseltine, C.W., O.L. Shotwell, W.F. Kwolek, E.B. Lillehoj, W.K. Jackson and R.J. Bothast. 1976. Aflatoxin occurrence in 1973 corn at harvest. II. Mycological studies. Mycologia 68:341-353.

11. Hobson, W., J. Baley, A. Kowalk and G. Fuller, 1977. The hormonal potency of zearalenone in human primates. *In* Mycotoxins in Human and Animal Health. Pathotox, Park Forest South, Illinois, USA. Pp. 365-390.

12. Hurd, R. 1977. Structure activity relationships in zearalenone. *In* Mycotoxins in Human and Animal Health. Pathotox, Park Forest South, Illinois, USA. Pp. 379-390.

13. Jackson, M.L. 1964. Soil clay mineralogical analysis. *In* Soil Clay Mineralogy, C.I. Rich and G.W. Kunze, eds. University of North Carolina Press, Chapel Hill, North Carolina, USA. Chapter 8.

14. Jimeno, M.C., N.B. Ruiz and N.E. Peña. 1980. Niveles de aflatoxina B en sorgo recolectado en dos zonas del pais, primera cosecha, 1978. Revista ICA (Colombia) 15(2):129-134.

15. Johnson, L.F., and E.A. Curl. 1978. Methods for Research on the Ecology of Soil-Borne Plant Pathogens. Burgos Publishing, Minneapolis, Minnesota, USA.

16. Kraybill, F. 1977. Newer approaches in assessment of environmental carcinogenesis. *In* Mycotoxins in Human and Animal Health. Pathotox, Park Forest South, Illinois, USA. Pp. 675-685.

17. Krogh, P. 1977. Ochratoxins. *In* Mycotoxins in Human and Animal Health. Pathotox, Park Forest South, Illinois, USA. Pp. 489-497.

18. Lee, L.S., E.B. Lillehoj and W.F. Kwolek. 1980. Aflatoxin distribution in individual corn kernels from intact ears. Cereal Chemistry 57(5):340-343.

19. Leistner, L., and J.I. Pitt. 1977. Miscellaneous *Penicillium* toxins. *In* Mycotoxins in Human and Animal Health. Pathotox, Park Forest South, Illinois, USA. Pp. 639-652.

20. Lillehoj, E.B., A. Ciegler and R.W. Detroy. 1970. Fungal toxins. *In* Mycotoxins: Essays in Toxicology, Frank Blood, ed. Academic Press, New York, N.Y., USA.

21. Lillehoj, E.B., W.F. Kwolek, D.I. Fennell and M.S. Milburn. 1975. Aflatoxin incidence and association with bright greenish-yellow fluorescence and insect damage in a limited survey of freshly harvested high-moisture corn. Cereal Chemistry 52:403-412.

22. Lillehoj, E.B., D.I. Fennell and W.F. Kwolek. 1976. *Aspergillus flavus* and aflatoxin in Iowa corn before harvest. Science 193:495-496.

23. Lillehoj, E.B., and C.W. Hesseltine. 1977. Aflatoxin control during plant growth and harvest of corn. *In* Mycotoxins in Human and Animal Health. Pathotox, Park Forest South, Illinois, USA. Pp. 107-116.

24. Marasas, W.F.O., F.O.N.P.J. Kriek, V.M. Wiggins, P.S. Steyn, D.K. Towers and T.J. Hastie. 1979. Incidence, geographic distribution, and toxigenecity of *Fusarium* species in South African corn. Phytopathology 69:1181-1185.

25. Marsh, P.B., M.E. Simpson, R.J. Ferretti, T.C. Campbell and G.V. Merola. 1955. Observations on certain fluorescent spots in raw cotton associated with the growth of microorganisms. Textile Research Journal 25:1007.

26. Marti *et al.* 1978. Determination of citrinin in corn and barley. Journal of the Association of Official Analytical Chemists 61(6):1352-1358.

27. Mirocha, J., V. Pathre and C. Christensen. 1977. Zearalenone. *In* Mycotoxins in Human and Animal Health. Pathotox, Park Forest South, Illinois, USA. Pp. 345-365.

28. Mislivec, P.B., and J. Tuite. 1970. Species of *Penicillium* occurring in freshly harvested and in stored dent corn kernels. Mycologia 62:67-74.

29. Mislivec, P.B., and J. Tuite. 1970. Temperature and relative humidity requirements of species of *Penicillium* isolated from yellow dent corn kernels. Mycologia 62:75-88.

30. Mislivec, P.B., C.T. Dier and V.R. Bruce. 1975. Effect of temperature and relative humidity on spore germination of mycotoxin species of *Aspergillus* and *Penicllium*. Mycologia 67(6):1187-1189.

31. Nesheim, S. 1969. Isolation and purification of ochratoxins A and B and preparation of their methyl esters. Journal of the Association of Analytical Chemists 52(5):975-979.

32. Onions, A.H. 1966. Descriptions of pathogenic fungi and bacteria. 10:91-100. CMI, Surrey, England.

33. Pabón, G., and L. Arango. 1981. Estudios preliminares sobre el problema de aflatoxinas en algunos granos de consumo humano y animal en la ciudad de Medellín. Gallescencia 1(1):79-88.

34. Rapers, R.B., and D. Fennell. 1965. The Genus *Aspergillus*. Willians and Wilkins, Baltimore, Maryland, USA.

35. Shigematsu, M.L., R.G. Osorio and A. Restrepo. 1976. Presencia de aflatoxinas y de *Aspergillus flavus* toxigénico en maíz almacenado. Antioquía Medica 26(4):289-300.

36. Stotzky, G., and R. Torrence. 1963. Soil mineralogy in relation to the spread of *Fusarium* wilt of banana in Central America. Plant and Soil 18(3):317-337.

37. Stotzky, G., and A.H. Post. 1967. Soil mineralogy as a possible factor in geographic distribution of *Histoplasma capsulatum*. Canadian Journal of Microbiology 13:1-7.

38. Ware, G.W., and Thorpe. 1978. Determination of zearalenone in corn by high pressure chromatography and fluorescence detection. Journal of the Association of Official Analytical Chemists 6(5):1058-1062.

39. Zuber, M.S., O.H. Calvert, E.B. Lillehoj and W.F. Kwolek. 1976. Preharvest development of aflatoxin B_1 in corn in the United States. Phytopathology 66:1120-1121.

49. Zuber, M.S., O.H: Calvert, W.F. Kwolek, E.B. Lillehoj and M.S. Kang. 1978. Aflatoxin B_1 production in eight-line diallel of *zea mays* infected with *A. flavus*. Phytopathology 68:1346-1349.

Aflatoxin in Ecuador

S. Espin de Rivera, Nutrition Department, National Agricultural Research Institute, Quito, Ecuador

The consumption of foods containing mycotoxins (toxic substances produced by microscopic fungi) can have serious effects on human health. Environmental factors, such as moisture and temperature, as well as the presence of specific fungal species, can result in the presence of mycotoxins in food. The amount of mycotoxin contamination in food products depends on geographical conditions, production and storage methods and types of food. The aflatoxins are produced by the fungi *Aspergillus flavus* and *A. parasiticus*, which are widely found and can cause high levels of contamination in agricultural commodities. Aflatoxin contamination is more prevalent in the tropics and subtropics than in temperate regions.

Nearly all plant materials can serve as hosts for fungal growth and the subsequent formation of mycotoxins; this can lead to the direct contamination of food products and also to indirect contamination of animal products when livestock eat contaminated feed. The animals not only suffer ill effects due to the toxin, but can also pass the contamination on to humans through milk and meat (1).

The Cultivation of Hard-Endosperm Maize in Ecuador

Over the past decade, the cultivation of hard-endosperm maize has assumed great importance in the Ecuadorian economy because of its demand in livestock feed. Annual production of hard-endosperm maize has increased from 120 tons in 1971 to 269 tons in 1984 (3) (Table 1).

Table 1. Area, production and yield of hard-endosperm maize, Ecuador, 1971-1984

Year	Area (ha)	Production (tons)	Yield (t/ha)
1971	110,740	120,528	1.09
1972	101,840	100,748	0.99
1973	140,580	153,346	1.09
1974	161,640	185,628	1.15
1975	166,038	203,392	1.22
1976	171,210	209,108	1.22
1977	163,000	164,100	0.99
1978	132,537	136,513	1.03
1979	170,371	182,329	1.07
1980	166,708	196,414	1.18
1981	184,729	232,620	1.26
1982	155,418	269,287	1.73
1983	153,020	257,350	1.68
1984	182,830	269,020	1.47

Source: Ministerio de Agricultura y Ganadería. Estimación de la superficie cosechada y de la producción agrícola del Ecuador (1971-1984). Dirección General de Planificación, Departamento de Estadísticas Agropecuarias, Quito, Ecuador

There has been a 20% increase in the amount of land planted to hard-endosperm maize in the lowlands of Ecuador. This increase results from banana, cacao and coffee growers changing to maize production because it offers a higher return, due to the great amount of grain needed by the poultry farms and the flourishing livestock industry.

Production areas

Principal production areas for hard-endosperm maize in Ecuador are shown in Figure 1. In 1984, there was a production of 98,846, 63,605 and 36,600 tons, respectively, in the Los Rios, Manabi and Guayas coastal areas. In the highlands, Loja was the main producer, with 27,500 tons (Table 2).

Growth of domestic requirements

The growing poultry industry, especially in the Manabi region, has led to an increased demand for hard-endosperm maize, the principal ingredient (55%) in poultry feed. In 1981, 9,575,000 chickens were produced in Manabi. Their feed requirements were 287,250 tons, of which 157,987 tons consisted of maize grain. That year only 37,380 tons of local maize was available to the industry in Manabi, or only 24% of demand (3). Despite the increase in maize production area in Ecuador, the high domestic demand still requires importing maize grain to satisfy domestic requirements.

Figure 1. Principal regions for the production of hard-endosperm maize, Ecuador

Table 2. Estimates of area, production and yield of agricultural products, Ecuador, 1984

Province	Area (ha)	Production (tons)	Yield (t/ha)
Highlands			
Loja	33,950	27,500	8.10
Pichincha	4,922	8,929	1.81
Total	47,662	47,619	9.91
Coast			
Esmeraldas	7,000	9,025	1.31
Manabi	46,734	63,605	1.36
Guayas	19,691	36,600	1.85
Los Rios	50,380	98,846	1.96
El Oro	2,500	3,462	1.38
Total	126,305	211,718	1.67
Country total	182,830	269,020	1.47

The domestic marketing system

Domestic marketing is regulated by the National Marketing and Storage Agency (ENAC), which has storage facilities located in various parts of the country. The organization is able to store 65,659 tons in silos and 34,000 tons in warehouses; the total equals only 32% of national production.

Aflatoxin Contamination in Stored Maize

This study was made using samples of 1985 winter-crop maize in ENAC silos and warehouses. Total production had been approximately 363,000 tons that season (2). The samples tested were found to have between 1 and 2% fungal contamination and 13% moisture. The average storage temperature was 27° ± 2°C. Samples were taken from the upper and lower sections in those silos that had shown a high percentage of fungi and impurities. The combined sample weighed approximately three kilos, from which a 40-g sample was utilized for analysis.

Method of analysis

For analysis, the sample was homogenized and treated with acidified acetone. The extract was then purified with 10% $(NH_2)_2SO_4$ and Celite 545, followed by a liquid separation of the purified extract, using toluene. The toluene fraction was concentrated to dryness and taken to volume with a benzene-acetonitrile mixture. Two-dimensional, thin-layer chromatography was used to identify and quantify aflatoxin.

Results

Of the 52 maize grain samples studied, 39 were found to be contaminated with aflatoxin B_1 at levels ranging from 5 to 50 μg/kg. The amount of aflatoxin detected in grain from each of the maize-producing provinces is shown in Table 3. Contamination values ranged from 40 to 50 μg/kg for 3.8% of the samples tested, 30 to 40 μg/kg for 7.6%, 20 to 30 μg/kg for 1.9%, 10 to 20 μg/kg for 28.8%, 5 to 10 μg/kg for 32.7% and less than 5 μg/kg for 13.5%; the remaining 11.6% of the samples showed no contamination. Aflatoxins B_2, G_1 and G_2 were not detected. The highest level of contamination detected (50 μg/kg) was found in samples from the province of Esmeraldas. Samples showing contamination levels between 5 and 20 μg/kg of aflatoxin B_1 represented a majority; 61.5% of the samples tested were within this range.

Table 3. Presence of aflatoxins in hard-endosperm winter maize, Ecuador, 1985

Province	No. of samples	Aflatoxin B_1 (μg/kg)
Coast		
Guayas	2	0
	9	5-25
	1	30-40
Los Rios	2	5
	8	5-20
	2	30-40
Manabi	2	0
	1	5
	4	5-20
Esmeraldas	2	5
	4	5-20
	1	50
Highlands		
Pichilingue	1	5
	7	10-30
Bolivar	2	5-10
	1	40
Loja	2	0
	1	5

Note: No aflatoxins B_2, G_1 or G_2 were found

There are regions in Ecuador where stored maize grain is less prone to contamination (Table 3). The provinces of Manabi and Loja appear to offer better storage conditions, thus reducing the risk of aflatoxin contamination.

Aflatoxins in Foodstuffs

No quality standards for mycotoxin contamination in foods have been established in Ecuador. Although the National Agricultural Research Institute is not a regulatory agency, producers of livestock and manufacturers and distributors of feed often bring samples to the Nutrition Department laboratories for aflatoxin testing (Figure 2).

Among these samples, poultry and swine feed, which is composed principally of maize grain, have levels of aflatoxin B_1 contamination ranging from 20 to 25 μg/kg. In samples of corn meal used in feed, aflatoxin B_1 ranges from 5 to 20 μg/kg. No aflatoxin contamination has been found in a number of other food products, including soybeans, soybean cakes, castor bean cakes, rice flour, imported rice, fish meal and dried cassava (Table 4).

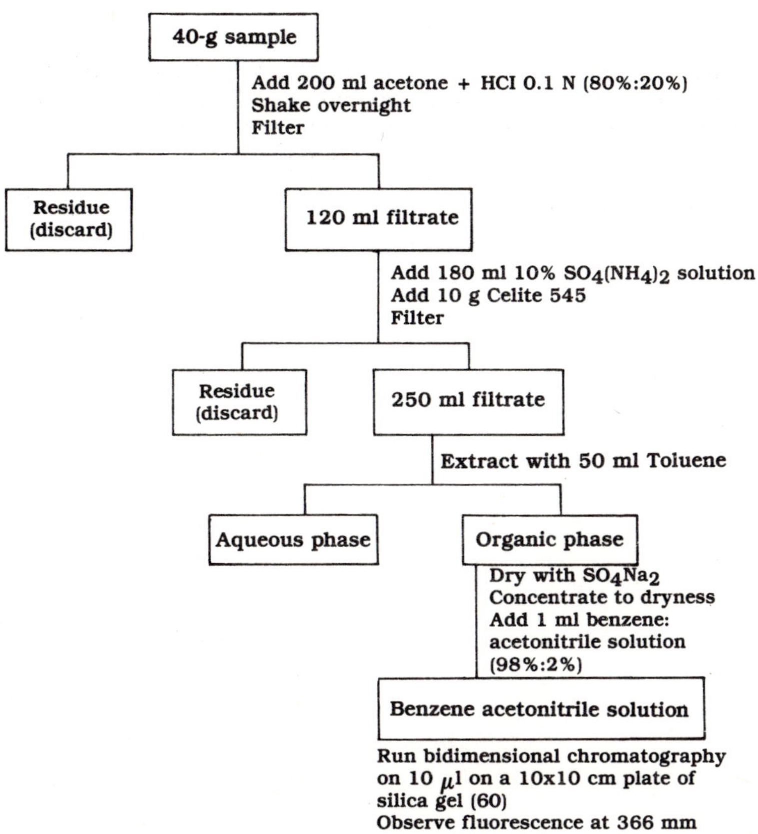

Figure 2. Flow chart, method used for aflatoxin testing of foods, Nutrition Department, National Agricultural Research Institute, Ecuador

Conclusions

Environmental conditions in Ecuador support the formation of aflatoxin B_1 in maize grain; 88.5% of test samples were contaminated. Research must be continued on the presence of aflatoxin and other mycotoxins to establish permissible levels. Further systematic research is needed on levels of contamination of mycotoxins in food and feed in the various regions of Ecuador, and on their effects on human and animal health. Studies must also be initiated on the factors that foster fungal growth and the formation of mycotoxins in grains under both pre- and post-harvest conditions, as well as in storage.

References

1. Critères d'hygiene de l'environnement 11. 1980. Micotoxines. World Health Organization, Geneva, Switzerland.

2. Departamento de Comercialización. Empresa Nacional de Comercialización y Almacenamiento, Quito, Ecuador.

3. Plan de investigación para cultivo de maíz duro. 1985. Programa de maíz, Estación Experimental Pichilingue. Instituto Nacional de Investigaciones Agropecuarias, Pichilingue, Ecuador.

Table 4. Presence of aflatoxins in maize tested at the Nutrition Department laboratories, National Agricultural Research Institute, Ecuador

Product	No. of samples	Aflatoxin B_1 ($\mu g/kg$)	Comments
Feed	1	25	Mortality in chickens
Concentrated pig feed	1	20	Mortality in swine, miscarriages, still births
Corn meal for feed	15	5-25	
Castor bean cakes	1	-	
Soybeans	2	-	
Soybean cakes	2	-	
Sorghum	2	-	
Rice flour	1	-	
Bran	1	-	
Rice	32	-	Imported from Thailand
Fish meal	1	-	
Dried cassava	6	-	
Hard-endosperm maize	3	5-40	

Aflatoxin in the Southeastern USA

T.E. Nichols, Jr., Department of Economics and Business, North Carolina State University, Raleigh, North Carolina, USA

Abstract

Relatively high moisture and temperature conditions in the southeastern USA favor the growth of molds and subsequent production of aflatoxin in foodstuffs. Occurrence of aflatoxin in grain varies from year to year and state to state. The level of field contamination appears to have stabilized or improved from earlier years, except for 1977, 1980 and 1983 when the maize crop was severely damaged by drought. The natural occurrence of aflatoxin in preharvest maize, and the cumulative effects of aflatoxin, which may increase progressively from field to animal rations, enhance the risks southeastern grain producers and livestock feeders normally face. Contaminated maize and feed cause economic losses to these producers and pose increasing concerns for animal and human health. Control of aflatoxin in foods and feeds is difficult. The most effective approach is to prevent mold growth at all levels of production, storage, processing and utilization. Greater efforts should be made to find resistance to aflatoxin in plant or animal species, and to find methods to detoxify contaminated grain and other feedstuffs. These research objectives should be given very high priority.

Resumen

Los niveles de humedad y temperatura relativamente altos que imperan en el sureste de Estados Unidos favorecen el crecimiento de moho y la subsecuente producción de aflatoxinas en los productos alimenticios. La presencia de aflatoxinas en los granos varía de un año a otro y de un estado a otro. En comparación con años anteriores, el nivel de contaminación en el campo parece haberse estabilizado o mejorado, con excepción de 1977, 1980 y 1983, años en que la sequía dañó gravemente la cosecha de maíz. La presencia natural de aflatoxinas en el maíz durante la fase de precosecha y los efectos acumulativos de las aflatoxinas, que suelen incrementar en forma progresiva de las plantas a los animales, aumentan los riesgos que tienen que encarar normalmente los productores de granos y los ganaderos del sureste de Estados Unidos. El maíz y el alimento contaminado ocasionan pérdidas económicas a estos productores y constituyen una amenaza cada vez mayor para la salud de los animales y los seres humanos. Resulta difícil controlar la presencia de aflatoxinas en los alimentos humano y animal. El método más eficaz consiste en evitar el desarrollo de moho en todos los niveles de producción, almacenamiento, procesamiento y utilización. Es preciso realizar grandes esfuerzos para encontrar resistencia a las aflatoxinas, tanto en especies vegetales como animales, así como descubrir métodos para eliminar las toxinas de granos y otros alimentos contaminados. Hay que dar una gran prioridad a estos objetivos de investigación.

Numerous studies report the types and incidence of toxic molds that contaminate food and agricultural commodities worldwide (1,6,16,19,26), indicating their ubiquitous distribution and the health risks associated with the mycotoxins that may subsequently be produced. Toxin-producing molds in foods and animal feeds include members of the genera *Aspergillus*, *Penicillium*, *Fusarium*, *Alternaria*, *Trichothecium*, *Cladosporium*, *Byssochlamys* and *Sclerotinia*. These microorganisms are capable of growth on a wide variety of substrates and under diverse moisture levels, pH and temperature conditions. Foods and feeds are susceptible to fungal invasion and growth during any stage of production, processing, transport and storage. Studies have shown that in the poultry industry, aflatoxin contamination can occur during the manufacture of feed and in the feed distribution system, with mold growth in such places as storage bins, troughs, pans and feeder bins (10). The types and level of molds present in a commodity are directly affected by the substrate composition, genetic properties of the mold strains, moisture, temperature and handling conditions. A more detailed account of these factors in the production of mycotoxins in cereals has been given by Hesseltine (8).

During the past several years, over 200 mycotoxins have been isolated from mold cultures. However, not all of the toxins have been shown to have a role in human or animal disease. Among those that are considered to pose the greatest potential hazard to animal and human health are ergot, aflatoxins, ochratoxins, sterigmatocystin, patulin, citrinin, fungal tremorgens and fusarium toxins, such as zearalenone, T-2 toxin, vomitoxin and other trichothecenes. Although all of these mycotoxins are important in animal agriculture and human health, this paper will address only aflatoxins.

Aflatoxins

Aflatoxins are produced by some strains of the fungus *Aspergillus flavus*, and by most strains of *Aspergillus parasiticus* (4). *Aspergillus flavus* occurs in most of the soils of the southern USA, where growth of the fungus is favored by high temperatures and moisture levels. It occurs on many types of organic materials in various stages of decomposition, including forages, cereal grains, and food and feed products. However, not all isolates of the organism produce aflatoxins. The presence of *A. flavus* on crop plants, grain or foodstuffs does not mean that aflatoxin will be present.

Aflatoxins have been shown to be a primary cause of liver cancer in certain animals. Aflatoxin B_1 is the most potent, naturally occurring, cancer-producing substance known (25). It has adverse biological effects, including liver damage, impaired feed efficiency, reduced growth rate and immunosuppression at levels approximating 1 ppm in the diets of most domesticated and experimental animals. There are also reports in the literature that associate aflatoxins with acute poisoning in humans (2). Reye's syndrome, an encephalitis-like disease that occurs mostly in children, is characterized by vomiting, hypoglycemia, convulsions, coma and usually death (18). Since the initial reports of the disease, *A. flavus* has been linked to several outbreaks of Reye's syndrome in New Zealand, Czechoslovakia, the USA and Thailand. The incidence of primary liver cancer varies throughout the world, with the highest incidences occurring in tropical regions and in those countries where large quantities of cereal grains, rice and nuts that may contain molds are consumed directly (12,20). In countries where meat is eaten, animals frequently serve as a means of screening for feed contaminated by aflatoxin. Because humans are on the higher end of the food chain, they can avoid some exposure to these mycotoxins.

In the USA, there is no evidence that aflatoxins are a significant cause of liver cancer or other diseases in humans. On a per capita basis, aflatoxin intake in the southeast is about nine times greater than for the country as a whole, but the proportion of deaths attributable to liver cancer is lower than that for the country as a whole.

Aflatoxin contamination of foods originating from animals is a major concern for animal and human health. Aflatoxin M_1, which is a hydroxylated metabolite of aflatoxin B_1, has been found in milk from animals fed grain or other feedstuffs containing aflatoxins. Aflatoxins have been found in commercial fluid milk in the USA, Germany and South Africa, and in samples of domestic and imported cheeses in the USA (21,23). Aflatoxins B_1 and M_1 have also been found in livers, kidneys and other tissues of pigs, and in the tissues of broiler chickens and eggs of laying hens fed aflatoxin-contaminated diets (10,22). Although aflatoxin levels in foods are generally low, studies indicate that indirect exposure to aflatoxins could occur from consumption of milk, meat and eggs if the animals are fed aflatoxin-contaminated feed.

Controlling aflatoxin in foods is a complex and difficult task. The most effective approach to eliminating food contamination is preventing mold growth at all levels of production, harvest, transport and storage. However, this is not as simple and straightforward as it may seem. Aflatoxin is cumulative, and once in food, it appears to persist and remain biologically active longer than other mycotoxins. Because of the potential dangers of human and animal exposure to mycotoxins, foodstuffs are closely monitored by the Food and Drug Administration (FDA) in the USA. The FDA treats aflatoxins as poisonous and deleterious substances and regulates them accordingly. At present, there is no established tolerance level for aflatoxin in any food, because toxicological data upon which such a level might be based are lacking. Instead, the FDA has set what it considers to be practical limits for levels of aflatoxin in foods and feeds; these are based primarily on the limitations of the detection and measurement techniques of present analytical methods, and to some extent on the ability of current agronomic and technological practices to prevent contamination (24). These limits have been set forth by the FDA as working guidelines for regulatory action, and apply to all products known to be susceptible to aflatoxin contamination. The guideline for total allowable aflatoxin in cornmeal, grits and peanut butter is 20 ppb (μg/kg), and in whole milk, skim milk and low fat milk, 0.5 ppb.

Exceptions to the guidelines have been made from time to time by the FDA when petitioned by individual states where agriculture has been affected by a severe aflatoxin problem (7). For example, in 1977 in the southeastern USA, aflatoxin occurred widely in maize at levels above 20 ppb and orderly marketing of the commodity was effectively halted in some states. Upon application by several states, the FDA raised the guidelines to 100 ppb in animal feed for intrastate use, as long as the contaminated maize or its products did not enter into human food and did not adversely affect animal health. This action, which applied only in 1977, was not extended to interstate commerce because the end use of contaminated maize could not be controlled effectively.

A second exception was issued in 1980, another crisis year for aflatoxin in maize in the southeastern US. Because of the severity of the problem, the guideline was raised from 20 to 200 ppb for intrastate use; for the first time, maize containing up to 100 ppb could cross state lines. However, movement of the commodity was closely monitored by USDA grain inspectors and policed by FDA field-workers. Maize containing up to 400 ppb could be blended with aflatoxin-free grain, provided that such material would not enter human foods and that it was fed to nonlactating and mature animals that normally would be unharmed by such concentrations of aflatoxin. Recently, the FDA responded to a crisis situation in Arizona by permitting cottonseed meal containing up to 300 ppb aflatoxin to be marketed, with the restriction that it be fed only to mature, nonlactating animals. Although the efforts of regulatory agencies such as the FDA are very important in preventing human and animal exposure to mycotoxins, the responsibility for actual control and prevention of this hazard lies with everyone in the agricultural and food industries.

Incidence of Aflatoxin in the Southeast

Preharvest

Aspergillus flavus growth and subsequent toxin production was once classified as primarily a storage-related problem, but subsequent studies have shown that the fungus can infect developing maize kernels in the field (3). *Aspergillus flavus* can colonize on maize silks and invade developing kernels in all maize-producing regions of the USA, but conditions in the southern part of the country are more favorable for its growth (15).

In a survey of 20 hybrids grown at diverse locations in 1980 and 1981, aflatoxin was routinely observed at levels of 20 ppb or higher in mature kernels from the southern USA and sporadically at lower concentrations in samples from the Corn Belt region. Aflatoxin-positive rates suggested that locations could be grouped into categories of high, medium and low occurrence. Incidence of the toxin in samples from the high-occurrence southeastern states (Florida, Georgia, North Carolina, South Carolina and Mississippi) varied from 90 to 100%, whereas in medium-occurrence locations (Tennessee and Kansas), incidence levels ranged from 30 to 45%. The low-occurrence locations (Indiana and Ohio) had detectable aflatoxin in 2 to 25% of the samples. No definite year-to-year variations in aflatoxin were noted in samples from locations with high and medium occurrence, but the incidence level was higher in 1981 than in 1980 in samples from the two low-occurrence locations in the Corn Belt.

An unusual pattern of aflatoxin concentration was observed in this study (Table 1). Samples from three locations (Georgia, South Carolina and Mississippi) uniformly produced mean toxin levels exceeding 100 ppb. Although a high incidence of aflatoxin was observed in samples from Florida and North Carolina, toxin levels were variable; Florida maize contained relatively low toxin concentrations in both years, whereas North Carolina samples varied from a low of 19 ppb of aflatoxin in 1980 to 301 ppb in 1981.

Table 1. Occurrence and mean aflatoxin levels in kernels of 20 maize hybrids grown at diverse locations in the USA, 1980 and 1981

Location	Occurrence of aflatoxin (%)			Aflatoxin B_1 and B_2 (ng/g)		
	1980	1981	Mean	1980	1981	Mean
Florida	98	90	94	34	33	34
Georgia	100	100	100	182	355	254
North Carolina	96	95	96	19	301	76
South Carolina	100	100	100	118	141	129
Mississippi	—	100	100	—	362	362
Indiana	2	25	14	33	3	10
Ohio	5	23	14	1	4	2
Tennessee	45	30	38	6	3	4
Kansas	—	33	33	—	11	11

Source: Lillehoj *et al.* (16)

In addition, the varied aflatoxin incidence noted in Corn Belt samples was not consistently linked to equivalent differences in toxin levels. The results of this survey substantiated earlier studies indicating that *A. flavus* infection was widespread in the south and sporadic in the Corn Belt.

Many environmental factors and cultural practices affect aflatoxin contamination in preharvest maize (12). Water stress or drought has been identified as a major factor. Drought conditions that prevailed over the southeastern USA during the summers of 1977, 1980 and 1983 were conducive to the growth of *A. flavus* on developing maize and subsequent production of aflatoxin. Average yields per acre were sharply reduced and aflatoxin concentration significantly increased (Figure 1).

A survey of commercial maize samples taken at North Carolina grain elevators from 1976 through 1980 indicates the level of field contamination (Table 2). In 1976, nearly two-thirds (64.2%) of the samples had less than 20 ppb of aflatoxin. Only 8% contained 100 ppb or more. In 1978, nearly 90% of the crop was aflatoxin-free, with only 1% containing 100 ppb or more. In sharp contrast, the 1980 survey showed that over 17%, twice the number of samples reported in 1976, had levels

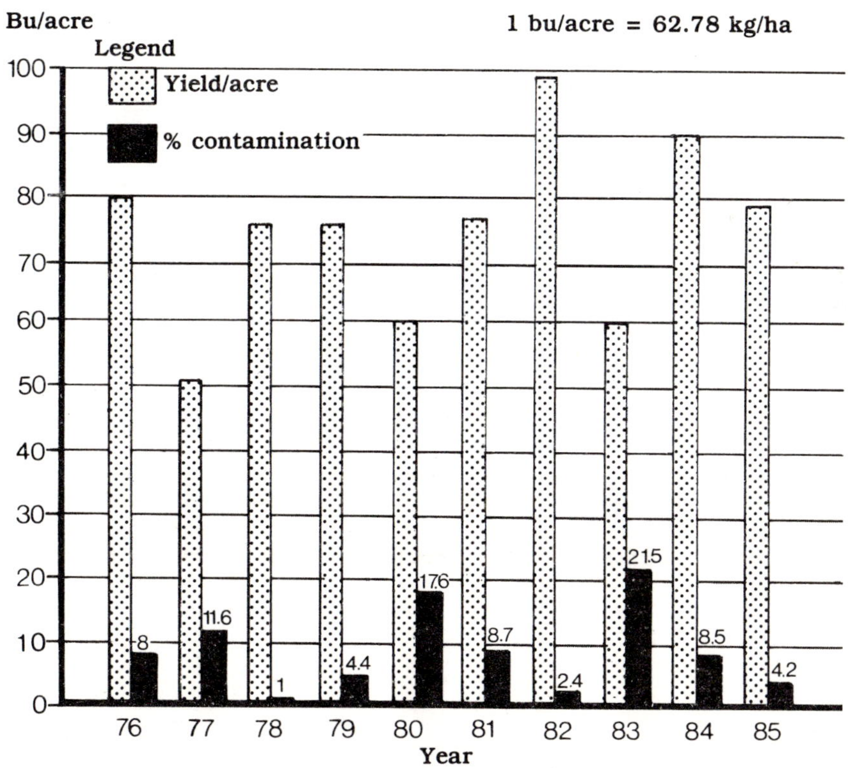

Figure 1. Relationship between average maize yield per acre and aflatoxin contamination, North Carolina, 1976 to 1985

Note: Figures represent percent of samples with aflatoxin concentrations above 10 ppb

exceeding 100 ppb; only one-third (34.3%) of the samples had less than 20 ppb aflatoxin.

Annual surveys have not been taken in the last five years, so the incidence of aflatoxin in North Carolina maize is not precisely known. However, estimates have been made using data submitted by farmers to field laboratories operated by the Food and Drug Protection Division of the North Carolina Department of Agriculture (Table 3). These data indicate that the level of field contamination has stabilized or improved from earlier years, except in 1983 when the maize crop was severely damaged by drought.

Postharvest

Economic losses caused by the outbreak of aflatoxin in 1977 and again in 1980 have made southeastern US grain and livestock producers more cautious. A greater effort is being made, at least by the major poultry and livestock feeders, to obtain feed or ingredients free of aflatoxin. This is much easier when a quality assurance laboratory exists, and now many of the large poultry and swine producing firms in the USA, particularly in the southeast, have such laboratories.

Nearly all grain elevators now apply some type of quality control test procedure to incoming maize and other cereal grains. Two tests routinely used are the blacklight test and the minicolumn test. The blacklight test involves a long-wave ultraviolet illumination of damaged maize kernels and visual identification of bright greenish-yellow fluorescence (BGYF). If one or more BGY-fluorescing particles is found in a 2- to 4.5-kg sample, a minicolumn test is performed to determine if the sample contains more than 20 ppb aflatoxin.

Some processors, particularly maize millers, require results from a thin-layer chromatography (TLC) test before purchase. Since TLC provides quantitative (precise level) readings, most buyers prefer this method over minicolumn or blacklight tests. The lack of commercial facilities equipped to service the grain and feed industries and the cost of conducting the TLC test have been limiting factors. However, a new assay technology will be marketed by at least one firm by May 1986 and may help solve this problem (D. Jackson, personal communication).

Table 2. Incidence of aflatoxin in North Carolina maize, 1976 to 1980[a]

Crop year	Aflatoxin level (ppb) 0-19 (%)	20-100 (%)	100 (%)
1976	64.2	27.7	8.0
1977	58.1	30.2	11.6
1978	87.0	12.0	1.0
1979	67.3	28.3	4.4
1980	34.3	48.1	17.6

a/ Based on nonrandom samples of maize taken from grain elevators across the state

Table 3. Incidence of aflatoxin in North Carolina maize, 1981 to 1984[a]

Crop year	Aflatoxin level (ppb) 0-19 (%)	20-100 (%)	100 (%)
1981	76.4	14.9	8.7
1982	91.8	5.8	2.4
1983	49.9	28.6	21.5
1984	79.6	11.6	8.8
1985	85.4	10.4	4.2

a/ Based on samples submitted by farmers to the N.C. Department of Agriculture field laboratories

The new system uses monoclonal antibodies having a high affinity for aflatoxin, and can be run in less than 10 minutes after a 50-g sample has been prepared. Furthermore, the assay, which can be conducted by less highly skilled personnel than the TLC method, will cost about US$ 6 per sample compared to $45 to $47 for the TLC test run in a commercial laboratory. If this new test is approved by the Association of Official Analytical Chemists (AOAC) and adopted by grain handlers and feed processors, it will revolutionize the surveillance techniques and quality control procedures necessary to protect human and animal health.

The Economic Impact of Aflatoxin

The natural occurrence of aflatoxin in maize and other feed ingredients in the southeastern USA increases the risks that grain producers and livestock feeders normally face. Contaminated maize causes economic losses to producers, handlers and feeders. Losses vary from year to year, depending on the occurrence and potency of the aflatoxin. Direct costs associated with lower yields, reduced market price, extra drying, handling and testing expenses and storage problems are fairly easy to document and measure.

Losses that result from using contaminated grain as feed are more difficult to measure. The effects of chronic or acute ingestion of aflatoxin on poultry, livestock and other animals are varied. Death from aflatoxicosis can often be readily diagnosed and measured, but the more subtle effects associated with contaminated feed that does not produce clinical symptoms of toxicity are difficult to document. These symptoms include reduced growth rate and feed efficiency, the infertility syndrome in swine and cattle, and loss of quality in animals and animal products because of abnormal bone and leg development, fatty livers, hepatic lesions and hemorrhaging. A complicating factor in evaluating aflatoxin effects in animals is that other mycotoxins can be found in the feed ingredients, and synergistic interactions between these mycotoxins are well documented (14).

Losses associated with aflatoxin in maize and other feed ingredients are not restricted to private individuals or firms; public costs are associated with the monitoring, sampling and analysis of feed ingredients, and the epidemiology, education and training that underlie and support risk assessment and regulatory control programs. Additional funds are used for research and extension efforts to minimize the aflatoxin problem and to help decision makers reduce losses until a solution is found. No recent studies have been made to assess the private and public costs of aflatoxins in the southeast, but in 1977 and 1980, years in which there were serious aflatoxin outbreaks, the costs were estimated at nearly US$ 200 and $138 million, respectively (11).

Future Research Needs

During the past decade, research has centered on understanding the epidemiology of aflatoxin formation by *A. flavus* and the biological effects of aflatoxin on animal health. These efforts should continue, but perhaps should focus more precisely on methods of preventing aflatoxin in the field. Laboratory and field studies have shown varied differences in maize resistance to *A. flavus* and production of aflatoxin. However, most scientists agree that it is more difficult to breed for plant resistance to *A. flavus* infection and to aflatoxin production than for plant resistance to other diseases. Another approach might be to find animal species that have natural immunities or resistance to aflatoxin-contaminated feed. Some strains of poultry have shown limited resistance, and there may be other animals that could be bred for this trait.

Equal emphasis should be placed on finding methods for detoxification and use of contaminated grains and other feed ingredients. The treatment of whole-kernel maize with sodium bisulfite has been found to effectively destroy aflatoxin B_1 and other mycotoxins in laboratory tests (7). Density segregation using a combination of brine and sucrose solutions also promises to be an effective method for reducing concentrations of these mycotoxins in grain (10). Further research is needed to adapt successful laboratory tests to practical on-farm or commercial applications.

Finally, additional research efforts should be made to study involvement of aflatoxins in human disease. Ways must be found not only to protect animals from contaminated feed but also to prevent the occurrence of mycotoxin residues in human foods of animal origins.

References

1. Allcroft, R. 1965. Aspects of aflatoxicosis in farm animals. *In* Mycotoxins in Foodstuffs, G.N. Wogan, ed. Massachussetts Institute of Technology Press, Cambridge, Massachussetts, USA.

2. Bullerman, L.B. 1979. Significance of mycotoxins to food safety and human health. Journal of Food Protection 42:65-86.

3. Council for Agricultural Science and Technology. 1979. Aflatoxins and other mycotoxins: An agricultural perspective. Iowa State University, Ames, Iowa, USA.

4. Diener, U.L., and N.D. Davis. 1969. Aflatoxin formation by *Aspergillus flavus*. *In* Aflatoxin, L.A. Goldblatt, ed. Academic Press, New York, N.Y., USA.

5. Goldblatt, L.A., ed. 1969. Aflatoxin. Academic Press, New York, N.Y., USA.

6. Hamilton, P.B. 1984. Determining safe levels of mycotoxins. Journal of Food Protection 47:570-575.

7. Hagler, W.M., Jr., J.E. Hutchins and P.B. Hamilton. 1983. Destruction of aflatoxin B_1 with sodium bisulfite: Isolation of the major product aflatoxin BS. Journal of Food Protection 46:295-300.

8. Hesseltine, C.W. 1969. Mycotoxins. Mycopathologia et Mycologia Applicata 39:371-383.

9. Huff, W.E., and W.M. Hagler, Jr. 1985. Density segregation of corn and wheat naturally contaminated with aflatoxin, deoxynivalenol and zearalenone. Journal of Food Protection 48:416-420.

10. Jarvis, B. 1976. Mycotoxins in food. *In* Microbiology in Agriculture, Fisheries and Food, F.A. Skinner and J.G. Carr, eds. Academic Press, London, England.

11. Jones, F.T., W.H. Hagler and P.B. Hamilton. 1982. Association of low levels of aflatoxin in feed with productivity losses in commercial broiler operations. Poultry Science 61:861-868.

12. Kraybill, H.F., and R.E. Shapiro. 1969. Implication of fungal toxicity to human health. *In* Aflatoxin, L.A. Goldblatt, ed. Academic Press, New York, N.Y., USA.

13. Lillehoj, E.B. 1983. Effect of environmental and cultural factors on aflatoxin contamination of developing corn kernels. Aflatoxin and *Aspergillus flavus* in Corn, U.L. Diener, R.L. Asquith and J.W. Dickens, eds. Southern Cooperative Series Bulletin 279. Alabama Agricultural Experiment Station, Auburn, Alabama, USA.

14. Lillehoj, E.B., and A. Ciegler. 1975. Mycotoxin synergism. *In* Microbiology, D. Schlessinger, ed. American Society of Microbiology, Washington, D.C., USA.

15. Lillehoj, E.B., M.S. Zuber, L.L. Darrah, W.F. Kwolek, W.R. Findley, E.S. Horner, G.E. Scott, A. Manwiller, D.B. Sauer, D. Thompson, H. Warren, D.R. West and N.W. Widstrom. 1983. Aflatoxin occurrence and levels in preharvest corn kernels with varied endosperm characteristics grown at diverse locations. Crop Science 23:1181-1184.

16. National Academy of Science (USA). 1979. Interactions of mycotoxins in animal production. Washington, D.C., USA.

17. Nichols, T.E. 1983. Economic impact of aflatoxin in corn. Aflatoxin and *Aspergillus flavus* in Corn, U.L. Diener, R.L. Asperquith and J.W. Dickens, eds. Southern Cooperative Series Bulletin 279. Alabama Agricultural Experiment Station, Auburn, Alabama, USA.

18. Riela, A.R., and E.S. Roach. 1983. Reye's syndrome: Twenty years in perspective. North Carolina Medical Journal 44:351-355.

19. Rodricks, J.V., and H.R. Roberts. 1977. Mycotoxin regulation in the United States. *In* Mycotoxins in Human and Animal Health, J.V. Rodricks, C.W. Hesseltine and M.A. Mehlman, eds. Pathotox, Park Forest South, Illinois, USA.

20. Shank, R.C. 1976. The role of aflatoxins in human disease. *In* Mycotoxins and Other Fungal Related Food Problems, J.V. Rodricks, ed. Advances in Chemistry Series 149. American Chemical Society, Washington, D.C., USA.

21. Stoloff, L. 1976. Occurrence of mycotoxins in foods and feeds. *In* Mycotoxins and Other Fungal Related Food Problems, J.V. Rodricks, ed. Advances in Chemistry Series 149. American Chemical Society, Washington, D.C., USA.

22. Stoloff, L. 1977. Aflatoxins: An overview. *In* Mycotoxins in Human and Animal Health, J.V. Rodricks, C.W. Hesseltine and M.A. Mehlman, eds. Pathotox, Park Forest South, Illinois, USA.

23. van Walbeek, W. 1973. Fungal toxins in foods. Canadian Institute for Food Science and Technology Journal 6:96.

24. Wessel, J.R., and L. Stoloff. 1973. Regulatory surveillance for aflatoxin and other mycotoxins in feeds, meat and milk. Journal of the American Veterinary Medicine Association 163:1284.

25. Wogan, G.N. 1966. Chemical nature and biological effects of the aflatoxins. Bacteriology Review 30:460.

26. Wyllie, T.D. and L.G. Morehouse, eds. 1978. Mycotoxic fungi, mycotoxins, mycotoxicoses. An encyclopedia handbook, Vol. II. Mycotoxicoses of domestic and laboratory animals, poultry and aquatic invertebrates and vertebrates. Marcel Dekker, New York, N.Y. USA.

Aflatoxin in Kenya

A. Manwiller, USAID, Nairobi, Kenya

Outbreaks of mycotoxicosis in Kenya received wide publicity in 1978 and again in 1984-85 when large numbers of dogs and poultry died. In 1978, the Government Chemists Department collected 336 samples of food and feed (2). Of these, aflatoxin levels in 52 exceeded 150 ng/g^{-1}, the highest being 3000 ng/g^{-1}, which was found in dog meal. None of the human food samples were contaminated. One of the reasons for the high levels may have been that a bumper crop was produced in the 1977-78 season, far in excess of proper storage facilities; 1984-85 was a famine period and maize was imported from several countries, including the USA. At least one shipload of incoming maize was ordered dumped in the ocean as aflatoxin made it unfit for use.

In western Kenya, 300 maize samples were collected and analyzed for aflatoxin during 1984-85 as part of an on-farm storage project (1). Only three samples tested positive with the bright greenish-yellow fluorescence (BGYF) color test; one tested positive with the Holaday-Velasco minicolumn method.

In general, the Kenya commercial maize crop seems to be monitored rather carefully. Maize produced on subsistence farms is always suspect during a drought period, but it is difficult to control the situation on these thousands of small units.

References

1. P.H. Giles, personal communication.

2. Muraguri, N., L.C. Omulkoola, G.M. Kenji and G.A. Condier. 1981. A survey of mycotoxins in human and animal foods: Part 1. East African Medical Journal, vol. 58, no. 7.

Aflatoxin in India: I

K.S. Bilgrami and K.K. Sinha, Post-Graduate Department of Botany, Bhagalpur University, Bhagalpur, India

It is well established that aflatoxins are a potential threat to human and animal health. Practically all tropical countries encounter the problem of aflatoxin contamination in one way or another. Climatic conditions, coupled with inadequate socioagronomic systems, are mainly responsible for the problem in India. A variety of agricultural and industrial commodities are affected by aflatoxins and the seriousness of the problem is reported from several parts of the country.

Fortunately, some active centers of research have arisen during the last ten years. Interest in the aflatoxin problem is also demonstrated by the fact that a number of institutions have recently submitted research proposals to various funding agencies. Indian work on mycotoxins was first summarized by Hesseltine in 1976 (22). Subsequently, the Indian Council of Medical Research, New Delhi, published a monograph entitled, *Health Hazards of Mycotoxins in India* (8). Aflatoxin problems in India were discussed in detail at a national symposium at Bhagalpur University (17), and the mycotoxin problem in India was the focus of the International Conference on Mycotoxins held in Cairo, Egypt, in 1983 (51).

The progress made by Indian workers on the mycotoxin problem in food has been reviewed by the authors and published in the Indian Review of Life Sciences (10). This paper is a status report on work being done on mycotoxins at the various research centers in India; emphasis is, however, on aflatoxin contamination in maize. Areas that need attention, as well as the approach and scope for collaborative research, are outlined.

Aflatoxin Incidence

The natural incidence of aflatoxins in food and feed is influenced by climate, the quantity of *Aspergillus flavus* conidia in the air and soil, and the nature of the substrate.

The potential of *A. flavus* isolates to produce aflatoxin varies widely. Initial reports from India show that less than 10% of *A. flavus* isolates produce toxin (38,46). However, Maggon and colleagues (31) recorded aflatoxin elaboration by all seven isolates of *A. flavus* obtained from soil of the Delhi region. These strains elaborated only aflatoxin B_1 and B_2. Aflatoxin B_1 production by all the isolates of *A. flavus* obtained from peanuts was reported from Punjab (25). More than 90% of *A. flavus* isolates were found to be toxigenic in an extensive study undertaken by Indulkar *et al.* (23), who screened 1800 isolates of cottonseed. Only one isolate produced all four aflatoxins (B_1, B_2, G_1 and G_2); the majority produced only B_1 and B_2. From the *tarai* (foothill) areas of Nainital, 162 isolates of *A. flavus* (from a total of 276 isolates) obtained from various agricultural commodites were found to be aflatoxin producers (34). Nearly 80% of the *A. flavus* isolates obtained from standing maize crops and from stored maize grain produced aflatoxins in studies made by a research team at Bhagalpur University (12,45). Other reports from the country also reflect a high frequency of naturally occurring toxigenic strains of *A. flavus* with varying qualitative and quantitative potentials.

No edible commodity is absolutely safe from toxigenic strains or aflatoxin contamination. Reports from different areas of India suggest that a wide range of food and feed materials are good substrates for aflatoxin contamination. The list includes cereals (maize, wheat, paddy rice and sorghum), pulses (mung and *urad*), oil seeds and their products (peanuts, cottonseed and mustard), dried fruits and nuts (coconut, almonds, cashew nuts, walnuts, raisins and *makhana*), spices (chile, coriander, cumin, black cumin, black pepper, cardamom, fennel, ginger, turmeric and red pepper), fruits (guava, banana, *Artocarpus lakoocha* and apples), vegetables (*Momordica charantia*), milk and poultry and cattle feed (10). This report considers only aflatoxin in maize.

Maize is grown throughout India, predominantly in Uttar Pradesh, Bihar and Rajasthan. In Bihar and Uttar Pradesh, the crop is grown throughout the year and is harvested in all of the three major seasons, monsoon, winter and summer. Surveys of aflatoxin contamination of maize in India are inadequate; fragmented reports, however, indicate that a serious problem exists. In 1974, a large number of human mortalities were recorded in the tribal belts of western India, due to an outbreak of acute hepatitis. A team of scientists from the National Institute of Nutrition, Hyderabad, surveyed the staple food (maize grain) of the affected tribal population in the Banswara district of Rajasthan and the Panchmahal district of Gujarat. The amount of aflatoxins in the contaminated maize grain was found to be very high, i.e., up to 15.6 ppm (26); in a 1975 survey, the amount was lower, not exceeding 0.1 ppm (27). Maize in the area was also analyzed by a team of workers of the All-India Institute of Medical Sciences, New Delhi (49). They detected aflatoxins in 13 out of 14 samples collected from the affected families. While mostly aflatoxins B_1 and G_1 were found, a few samples had traces of M_1 also. The concentration of aflatoxin B_1 in most of the samples was in the range of 100 to 600 ppb, but two samples had 900 and 1100 ppb. Scientists from the Indian Agricultural Research Institute, New Delhi, recorded aflatoxin in 15 of 22 maize samples from the affected areas (24). Seven of these samples were collected from families with no recorded deaths of humans or dogs. Concentration of aflatoxin B_1 in the contaminated samples was appreciably high, up to 8.5 ppm.

From the *tarai* areas of Nainital, Misra (34) reported aflatoxin contamination in 35.55% of 135 maize samples collected during 1976-77. The quantity of aflatoxin B_1 in the contaminated samples ranged from 8 to 1850 ppb. Few samples showed the presence of all four aflatoxins (B_1, B_2, G_1 and G_2). Other regional reports of aflatoxin contamination in stored maize are from Lucknow (53), Madras (4) and Coimbatore (5). High incidence of aflatoxins was also recorded in stored maize, which was collected from 1978 to 1980 in the tribal belts of Santhal Pargana, Bihar (18). Aflatoxin B_1 levels in some of the positive samples reached 3.14 ppm. Maize samples from Bhagalpur district also exhibited 35% aflatoxin incidence during the survey year 1978-79 (45). The amount of aflatoxin B_1 was in the range of 80 to 2288 ppb. A research group of Ujjain University surveyed stored maize grain for three years (1979 to 1981) and found 26 of 125 samples to be aflatoxin positive with a concentration of 20 to 750 ppb (32).

In addition to stored materials, aflatoxin is quite frequently associated with kernels of the standing maize crop in some parts of the country. In an extensive survey conducted for three consecutive years (1977 to 1980) 615 maize samples from different areas of Bihar were collected (12,13); of these, 170 samples (27.6%) had aflatoxin contamination. In most of the samples, aflatoxins B_1 or a combination of B_1 and B_2 (81 and 70, respectively) were present. Only three samples contained all four aflatoxins, B_1, B_2, G_1 and G_2. The amount of aflatoxin B_1 in the contaminated samples ranged from 8 to 1640 ppb. A majority of the samples contained aflatoxin B_1 at levels above 20 ppb.

Storage of maize ears within an Indian farm home

Factors Affecting Aflatoxin Contamination

High incidence of aflatoxins in maize may be attributed to the prevailing climate, as well as to nonscientific agricultural and storage methods practiced in India. It has been established that aflatoxin-producing fungi require a warm, humid climate for growth and metabolism. Most of the Indian subcontinent provides such conditions. Moreover, unseasonal rains and flash floods, which are very frequent, result in increased kernel moisture. Sometimes the standing maize crop is totally submerged in floodwater and harvesting is done under very humid conditions. The crop

Typical maize storage bins of mud over bamboo strips with paddy straw roofs

is heaped on roadsides while ear moisture is very high. When it is cloudy, drying can be prolonged, and the ears are ideal targets for fungal invasion. In some cases, many fungal colonies are distinctly visible after removal of husks. Due to the high frequency of toxigenic conidia of *A. flavus* in the air, the kernels are infected by the fungus before they attain a safe moisture level.

Conditions for storage in the countryside are unsatisfactory. Storage structures are usually made of mud, bamboo strips, *Cajanus cajan* reeds, palm leaves or paddy straw. Earthenware containers of different shapes and sizes are also used for storage. In parts of northern and western India, unshelled maize ears are stored on rooftops. These ears serve as seed stock for the next crop. Besides these traditional methods, food grain is also stored in metal pots, polyethylene bags and jute bags.

Effect of Aflatoxins on Animal Systems

Aflatoxin-induced abnormalities in animals have received considerable attention in India. Different laboratory animals, poultry and dairy cattle have been reported to be severely affected by aflatoxin poisoning.

One of the earliest reports on aflatoxicosis in dairy cattle was by Sastry *et al.* (42) from Andhar Pradesh, who recorded loss of appetite, diarrhea, dullness, ascites, emaciation and icterus in 24 Murrah buffalo. Histopathological studies revealed centrilobular hepatic cell necrosis, bile duct proliferation and central vein occlusion. An outbreak of aflatoxicosis in dairy cattle was also reported from Karnataka in which mortality of more than 58 crossbred cattle was recorded (19). Mohinddin and Ali (35) also recorded suspected cases of aflatoxin poisoning in female buffalo, leading to abortions in most of the animals.

The research group of Kerala Veterinary College and Research Institute recorded identical features of aflatoxicosis among ducklings of the government duck farm at Niranam, Kerala, where large-scale mortality was reported (8). Madhavan and Rao (28) also observed hepatic infarction and bile duct proliferation in ducklings due to aflatoxin poisoning. More than 2219 chicks died in an outbreak of aflatoxicosis in fowls in Karnataka (20). The symptoms were severe anorexia, loss of weight, staggering gait and convulsive movements.

Mehrotra and Khanna (33) made a detailed survey in the Kulu Valley, where more than 4000 rabbits died due to consumption of aflatoxin-contaminated pellets. They observed characteristic changes in the liver of affected animals, for example, proliferation of connective tissue and hyperplasia of the bile duct. Aflatoxicosis was also recorded in swine from a breeding farm at Mannuthy, Kerala (8). Large-scale mortality of dogs that fed on food remnants containing high levels of aflatoxin was reported in western India in 1974 (26,27).

The effect of aflatoxins on primates was demonstrated for the first time at the National Institute of Nutrition, Hyderabad (52). Monkeys receiving as little as 0.5 mg of aflatoxins for three to four weeks developed fatty livers. Research by Madhavan and Gopalan (30) revealed that low levels of dietary protein had an inhibitory effect on aflatoxin-induced carcinogenesis in rats. Hepatocellular carcinoma of the giant cell type was induced in a male monkey by feeding it aflatoxin for 5.5 years (21). Aggravation of the toxic effects of aflatoxins due to vitamin A deficiency was also established in rats (40).

In guinea pigs, renal changes, such as tubular epithelial reflux, were observed due to aflatoxin poisoning (29).

Aflatoxin B_1 was shown to inhibit the increment in liver weight and to suppress hepatocellular replication in rats (50). The quantity of aflatoxin M_1 excreted in the urine of albino rats, guinea pigs and monkeys in relation to administered radioactive aflatoxin was determined at the Central Food and Technological Research Institute, Mysore (43). Nearly 80% of the aflatoxin M_1 was excreted within six hours after toxin administration. The effect on the reproductive behavior of albino rats fed aflatoxin B_1 was also studied (36). Daradhiyar (18) recorded liver discoloration, change in texture, enlargement of the lobes and development of necrotic lesions on the liver surface in rats fed with aflatoxin-contaminated maize grain. The synergistic effects of aflatoxin B_1 and ochratoxin A were also observed in rats (39). Singh and Chauhan (44) studied the biochemical alterations in albino rats fed *Aspergillus*-infested maize diets. Changes in body weight, hemoglobin percentage, total tissue proteins, level and activity of Glutamic-Oxaloacetic Transaminase (GOT) and Glutamic Pyruvic Transaminase (GPT) were recorded in the liver, brain, kidney and heart tissues of the animals.

At Bhagalpur University, research has also identified several consequences of aflatoxin poisoning in laboratory animals, such as Swiss mice and guinea pigs. In one set of experimental animals, a small tumor was observed in a mouse after 30 weeks of aflatoxin feeding; this was a secondary carcinoma (9). Some of the central veins of the affected liver were found to be dilated and were surrounded by a large number of polymorphs and mononuclei. Remarkable microscopic changes in the hepatocytes of liver were also recorded. Mononuclear infiltrations were quite frequent in focal areas of interstitia of the affected kidney.

Skin of guinea pigs was also affected by aflatoxin poisoning, with loss of hair. The sebaceous and sweat glands also revealed atrophic changes. Aflatoxins caused significant changes in the hematological features of the experimental guinea pigs. Mutagenic effects of aflatoxins are also being investigated in *Drosophila* and Swiss mice. Aflatoxins have been found to act as mitotic inhibitors, with increases in the proportion of prophase cells.

Effect of Aflatoxins on Human Beings

There is no direct correlation of aflatoxin ingestion with a defined aflatoxicosis in humans, but circumstantial evidence suggests the involvement of the toxin in human disease and deaths. Indian childhood cirrhosis (ICC), a serious disorder of the liver in children, is confined mostly to the Indian subcontinent. Although the etiology of this disease is not clear, Robinson (41) suggested the role of aflatoxins in the development of the disease. According to Amla *et al.* (1,3), there is sufficient circumstantial evidence to show that children exposed to aflatoxins through mothers' milk and foods such as parboiled rice and unrefined peanut oil may acquire ICC. They also reported that malnourished children who had consumed aflatoxin-contaminated protein flour developed hepatic lesions similar to that of ICC (2). Yadgiri *et al.* (54) analyzed the presence of aflatoxin-like compounds in urine and liver extracts of children suffering with ICC. Subsequently, presence of aflatoxins was confirmed in 7% of the urine samples of 332 children examined (37). However, later phases of research by the scientists of RNT Medical College, Udaipur, strongly contradicted such claims (6,7). They established that ICC was not due to aflatoxin but to accumulation of excess copper in the liver.

The first report of aflatoxicosis in humans was made by the scientists from the National Institute of Nutrition,

Hyderabad (26). They undertook detailed investigations on the outbreak of acute hepatitis in the tribal belts of western India in 1974. The illness had a subacute onset and was characterized by high fever, rapidly progressing jaundice, ascites and portal hypertension. More than 990 individuals (2:1 male, ages 5 to 14 and over 30 years) were affected and about 100 died. In an autopsized human liver, bile duct proliferation and multinucleate giant cells were observed. Based on a combination of epidemiological, mycological, mycotoxic and histopathological studies, it was concluded that the disease was caused by the consumption of maize grains heavily contaminated with aflatoxins, i.e., up to 15.6 ppm (27). The same affected areas were investigated by the scientists of the All-India Institute of Medical Sciences, New Delhi. Initially they were doubtful about the role of aflatoxin as the sole factor responsible for the outbreak (48,49); subsequent studies, however, confirmed its role in the etiology of the disease (47).

Areas Needing Further Research

Despite frequent reports about the health hazards caused by aflatoxins, it has not been possible to make a proper assessment of the magnitude of this problem, especially in rural India, the tribal belts and the slums. Therefore, there is an urgent need for regular monitoring of human and livestock populations, especially in the areas of high temperature and high humidity where chances of aflatoxin contamination are high. Since aflatoxins can also be present in standing crops, there is need to assess crops before and after harvest for aflatoxin contamination. Factors affecting the natural formation of aflatoxins in field crops also require attention, because of the varying geographical and environmental conditions on the Indian subcontinent. To correctly evaluate the toxic and mutagenic effects of aflatoxins on humans, it is essential that test systems be suitably selected, and that the parameters and doses are similar to those the populations encounter under natural conditions. Aflatoxins have already been found to be mutagenic for prokaryotes and cells in culture; therefore, such effects on human populations cannot be ruled out, especially when persons have been exposed to aflatoxin contamination over long periods of time.

In view of the carcinogenic properties of aflatoxin, emphasis should be given to developing effective and economic control measures against aflatoxin elaboration. Cooperative research on this issue can be developed among scientists of different countries. Of the three basic strategies, i.e., prevention, inactivation and detoxification, prevention provides the best solution. Identification of the genotypes of resistant crop varieties, as well as the use of effective chemical compounds against *A. flavus* and aflatoxin development, should be beneficial. Several years ago, 15 common maize varieties were screened against aflatoxin formation under laboratory conditions, and Him-123, Ganga-2 and Ganga-5 exhibited resistance (15). Prevention of aflatoxin formation in some cereals, including maize, has also been achieved through the use of certain plant extracts and phenolic compounds (11,14,16). Mycotoxin workers at Bhagalpur University are prepared to cooperate in collaborative research programs to develop inactivation and detoxification processes to minimize the extent of aflatoxin hazards.

References

1. Amla, I., K. Shymala, V. Sreenivasamurthy, A.P. Jayaraj and H.A.B. Parpia. 1970. Role of aflatoxin in Indian Childhood Cirrhosis. Indian Pediatrician 7:262-270.

2. Amla, I., C.S. Kamala, G.S. Gopalakrishna, A.P. Jayraj, V. Sreenivasamurthy and H.A.B. Parpia. 1971. Cirrhosis in children from peanut meal contaminated by aflatoxin. American Journal of Clinical Nutrition 24:609-614.

3. Amla, I., V. Sreenivasamurthy, A.P. Jayaraj and H.A.B. Parpia. 1974. Aflatoxin in Indian Childhood Cirrhosis: Review. Journal of Tropical Pediatrics. 20:28-33.

4. Anonymous. 1973. Studies on the toxicities of metabolites of fungi other than *A. flavus* to obtain basic information needed to protect existing markets of wheat and rice. Final Technical Report, Project PL-480. Biochemical Laboratory, University of Madras, Madras, India.

5. Anonymous. 1980. Survey for the presence of mycotoxins with special reference to patulin, sterigmatocystin, aflatoxin and penicillic acid in foods, food products, animal feeds and concentrates available in the Tamil Nadu region of India. Final Technical Report, DST. Department of Food Technology, Tamil Nadu Agricultural University, Coimbatore, India.

6. Bhandari, B., and M.D. Sharda. 1984. Indian Childhood Cirrhosis: A search for cause and remedy. Indian Journal of Pediatrics 51:135-138.

7. Bhandari, B., and B. Sharda. 1984. Indian Childhood Cirrhosis (ICC): Recent observations. Indian Pediatrician 21:569-573.

8. Bhat, R.V., V. Nagarajan and P.G. Tulpule. 1978. Health Hazards of Mycotoxins in India. Indian Council of Medical Research, New Delhi, India. P. 58.

9. Bilgrami, K.S. 1985. Mycotoxins in dry fruits and spices. Final technical report, UGC Project. Bhagalpur University, Bhagalpur, India. P. 80.

10. Bilgrami, K.S., and K.K. Sinha. 1984. Mycotoxin contamination in food and its control. Indian Review of Life Sciences 4:19-36.

11. Bilgrami, K.S., R.S. Misra, K.K. Sinha and P. Singh. 1979. Prevention of aflatoxin production in liquid and solid substrates through some plant extracts. Indian Journal of Experimental Biology 17:1151.

12. Bilgrami, K.S., T. Prasad, R.S. Misra and K.K. Sinha. 1980. Survey and study of mycotoxin producing fungi associated with grains in the standing maize crop. Final technical report, CAR Project. Bhagalpur University, Bhagalpur, India. P. 98.

13. Bilgrami, K.S., T. Prasad, R.S. Misra and K.K. Sinha. 1981. Aflatoxin contamination in maize under field conditions. Indian Phytopathology 34:67-68.

14. Bilgrami, K.S., K.K. Sinha and P. Singh. 1981. Inhibition of aflatoxin production by ferulic acid on some cereals and oil-seeds. Current Science 50:997-998.

15. Bilgrami, K.S., R.S. Misra, T. Prasad and K.K. Sinha. 1982. Screening of different varieties of maize for aflatoxin production by *Aspergillus parasiticus*. Indian Phytopathology 35:376-378.

16. Bilgrami, K.S., K.K. Sinha and P. Singh. 1982. Prevention of aflatoxin production on some cereals and oil-seeds by O-vanillin. Current Science 51:138.

17. Bilgrami, K.S., T. Prasad and K.K. Sinha, eds. 1983. Mycotoxins in Food and Feed. Allied Press, Bhagalpur, India. P. 315.

18. Daradhiyar, S.K. 1980. Studies on aflatoxins associated with storage and standing crop of maize, poultry feed and cattle cakes in the tribal belt of Santhal Pargana District. PhD Thesis. Bhagalpur University, Bhagalpur, India.

19. Gopal, T., S. Zaki, M. Narayanaswamy and S. Premlata. 1968. Aflatoxicosis in dairy cattle. Indian Veterinary Journal 45:707-712.

20. Gopal, T., S. Zaki, M. Narayanaswamy and S. Premlata. 1969. Aflatoxicosis in fowls. Indian Veterinary Journal 46:348-349.

21. Gopalan, C., P.G. Tulpule and D. Krishnamurthi. 1972. Induction of hepatic carcinoma with aflatoxin in the rhesus monkey. Food and Cosmetics Toxicology 10:519-521.

22. Hesseltine, C.W. 1976. Mycotoxin research in India. Mycopathologia 58:157-161.

23. Indulkar, A.S., S. Vedanayagam and S.R. Rao. 1971. Seed mycoflora and aflatoxins in cotton-seed: A preliminary survey. Abstract. *In* the Second International Symposium on Plant Pathology, IARI, New Delhi, India. Pp. 31-32.

24. Kandhari, J., M.M. Payak, S.L. Mehta and R.C. Sharma. 1980. Investigations on microbial toxins in stored maize samples from Rajasthan. *In* the Symposium on Environment Pollution and Toxicology. Pp. 319-326.

25. Kang, M.S. 1970. Pathogenesis in groundnut (*Arachis hypogaea* L.) by *Aspergillus flavus* Link ex Fries. PhD Thesis. Punjab Agricultural University, Ludhiana, India.

26. Krishnamachari, K.A.V.R., R.V. Bhat, V. Nagarajan and T.B.G. Tilak. 1975. Hepatitis due to aflatoxicoses: An outbreak in western India. Lancet i:1061-1063.

27. Krishnamachari, K.A.V.R., R.V. Bhat, V. Nagarajan, T.B.G. Tilak and P.G. Tulpule. 1977. The problem of aflatoxic human diseases in parts of India: Epidemiological and ecological aspects. Annales de la Nutrition et de l'Alimentation 31:991-996.

28. Madhavan, T.V., and K.S. Rao. 1966. Hepatic infarction in ducklings in afatoxin poisoning. Archives of Pathology and Laboratory Medicine 81:520-524.

29. Madhavan, T.V., and K.S. Rao. 1967. Tubular epithelial reflux in the kidney in aflatoxin poisoning. Journal of Pathology and Bacteriology 93:329-331.

30. Madhavan, T.V., and C. Gopalan. 1968. Effect of dietary protein on carcinogenesis of aflatoxin. Archives of Pathology and Laboratory Medicine 85:133-137.

31. Maggon, K.K., L. Vishwanathan, T.A. Venkitasubramanian and K.G. Mukerji. 1969. Aflatoxin production by some Indian strains of *Aspergillus flavus* Link ex Fries. Journal of General Microbiology 59:119-124.

32. Mall, O.P., K.K. Ranawat and S.K. Chauhan. 1983. Mould flora and aflatoxin contamination in maize kernels. *In* Mycotoxins in Food and Feed, K.S. Bilgrami, T. Prasad and K.K. Sinha, eds. Allied Press, Bhagalpur, India. Pp. 37-44.

33. Mehrotra, M.L., and R.S. Khanna. 1973. Aflatoxicosis in Angora rabbits. Indian Veterinary Journal 37:31-34.

34. Misra, R.S. 1977. Aflatoxin contamination of some agricultural commodities in Tarai and biochemical effects of aflatoxin B_1 on maize seeds. PhD Thesis. G.B. Pant University of Agriculture and Technology, Pantnagar, India.

35. Mohiuddin, S.M., and Z.M. Ali. 1973. Suspected abortion due to aflatoxin in she-buffalo. Indian Veterinary Journal 50:623-625.

36. Panda, P.C., V. Sreenivasamurthy and H.A.B. Parpia. 1970. Effect of aflatoxin on reproduction in rats. Journal of Food Science and Technology (Mysore) 7:20-22.

37. Parpia, H.A.B., V. Sreenivasamurthy, T. Shantha, A. Shantha, A. Shankaramurthi, S. Srikantia and I. Amla. 1972. Urinary aflatoxin B and effects of preservation on its detection. American Journal of Clinical Nutrition 25:13-15.

38. Rao, K.S., T.V. Madhavan and P.G. Tulpule. 1965. Incidence of toxigenic strains of *Aspergillus flavus* affecting groundnut crop in certain coastal districts of India. Indian Journal of Medical Research 53:1196-1202.

39. Rati, E.R., S.C. Basappa, V. Sreenivasamurthy, H.P. Ramesh, B.S. Ramesh and G.B. Singh. 1981. The synergistic effect of aflatoxin B_1 and ochratoxin-A in rats. Journal of Food Science and Technology 18:176-179.

40. Reddy, G.S.R., T.B.G. Tilak and D. Krishnamurthi. 1973. Susceptibility of vitamin deficient rats to aflatoxin. Food and Cosmetics Toxicology 11:467-470.

41. Robinson, P. 1967. Infantile cirrhosis of the liver in India with special reference to probable aflatoxin etiology. Clinical Pediatrics 6:57-62.

42. Sastry, G.A., J.V. Narayana, P.R. Rao, K.J. Christopher and K.R. Hill. 1965. A report on the groundnut toxicity in Murrah buffaloes in Andhra Pradesh, India. Indian Veterinary Journal 42:79-82.

43. Shantha, T., V. Sreenivasamurthy and H.A.B. Parpia. 1970. Urinary metabolites of ^{14}C aflatoxin in some laboratory animals. Journal of Food Science and Technology 7:135-138.

44. Singh, A., and R.K.S. Chauhan. 1983. Studies on biochemical alterations in albino rats due to feeding of *Aspergillus* infested maize seeds. *In* Mycotoxins in Food and Feed, K.S. Bilgrami, T. Prasad and K.K. Sinha, eds. Allied Press, Bhagalpur, India. Pp. 109-116.

45. Sinha, K.K. 1980. Survey and study of aflatoxin producing isolates of *Aspergillus flavus* associated with maize grains in storage and standing crop. PhD Thesis. Bhagalpur University, Bhagalpur, India.

46. Sreenivasamurthy, V., A. Jayaraman and H.A.B. Parpia. 1965. Aflatoxin in Indian peanuts: Analysis and extraction. *In* Mycotoxins in Feedstuffs, G.N. Wogan, ed. Massachussetts Institute of Technology Press, Cambridge, Massachussetts, USA. Pp. 251-260.

47. Tandon, H.D., and B.N. Tandon. 1983. Clinical and pathological study of an human out-break of aflatoxicosis. *In* Proceedings of the International Symposium on Mycotoxins, Cairo, Egypt. March 19-24, 1983. Pp. 271-276.

48. Tandon, B.N., L. Krishnamurthy, A. Koshy, H.D. Tandon, V. Ramalingaswami, J.R. Bhandari, M.M. Mathur and P.D. Mathur. 1977. Study of an epidemic of jaundice, presumably due to toxic hepatitis, in northwest India. Gastroenterology 72:488-494.

49. Tandon, H.D., B.N. Tandon and V. Ramalingaswami. 1978. Epidemic of toxic hepatitis in India of possible mycotoxic origin. Archives of Pathology and Laboratory Medicine 102:372-376.

50. Tilak, T.B.G., and D. Krishnamurthy. 1973. Liver regeneration following partial hepatectomy. Archives of Pathology 96:18-20.

51. Tulpule, P.G., and V.K. Mehan. 1983. Research on mycotoxin problems in India. Paper presented at the International Conference on Mycotoxins, Cairo, Egypt, March 19-24, 1983.

52. Tulpule, P.G., T.V. Madhavan and C. Gopalan. 1964. Effect of feeding aflatoxin to young monkeys. Lancet i: 962-963.

53. Vora, V.C. 1978. A survey of toxin producing fungi and mycotoxins associated with post-harvest deterioration of field crops grown for human or animal consumption. Final technical report, PL-480 Project. CDRI, Lucknow, India.

54. Yadgiri, B., V. Reddy, P.G. Tulpule, S.G. Srikantia and C. Gopalan. 1970. Aflatoxin and Indian Childhood Cirrhosis. American Journal of Clinical Nutrition 23:94-98.

Aflatoxin in India: II

M.M. Payak, R.C. Sharma and S.R. Rao, Division of Mycology and Plant Pathology, Indian Agricultural Research Institute, New Delhi, India

Maize in India is grown in 17 states and six union territories under diverse conditions, from temperate to subtropical to exclusively tropical, in latitudes from 12° to 34°N and between elevations of 49 and 2600 meters. During 1983-84, maize was grown on 5,888,300 hectares, with a grain production of 7,923,600 tons (11). The major maize-producing states, in order of production, are Uttar Pradesh, Bihar, Rajasthan, Madhya Pradesh, Punjab, Andhra Pradesh, Gujarat, Himachal Pradesh, Jammu and Kashmir, Orissa, Karnataka, Maharashtra and Haryana. More than 95% of Indian maize production takes place in these states.

In northern India, there are two well-defined growing seasons, the summer/monsoon (*kharif*) and the winter (*rabi*). In peninsular India, the seasons are not so well defined, and maize can be grown practically throughout the year. In the north, maize is mainly a *kharif* season crop. Maize cultivation has started in the *rabi* season in the state of Bihar.

Maize is a staple food of the large population of the Himalayan region, which includes Arunachal Pradesh, Assam, Manipur, Meghalaya, Mizoram, Nagaland, the Sikkim and the sub-Himalayan regions, West Bengal in the northeast and Himachal Pradesh, Jammu and Kashmir in the northwest (11). Farmers, mostly of the Bhil/Meena tribe in the districts of Udaipur, Chittorgarh, Banswara and Durgapur in Rajasthan, Jhabua and Dhar in Madhya Pradesh and Panchmahals in Gujarat, depend almost exclusively on maize for food. This is true to the extent that the tribal members suffer from the nutritional disorder, pellagra (11). In the districts of Banswara and Panchmahals, aflatoxicosis among humans was first reported in 1975. In the two major maize-growing states of Uttar Pradesh and Bihar, maize is a major food cereal of the rural poor.

Almost 90% of the maize grain produced in India is for home consumption. Indeed, not more than 5% of the total maize crop reaches the market. It is in direct use of the grain that the problem of aflatoxin contamination becomes of greatest concern.

The presence of as many as 69 fungal species has been found on or in maize seed (Table 1). Some cause preharvest diseases in the crop (seed or kernel rot, seedling blight, stalk rot), whereas others develop after harvest and cause storage rots and/or seed deterioration. *Aspergillus flavus* Link ex Fries, the fungus mainly responsible for aflatoxin contamination, occurs both in pre- and postharvest stages. Another mold, *A. parasiticus* Speare, also known to produce aflatoxins, has so far not been recorded on maize in India.

The problem of aflatoxicosis in humans in India was first reported by Krishnamachari *et al.* (5); about 100 Bhil tribal members residing in the adjacent districts of Banswara in Rajasthan and Panchmahals in Gujarat died of hepatitis, which affected humans and dogs and was characterised by jaundice, rapidly developing ascites and portal hypertension. It was recorded that the people in question had consumed maize heavily contaminated with *A. flavus*. Approximately 400 persons were affected by the epidemic.

In its Delhi edition of March 25, 1975, the national daily *Indian Express* featured a major news item about this event. The Director General of Health Sciences, in a circular of May 15, 1975 addressed to all food and health authorities of states and union territories, issued a warning regarding outbreaks of hepatitis and measures to prevent its recurrence. A copy of that letter follows this paper.

The epidemic of hepatitis started in September, 1974, reached its peak in December, abruptly declined and finally ceased altogether in January, 1975. The late unseasonal rains during and after harvest undoubtedly contributed to a higher incidence of maize grain mold.

In response to the attention attracted by the epidemic, the All-India Institute

Table 1. Fungi associated with maize seed in India[a]

Acremonium kiliense	*Fusarium acuminatum*
A. strictum	*F. equiseti*
Actinomucor repens	*F. graminearum*
Alternaria tenuis	*F. moniliforme*
(=A. alternata)	*F. moniliforme* var. *subglutinans*
A. tenuissima	*F. oxysporum*
Aspergillus amstelodamii	*F. semitectum*
A. candidus	*Macrophomina phaseolina*
A. carbonarius	*Macrophomina* spp.
A. chevalieri	*Melanospora* spp.
A. flavus	*Memnoniella echinata*
A. niger	*Mucor* spp.
A. ochraceous	*Nigrospora oryzae*
A. ruber	*N. sphaerica*
A. sejunctus	*Penicillium citrinum*
A. sydowii	*P. chrysogenum*
A. tamarii	*P. frequentans*
A. terreus	*P. funiculosum*
A. terricola	*P. gunicolosum*
A. ustus	*P. implicatum*
A. versicolor	*P. islandicum*
Botryodiplodia theobromae	*P. oxalicum*
Chaetomium globosum	*P. purpurogenum*
C. indicum	*P. rugulosum*
Chrysosporium pruinosum	*P. viridicatum*
Cladosporum herbarum	*Penicillium* spp.
Cochliobolus spicifer	*Phoma glomerata*
Curvularia clavata	*Phoma sorghina*
C. lunata	*Pythium aphanidermatum*
Cystophaera mangifera	*Rhizopus nigricans*
Drechslera australiensis	*Rhizoctonia zeae*
D. carbonum	*Rhizoctonia* spp.
D. maydis	*Sclerotium rolfsii*
Epicoccum purpurescens	*Stachybotrys atra*
Exserohilum halodes	*Stenocarpella (=Diplodia) macrospora*
E. rostratum	*S. maydis*
E. turcicum	*Syncephalastrum racemosum*

[a] Fungi are listed in alphabetical order

of Medical Sciences formed a team of epidemiologists led by B.N. Tandon, a specialist in gastroenterology. The team collected samples of foods and grains from affected households (those that reported human mortalitiy) and nonaffected households. Most samples were of maize, while the remainder were wheat grain, wheat flour and sorghum grain. These samples were analyzed by Kandhari *et al.* (4). *Aspergillus flavus* was found in 90.3% of the samples and aflatoxins ranging from trace (less than one ppm) to 8.5 ppm were detected in 15 of 22 samples analyzed.

The incidence of aflatoxin-producing fungi in maize has been recorded to date from Gujarat, Tamil Nadu, Rajasthan, Uttar Pradesh, Bihar and Delhi. Undoubtedly, it also exists in other areas, but authentic reports are lacking. Bilgrami and associates (1,16) have also studied the incidence of *A. flavus* in preharvest maize. In the *rabi* season, 17% of the samples showed *A. flavus* contamination, whereas in the *kharif* season the incidence was as high as 72%. In stored maize samples the incidence was 35%.

Siradhana *et al.* (17) collected 39 maize samples from the Banswara district in Rajasthan and listed the fungi detected in them (including *A. flavus*), but data on frequency or proportion were not given. Om Prakash and Sirdhana (13, 14) studied the factors affecting aflatoxin B_1 formation in grains of hybrid Ganga-5. They also reported that four *Aspergillus* spp. can cause kernel rot in hybrid Ganga-5 (12). *Aspergillus flavus* induced a greenish type of kernel rot, and *A. tamarii* caused a dark brown rot; kernels infected by *A. niger* were black, whereas *A. terreus*-infected grains were stunted.

As reported by Kandhari *et al.* (4), the Bhil tribesmen follow a postharvest practice which in the local language is known as *ogawa*. Almost all of the maize plant, including the ear, is chopped. The chopped pieces are piled in bundles and left outside the houses. The ears are not husked or shelled, but are brought inside and shelled as they are needed. The grain is ground by dry milling and used to make thick *chapatis* (round, flat pancakes). This practice, even in normal seasons, favors the growth of grain molds. Proper education could in this instance eliminate one major source of mycotoxin contamination.

In spite of strong circumstantial evidence that aflatoxicosis was responsible for the deaths of the Adivasis in a three-month period of 1974-75, categorical cause and effect associations were not identified. There have been few follow-up studies. The team from the All-India Institute noted that there were some deaths due to cardiac arrest. In direct analysis, 16 of 40 samples showed the presence of hydrocyanic acid (HCN) in concentrations of 0.5 to 1.0 ppm (4). Stoloff studied the survivors of the hepatitis epidemic in Rajasthan and has not yet observed any long-term adverse effects (8).

Kandhari *et al.* (4) had tested the ability of propionic acid to prevent mold in maize. In two trials, spray application of cv. Basi on husked ears reduced *A. flavus* infestation to 6.7% and 0%, as compared to 30.7% and 76% in controls. Dhanraj *et al.* (3) had determined the effectiveness of Luprosil (a BASF product containing 99% propionic acid) in stored maize. The use of this acid in concentrations up to 340 ppm in grain destined for human consumption is internationally accepted. However, more work is needed on feasibility and cost effectiveness before recommending it for general use. Lal and Kapoor (7) have determined that, apart from propionic acid, boric acid and sorbic acid also can effectively reduce incidence of such storage fungi as *A. amstelodamii, A. chevalieri, A. lavus, A. niger, A. ruber* and *A. sydowii*.

Nagarajan and Bhat (10) tested maize cultivars (including an opaque composite, Shakti) with a rice strain of *A. flavus* and found that aflatoxin produced in the opaque composite sample was the lowest when compared to the hybrids Deccan and Ganga Safed-2. This study has shown that genetic differences do occur and must be investigated more vigorously.

Certain physical measures, such as reducing seed moisture to safer levels, are feasible. Singh (15) has devised a solar dryer in which the grain is exposed to a temperature of 100°C for two minutes; it has a capacity of 30 to 35 kg of grain per hour. This short exposure does not change the nutritional status of the treated grain. Similarly, storage bins of iron and stone have been made. Their cost, for a capacity of 260 kg, is 50 to 200 rupees (US$ 5-18). The work was carried out under the sponsorship of IDRC/ICAR as an Operation Research Project. The adoption of such small-scale technology should be encouraged among the Bhil tribal people.

The areas of research that need to be pursued vigorously can be divided into two categories, preharvest and postharvest. In the former, a study of genetic differences in maize germplasm for low or zero levels of fungal or ear infections, resistance to insect attack, tight and extended husks and drooping shanks should be rewarding. Tests on the application of fungitoxicant chemicals within the first 30 days of grain formation also need to be done.

In tropical countries, solar drying of ears and the use of low-cost solar dryers for shelled grain can prevent insect infection as well as mold after harvest. The effectiveness of storage in bins made from local materials also needs study. Chemicals that prevent aflatoxin contamination may not provide an economical approach, but research in this area is required. One example is the work of Codifer *et al.* (2), who found that HCHO (formaldehye) and $Ca(OH)_2$ (calcium hydroxide) applied to peanut meal reduced aflatoxin levels. Other approaches for postharvest detoxification also need to be followed.

Since 1975, a number of laboratories have become active and facilities for aflatoxin analysis built. A Mycotoxin Research Group Newsletter has been published by R.V. Bhat of the National Institute of Nutrition, Hyderabad, since 1979. In the same year, a monograph on health hazards of mycotoxins in India was published by the same institute. In March, 1980, the newsletter announced that a food contamination monitoring project, sponsored by FAO, had been initiated in India, Nepal, Pakistan and Sri Lanka. The objective of the project was the collection of data on the occurrence of aflatoxin, heavy metals and pesticide residue in staple food items. The Indian institutions collaborating in the project are the Central Food Laboratory, Calcutta, the Central Food Technological Research Institute, Hapur, the Industrial Toxicology Research Centre, Lucknow, the National Institute of Nutrition, Hyderabad, the Public Health Laboratory, Pune, and the Vallabhai Patel Chest Institute, New Delhi. The coordinating agency is the Directorate General of Health Services in New Delhi.

References

1. Bilgrami, K.S., T. Prasad, R.S. Misra and K.K. Sinha. 1981. Aflatoxin contamination in maize under field conditions. Indian Phytopathology 34:67-68.

2. Codifer, C.P., Jr., G.E. Mann and F.G. Dollear. 1976. Aflatoxin inactivation: Treatment of peanut meal with formaldehyde and calcium hydroxide. Journal of the American Oil Chemists' Society 53:204-206.

3. Dhanraj, K.S., J.S. Chohan, M.S. Sunar and S. Lal. 1973. Preliminary studies in the use of Luprosil in the prevention of microbial spoilage of wet maize during storage. Indian Phytopathology 26:63-67.

4. Kandhari, J., M.M. Payak, S.L. Mehta and R.C. Sharma. 1980. Investigations on microbial toxins in stored maize samples from Rajasthan. *In* Proceedings of the Symposium on Environmental Pollution and Toxicology, S.P. Raychaudhuri, ed. Today and Tomorrow Press, New Delhi, India. Pp. 319-328.

5. Krishnamachari, K.A.V.R., R.V. Bhat, V. Nagarajan and T.B.G. Tilak. 1975. Hepatitis due to aflatoxicosis (an outbreak in western India). Lancet, May 10, 1975, issue. P. 1061.

6. Krishnamachari, K.A.V.R., R.V. Bhat, V. Nagarajan and T.B.G. Tilak. 1975. Investigations into an outbreak of hepatitis in parts of Western India. Indian Journal of Medical Research 63:1036-1048.

7. Lal, S.P., and J.N. Kapoor. 1980. Efficacy of certain food preservatives in the control of species of *Aspergillus* in stored maize. Indian Phytopathology 33:336-338.

8. Mannon, J., and E. Johnson. 1985. Fungi down on the farm. New Scientist 1445:12-17.

9. Misra, R.S. 1977. Aflatoxin contamination of some agricultural commodities in *Terai* and biochemical effects of aflatoxin B_1 on maize seeds. PhD Thesis. G.B. Pant University of Agriculture and Technology, Pantnagar, India.

10. Nagarajan, V., and R.V. Bhat. 1972. Factors responsible for varietal differences in aflatoxin production in maize. Journal of Agricultural and Food Chemistry 20:911-914.

11. Payak, M.M., and R.C. Sharma. 1985. Maize diseases and approaches to their management in India. Tropical Pest Management 31:302-310.

12. Prakash, O., and B.S. Siradhana. 1978. Ear rot of maize caused by four *Aspergillus* species. Indian Phytopathology 31:115-116.

13. Prakash O., and B.S. Siradhana. 1978. Factors affecting formation of aflatoxin B_1 by *Aspergillus flavus* on Ganga-5 maize hybrid. Current Science 47:695-696.

14. Prakash O., and B.S. Siradhana. 1978. Aflatoxin B_1 formation by *Aspergillus flavus* on the grains of some common cultivars. Current Science 47:783.

15. Singh, J. 1982. The post-harvest problems in maize production. *In* Processing, Utilization and Marketing of Maize, M.S. Swaminathan, E.W. Sprague and J. Singh, eds. ICAR/CIMMYT, New Delhi, India. Pp. 221-227.

16. Sinha, K.K. 1983. Aflatoxin problem in storage and standing maize crops. *In* Proceedings of the Symposium on Mycotoxins in Food and Feed, K.S. Bilgrami, ed. Allied Press, Bhagalpur, India. Pp. 23-36.

17. Siradhana, B.S., S.R.S. Dange, R.S. Rathore and S.D. Singh. 1978. Seed mycoflora of maize with reference to mycotoxins. Current Science 47:59-60.

Copy of letter 17-13/75-PH(F&N) dated May 15, 1975, from the Director General of Health Services, New Delhi, to all food and health authorities of states and union territories

Subject: Outbreak of hepatitis in Rajasthan—regarding

A news report appeared in the Indian Express of 16th March, 1975 that more than a hundred Adivasis have died after eating rotten maize in scarcity-ridden districts of Banswara in Rajasthan and Panchmahals in Gujarat.

The clinical features of the condition are low-grade fever and general uneasiness as the earliest manifestations, jaundice ascites and oedema in feet follows in quick succession within a few days, in some cases death was sudden and the precise cause of death could not be established. . . .

According to preliminary survey reports conducted by the National Institute of Nutrition and others it has been suggested that the condition is likely to be due to consumption of maize heavily infected with *A. flavus* leading to aflatoxicity.

It has been suggested that a combination of several factors may contribute to the outreak:

1. Unseasonal and heavy rains affect the maize crop;
2. Storage of maize under conditions which promoted contamination with *Aspergillus flavus*; and
3. Selective consumption of the contaminated maize by members of the affected households.

To prevent a recurrence, it is recommended that people growing maize shold be adequately educated about the proper methods of storing it and advised not to consume visibly contaminated maize. The consumers may also be informed to clean the grain by hand-picking before the consumption as an extra measure of safety as it is not possible to hand-pick contaminated maize on commercial basis. These grains, if possible, may be further diluted with equal quantity of uncontaminated grains so as to lower the level of contamination. The Public Health Authorities should also be advised to be on their guard and bring to the notice of the State Government concerned, any intoxication symptoms as a result of consumption of food grains.

Aflatoxin in the Philippines

R.P. García, National Crop Protection Center, and L.L. Ilag, Department of Plant Pathology, both of the University of the Philippines at Los Baños, Laguna, Philippines

In the Philippines, aflatoxin contamination is a particular problem in maize, peanuts, copra and animal feeds. The risk potential from aflatoxin in tropical countries is especially alarming because the prevailing high temperatures and humidity favor fungal growth. Furthermore, in the Philippines heavy rain normally coincides with the peak harvest months of staple cereals, particularly maize; this results in high grain moisture and handling problems, and leads to to accelerated fungal growth and subsequent mycotoxin formation.

Of the mycotoxins, aflatoxins have received the most attention in the Philippines. Aflatoxin is an extremely potent carcinogen, affecting several animal species, and is a common contaminant of many agricultural commodites.

Studies on aflatoxin in the Philippines began in 1967, when the Food and Nutrition Research Institute (FNRI) in Manila conducted a survey of the aflatoxin content of various food items. Unpublished data taken from 1967 to 1982 show that maize and peanuts, under natural conditions, are the two commodities that contain highly toxic levels of aflatoxin (Table 1).

Another survey (Tables 2, 3 and 4) showed that maize and several agricultural commodities and their by-products were highly contaminated with the toxin (13). Again, maize and peanut samples commonly contained toxin. Although some samples showed no visible signs of the fungus *Aspergillus flavus*, they contained high levels of aflatoxin. Conversely, other samples yielded the fungus in agar plates, but yielded little or no toxin. The survey further reported that maize grown in Visayas and Mindanao had more of the toxin than maize grown in northern Luzon. *Palay* (unpolished rice), soybean and millet were noted to be poor substrates for aflatoxin production. The 1985 annual report of the National Crop Protection Center (NCPC) (11) showed that species of

Table 1. Incidence and range of aflatoxin in maize and peanuts, Food and Nutrition Research Institute, Manila, Philippines, 1967-1982

Sample	Incidence (%)	Aflatoxin range (ppb)
Maize:		
Raw, whole and dried	88(360/407)[a]	<3 - 1491
Raw, ground (grits)	95(305/322)	<3 - 510
Other products	59(166/279)	<3 - 735
Peanuts:		
Raw, fresh and dried	64(119/187)	<3 - 3321
Peanut butter	94(640/682)	<3 - 8600
Other products	64(477/742)	<3 - 2153

[a] Number of aflatoxin-positive samples/total samples analyzed

Table 2. ***Aspergillus flavus*** **infection and aflatoxin content of maize and its by-products, Visayas and Mindanao, Philippines**

Location[a/]	Sample	Number of units plated[b/]	Units yielding *A. flavus* (%)	Aflatoxin B_1 (ppb)
B-2	Whole maize, shelled, in drying yard	20	55	107
C-12	Whole maize on cobs, drying on cement, rural areas	20	70	1074
C-3	Maize gluten meal	25	20	333
C-11	Whole maize on cob, standing in field	21	70	929
C-14	Whole maize for animal feed, Cebu market	21	90	817
C-16	Trash from maize prior to milling (seedcoats, cob fragments, broken kernels, dust)	25	36	268
D-1	Whole maize for human consumption, shelled, ready for milling	20	30	503
D-2	Maize grits, taken from mill at bagging spout	20	25	86
D-3	Maize germ, for oil milling	21	47	233
D-4	Maize tiki-tiki for animal feed (dust and fine debris)	20	90	343
D-11	Whole maize on cob, wet, in sacks	20	35	948
I-6	Maize germ and bran from milling	18	100	367

[a/] B = Bacolod, C = Cebu, D = Davao City, I = Iloilo
[b/] For pulverized materials, clumps of 0.25-1.0 cc were used
Source: Santamaria *et al.* (13)

Aspergillus were the most frequently isolated organisms in various varieties of rice and maize, followed by *Penicillium* spp. Of 102 *Aspergillus* isolates analyzed for aflatoxin, 17 isolates were positive for aflatoxin B_1; only six isolates produced both aflatoxins B_1 and G_1.

The study at FNRI (Table 5) established the possible relationship between the consumption of aflatoxin-contaminated maize in the Philippines and the development of primary liver cancer (10). The incidence of liver cancer in the country has been found to be higher in maize-consuming

Table 3. *Aspergillus flavus* infection and aflatoxin content of peanut samples, Manila, northern Luzon, Visayas and Cotabato, Philippines

Sample[a]	Units yielding *A. flavus* (%)	Aflatoxin B_1 (ppb)
Freshly dug peanuts	5	14
Farmer stock peanuts for commerce, in bulk storage, dry	1	257
Shelled peanuts from farmer stock	45	964
Whole peanuts in shell	25	0
Large segregated peanuts separated from smaller discards, used for peanut products	90	114

a/ Samples consisted of 20 peanuts each, plated
Source: Santamaria *et al.* (13)

Table 4. *Aspergillus flavus* infection and aflatoxin content of various agricultural commodities, Philippines

Sample	Number of units plated[a]	Units yielding *A. flavus* (%)	Aflatoxin B_1 (ppb)
Crude copra, moldy, in pile	20	100	314
Sorghum, in pile	44	25	208
Sorghum, in storage, dry	30	0	0
Rice, unmilled, moldy, from storage	30	0	0
Rice, 2 years old, in storage	40	2	5
Soybean, drying on mat	25	16	21
Gabi, moldy, from public market	20	15	0
Coffee beans, in sacks	27	41	150

a/ For pulverized materials, clumps of 0.25 - 1.0 cc were used
Source: Santamaria *et al.* (13)

regions than in rice-consuming regions. The effect of aflatoxin-contaminated maize when ingested was synergistically aggravated by alcohol consumption.

Physiological studies conducted at UPLB showed that isolates of *A. flavus* obtained from various substrates differed in their temperature requirements (4). The most favorable temperature for aflatoxin formation by isolates inoculated in their respective substrates was 30°C for rice and maize; 25°C for peanut and copra isolates; and 20°C for soybean isolates.

In copra, the highest aflatoxin content (247 ppb B_1) was recorded in samples with 12.3% moisture (15). Copra with 7.8% and 6.8% moisture content had much less aflatoxin (15 ppb). No aflatoxin was obtained in samples incubated at 20°C, even though 100% mold infection was attained.

Aflatoxin-forming fungi (*A. flavus* and *A. parasiticus*) were found to infect maize in the field, especially when the ears were damaged prior to inoculation (7). Postharvest infection was higher in maize samples that were dried and/or stored as shelled grain and lower in maize grain on the husked cob; it was minimal in maize grain with intact husks. The degree of infection was directly related to the amount of *A. flavus* inoculum in the air, which was least in maize fields, higher in drying areas and highest in storage areas.

Although aflatoxin contamination may be an essentially postharvest problem, it may also occur in the field, particularly in insect-infested grains (13). Insects cause damage, which permits mold to penetrate the grain. Moreover, the insects serve as vehicles for the dissemination of molds to other commodities or to other parts of the storage areas. Also, the metabolic activities of insects produce enough heat and moisture (especially during long-term storage) to allow *Aspergillus* growth. Unfortunately, conditions in the Philippines are ideal for insects, thus making efforts to control mold growth more difficult.

Studies on rice indicate that under current handling practices, *palay* contains low levels of aflatoxin or none at all. However, milled or brown rice, when inoculated with toxigenic strains of *A. flavus* and *A. parasiticus*, support very high levels of aflatoxin. It was found that in brown rice aflatoxin B_1 was concentrated in the bran layers; the polished rice contained no toxin or only traces of aflatoxin (8).

Table 5. Primary liver cancer index (PLCI) and maize consumption by region, Philippines

Region	PLCI	Daily per capita maize consumption (g)
Eastern Visayas	8.5	147.2
Southwestern Mindanao	7.1	102.0
Northeastern Mindanao	6.4	84.6
Western Visayas	5.6	77.7
Southern Tagalog	4.1	2.1
Bicol	3.7	33.5
Ilocos	3.3	5.5
Cagayan Valley	2.9	71.8
Metropolitan Manila	6.6	1.6

Source: Jayme *et al.* (10)

Preventive measures are the ideal solution for controlling fungal invasion and subsequent aflatoxin formation during commodity handling and storage. Fungal growth can be prevented by drying the commodity immediately after harvest to a moisture level that will not allow molds to grow. After drying, it is important to maintain the low moisture level; otherwise, a flush of mold will appear as soon as the moisture content becomes favorable for growth.

Various means have been tested to prevent aflatoxin contamination in different commodities or to destroy the toxin when preventive measures fail. Maize grain treated with 0.1% thiabendazole at the time of inoculation with *A. flavus*, and grain treated one or two days prior to inoculation, was free of infection and aflatoxin contamination two weeks post-incubation. Appreciable levels of aflatoxin were produced when the fungicide was applied one or two days after inoculation (9). Sevin and Thiodan afforded some protection to maize in the field, but treated maize grain stored up to two months appeared to be a better substrate for fungal growth than untreated grain (3). The NCPC annual report (11) showed that all *Aspergillus* and *Penicillium* isolates from rice and maize grain exhibited no apparent growth in agar infused with 10 ppm Benlate (6). The growth of both *Aspergillus* and *Penicillium* isolates was progressively inhibited by increasing concentrations of Benlate. Captan also inhibited *Aspergillus* and *Penicillium* spp. growth but required a much higher concentration.

To protect fresh coconut meat to be used for copra production from fungal invasion, 47 chemical preservatives and fungicides were tested. Of these, 13 showed some promise against various fungi affecting copra under laboratory conditions. When fresh treated coconut meat was allowed to air dry in a copra warehouse, the chemicals AgriCaptafol, Diflatan and Dyrene allowed only mild infection, whereas the other chemicals tested allowed heavy mold infection (5).

A 10% calcium hydroxide coating on coconut meat gave protection against insects and rodents, and with proper ventilation it was found to be quite effective against molds (15). Treatment with a mixture of 5% sulfuric acid and 7% acetic acid on fresh coconut meat resulted in clean copra when dried within four to five days (14).

Chemical detoxification procedures with sodium hydroxide, ammonia, ammonium hydroxide, methylamine and propionic acid were tested on aflatoxin-contaminated whole and cracked maize, peanuts and sorghum. Propionic acid was found to be the best detoxifying agent; ammonia gave the least reduction in aflatoxin content. Although the chemically treated samples were less toxic than the untreated one, they were still lethal to chicken embryos (2). The washing of maize in lime for *binatog* preparation was found to cause a 68% reduction in aflatoxin B_1 and G_1. Boiling the *binatog* for ten minutes caused a 20% reduction in aflatoxin B_1 and a 22% reduction in G_1 (1).

Heat treatments were found to reduce aflatoxin levels. Autoclaving at 121°C was found to be more effective than boiling at 100°C for reducing the aflatoxin content in raw and sterilized inoculated mungbean seeds. More aflatoxin was removed as the heat treatment was prolonged (12). Popping corn at 130° ± 5°C for 3.5 minutes caused up to 97% less toxin (1).

The researchers at the Institute of Plant Breeding of the University of the Philippines at Los Baños (UPLB) have found some differences in the resistance of peanut cultivars to invasion by *A. flavus*. The resistance was found to be related to the tannin content of the seed coat. The ultimate goal of this work was to develop a peanut variety that was resistant to invasion by the toxic fungus and/or a variety that does not support toxin formation.

Current Research and Facilities

Currently, only a few studies on aflatoxin contamination and related problems are being pursued in the Philippines. This situation is probably caused by the lack of funds to procure expensive equipment and chemicals for aflatoxin analysis, the dearth of trained personnel in this research area and the absence of facilities to conduct safe research on aflatoxin.

Only the Institute of Chemistry at UPLB has an aflatoxin analytical laboratory that is fully equipped with thin-layer chromatography (TLC) facilities. This laboratory began operation in 1973 to handle the great bulk of samples from aflatoxin research conducted at the university, but it is mainly used as a service laboratory. At present, each scientist interested in pursuing aflatoxin research is trying to develop his own laboratory.

The National Postharvest Institute of Research and Extension (NAPHIRE) is doing some aflatoxin-related work. These studies include maize deterioration at both on- and off-farm levels of operation and long-term storage of grain under plastic cover.

At the Institute of Plant Breeding (IPB), UPLB, research on selection for field and postharvest resistance of peanut cultivars to infection by *A. flavus* and aflatoxin contamination is being conducted. At the NCPC, two projects, storage fungi associated with rice and maize in the Philippines and chemical control of storage fungi associated with maize, are being conducted, and at the Food Science Institute studies on modified atmosphere storage of peanuts in plastic bags are being carried out to determine the levels of CO_2 that either deter or prevent aflatoxin formation.

As this review shows, aflatoxin surveys have been conducted in the Philippines, and the commodities that need attention have been identified. However, little research on control measures has been done. To solve the potential hazards of aflatoxin contamination in maize and other agricultural commodities, a systems approach for developing an integrated program aimed at prevention and control of toxigenic fungi should be a prime objective.

Preventive measures at the farmers' level constitute the best approach to controlling contamination of commodities by toxigenic fungi. Research should be conducted at three levels and should be interdisciplinary. At the macrolevel, researchers should view the physical, social and economic environments in which the problem exists and where potential control strategies might be adopted. At the farmers' level, researchers must study the specific situation and context of the problem, and at the biological level, the implications of nutritional toxicology for humans and animals should be investigated.

In cooperative research, Philippine researchers will be able to screen germplasm of agricultural crops for resistance to mold infection and aflatoxin contamination, to study chemical preventive measures as well as decontamination and/or detoxification, and investigate biological control of fungal growth and aflatoxin formation.

References

1. Celino, L.P., and L.S.P. Madamba. 1984. Effect of processing on aflatoxin in contaminated maize. *In* Trichothecenes and Other Mycotoxins, Proceedings of the International Mycotoxin Symposium, Sydney, Australia. Pp. 185-192.

2. Dalmacio, I.F. 1980. Chemical detoxification of aflatoxin contaminated agricultural products. I. Corn, peanut and sorghum. NSDB-UPLB Project report. University of the Philippines at Los Baños, Laguna, Philippines.

3. Dalmacio, I.F., M.G. Toledo, F.T. Begonia and R.B. Aspiras. 1978. The effect of Sevin and Thiodan on the growth of *Aspergillus* and *Penicillium* during storage of corn and sorghum. NSDB-UPLB project report. University of the Philippines at Los Baños, Laguna, Philippines.

4. Fandialan, I.F., and L.L. Ilag. 1973. Aflatoxin production of *Aspergillus flavus* isolates from rough rice, corn, soybean, peanut and copra. Philippine Agriculture 57:254-263.

5. Fernandez, W.L. 1980. Chemical preservatives and fungicides against molds on copra. II. Copra warehouse storage. PCRDF-UPLB project report. University of the Philippines at Los Baños, Laguna, Philippines.

6. Garcia, R.P. 1985. Chemical control of storage fungi associated with corn. *In* Annual Report, National Crop Protection Center, University of the Philippines at Los Baños, Laguna, Philippines.

7. Ilag, L.L. 1973. *Aspergillus flavus* infection of preharvest corn, drying corn and stored corn in the Philippines. Philippine Phytopathology 9:37-41.

8. Ilag, L.L., and B.O. Juliano. 1982. Colonization and aflatoxin formation by *Aspergillus* spp. on brown rices differing in endosperm properties. Journal of Scientific Food and Agriculture 33:97-102.

9. Ilag, L.L., L.S.P. Madamba, E.M. Luis, L.P. Celino and A.H. Lapitan. 1974. Thiabendazole as seed treatment to control infection by *Aspergillus flavus* Link ex Fries and aflatoxin contamination in corn. Philippine Phytopathology 10:5-10.

10. Jayme, J.B., E.M. Almero and L.A. Salamat. 1976. Epidemiology of primary liver cancer in the Philippines with special considerations of a possible aflatoxin factor. Journal of the Philippine Medical Association 52:129-150.

11. National Crop Protection Center. 1985. Annual report. University of the Philippines at Los Baños, Laguna, Philippines. (Unpublished.)

12. Perez, R.A., and L.L. Ilag. 1977. Various heat treatments for aflatoxin decontamination in mungbean (*Vigna radiata* L. Wilczek). Philippine Phytopathology 13:24-31.

13. Santamaria, P.A., A.C. Pizarro and C.R. Jackson. 1972. Aflatoxin contamination on raw agricultural crops and their by-products in the Philippines. Philippine Phytopathology 8:12-20.

14. Subrahmanyan, V., I.C. Abdon, E.M. Payumo, J.G. Palad, E.C. Eusebio, P.L. Maniquez, C.E. Lotho and I. Capiral. 1966. A simple and inexpensive method of controlling infection during sun-drying of coconut kernel and the utilization of the product for oil, flour and protein. Philippine Journal of Nutrition 19:164-172.

15. Subrahmanyan, V., V.A. Paterno and D. Salon. 1964. Dehydrated coconut: Possibility of producing an edible product in the open and at ordinary temperature. Philippine Journal of Nutrition 17:289-301.

16. Tuason, M.A., and L.S.P. Madamba. 1980. Aflatoxin production in copra by *Aspergillus flavus*. Philippine Agriculture 63:189-196.

Aflatoxin in Thailand

D. Buangsuwon, Division of Plant Pathology and Microbiology, Department of Agriculture, Bangkhen, Bangkok, Thailand

Aflatoxin research in Thailand is now concentrated on toxin control in maize. Maize is second only to rice as a Thai cash crop; in 1985, 2.77 million tons were exported, with a value of US$ 307.47 million. Aflatoxin content has become a major factor affecting the export of maize and most importers have set an aflatoxin limit, usually in the range of 20 to 100 ppb. Aflatoxin restrictions and the world surplus of maize have made markets increasingly difficult to find, and prices are tending to fall. Thailand, therefore, has assigned top priority to research on aflatoxin control in maize. This work, coordinated by a national committee, has made rapid progress and many of the aflatoxin control measures that have been devised are being implemented and/or evaluated on a commercial scale.

Incidence of Aflatoxin Contamination in Thailand

Maize

Based on the testing of 90 samples (3), preharvest aflatoxin contamination of maize was found to be negligible in five Thai provinces from 1980 to 1983. A recent joint Department of Agriculture-UK study (2), based on 46 maize samples collected from two provinces during the major harvest period in the 1985 rainy season, confirms this low incidence of preharvest contamination, finding a mean total aflatoxin content of less than 4 ppb and a range of 0 to 27 ppb. Interestingly, most of the 46 batches monitored at harvest had been dried in the field from one to four weeks, but this did not result in any significant aflatoxin contamination.

Storage of undried ears, especially temporary farm storage for one to six weeks before the crop is sold, has been identified as the significant source of initial contamination. A mean aflatoxin content of 45 ppb with a range of 10 to 95 ppb was found in ten samples of freshly shelled maize in 1984, and an aflatoxin level of 74 ppb, with a range of 3 to 299 ppb, was found in 19 samples from two provinces in 1985. Effective aflatoxin control measures are needed soon after harvest, preferably within 48 hours. Although the critical initial contamination occurs during ear storage, high levels of aflatoxin contamination are usually associated with delays in drying shelled maize. Such delays can be due to rain during sun drying, and to transportation or storage of either undried or partially dried grain.

Peanuts

Incidence of preharvest aflatoxin contamination in peanuts has been found to be very low, with only 3 out of 80 samples being positive, and these at relatively low levels of 20 and 30 ppb (1). Low aflatoxin content was also found in 20 samples of unshelled peanuts collected from commercial warehouses after one week of storage (9 ppb, 20% incidence) and one month of storage (6 ppb, 15% incidence). After one year of storage, the incidence of aflatoxin-positive readings from ten samples was 40% and the mean total aflatoxin content 21 ppb. Higher incidence of aflatoxin contamination was found in shelled peanuts, especially in samples of damaged kernels, where 9 out of 15 samples had aflatoxin contamination ranging from 0 to 350 ppb, and in samples of damaged kernels in which all 20 samples had readings of 75 to 1350 ppb. Large, intact kernels had a low mean total aflatoxin content of 4 ppb, with only 3 out of 25 samples being positive.

Aflatoxin Research

All new aflatoxin research is coordinated by the National Committee on Mycotoxin Control in Agricultural Commodities and its three subcommittees, which cover the areas of research and development, extension and marketing. The committee is composed of representatives drawn from both the private sector and the government, including the Cooperatives Promotion Department, Office of Agricultural Economics, Department of Agriculture, Department of Agricultural Extension, Department of Foreign Trade, Board of Trade of Thailand, Bank of Agriculture and Agricultural Cooperatives and the National Economic and Social Development Board.

The duties and responsibilities of the committee are to:

- Publicize the nature and importance of aflatoxin, especially with regard to export commodities;
- Devise and evaluate effective aflatoxin control measures applicable to each stage in the marketing chain, and then vigorously promote these techniques through radio, television, newspaper advertisements, leaflets and posters;
- Coordinate aflatoxin research and maximize collaborative studies to increase efficiency and reduce excessive duplication of projects;
- Encourage mechanical drying of maize upcountry to minimize delays in drying wet grain to a safe moisture content, and hence reduce the risk of aflatoxin contamination; and
- Regularly monitor the aflatoxin content of maize and peanuts intended for export, so that any problems can be recognized early and remedial action can be taken promptly.

Chemical treatment

The Division of Plant Pathology and Microbiology of the Department of Agriculture has recently completed screening seven reagents in the laboratory for effectiveness in preventing or reducing aflatoxin contamination of maize. Only three of the reagents were found to be effective, sodium bisulphite, ammonia, and propionic acid:ammonium bis propionate at a ratio of 9:1. Sodium bisulphite and ammonia treatments both result in grain with a strong residual odor; the ammonia treatment also produces darker grain. The most promising reagent is the propionic acid-based fungicide formulation, which has been shown to effectively control both mold growth (*A. flavus*) and aflatoxin formation, while not adversely affecting the physical quality of the grain. The cost of the fungicide treatment should be offset by higher prices for better quality grain. Future work aims to reduce costs by minimizing the inclusion rate and improving the application method.

Mechanical drying

The UK-Thai Aflatoxin in Maize Project (1) has identified a set of criteria, called the UK-Thai Project (UTP) System, which has been shown to reliably produce low aflatoxin-content maize during the rainy season. With the UTP system, maize is first field dried for one to two weeks before harvesting to reduce moisture content to 18 to 22%. It is next shelled within 24 to 48 hours of harvest, and load shelled into a drier within at least 12 hours of shelling. Then, within 48 hours, it is dried to 14% moisture content, with no part exceeding 15%. Aflatoxin content is monitored by a special adaptation of the bright greenish-yellow fluorescence (BGYF) test.

Maize dried to 14% moisture content by the UTP system can be safely stored for a minimum of two months with no increase in aflatoxin content. With this system, 25 three-ton batches of maize were successfully produced with a mean total aflatoxin content of 2.5 ppb and a range of 0 to 16 ppb at drying sites in two provinces. The system is now being used commercially for about 50,000 tons of low-aflatoxin content maize.

Improved farm storage

A USAID-funded project, conducted by the Department of Agriculture, was begun in 1985. The project aims to develop and evaluate improved farm storage and drying methods. In 1985, crib stores of three sizes (0.5, 1.0 and 2.5 meters) were tested, as was a solar drier developed by the Asian Institute of Technology (AIT).

Mycological studies

A collaborative study, involving the Divison of Plant Pathology and Microbiology, the Department of Agriculture and the Tropical Agricultural Research Centre (TARC) of Japan, is being conducted on the incidence and occurrence of *Aspergillus flavus*. A very high incidence of *A. flavus* has been found in soil samples, especially in soil around drying facilities and warehouses. No *A. flavus* spores were detected in the atmosphere in maize fields, but high levels of spores were found in the air in warehouses used for maize storage.

A mycological study of maize was also done by the United Nations Development Programme/FAO in collaboration with the Thai Department of Agriculture in January and February, 1985. The work confirms low concentrations of *A. flavus* spores in the air in maize fields during the dry season, as well as the high concentration of spores in warehouses. *Aspergillus flavus* contamination in stored maize was found to be closely associated with weevil infestation (*Sitophilus zeamais*); the insects carried extremely high concentrations of *A. flavus* spores. Virtually no *A. flavus* was found before harvest in the second, dry-season crop of maize, but concentrations increased slowly during temporary storage of ears and grain.

Quality control methods for aflatoxin

The UK-Thai project has amassed data which strongly indicate that an adaptation of the bright greenish-yellow fluorescence (BGYF) test (2) can be used in Thailand to identify the level of aflatoxin presence in maize. Batches were classified according to the number of observed BGYF particles (e.g., 0, 3, 5, 10 counts). Correlation between these BGYF counts and the mean total aflatoxin of all batches within each classification was found to be excellent (correlation coefficient, $r = 0.92$). Sampling was found to be a critical factor when working toward a 20 or 30 ppb aflatoxin limit. The good correlations were only found when a 10-kg representative sample was coarsely ground with a hammer mill fitted with a 6-mm screen and the sample subdivided into four 125-g subsamples. Aflatoxin quality-control procedures based on the BGYF test have been devised for monitoring the production of low-aflatoxin content maize, and for assisting grain management at regional and export storage facilities. Monitoring is best done in conjunction with minicolumn testing to minimize consumer risk.

Analytical Services

Aflatoxin analysis is routinely done at a number of laboratories in Thailand. Unfortunately, sampling methods, sample preparation and analytical methods vary widely, although efforts are being made to standardize them.

Inspection companies offer an aflatoxin analysis service that is predominantly semiquantitative, based on minicolumn determination which is sometimes linked to a fluorotoxin meter. Fully quantitative aflatoxin determination is mainly performed in government laboratories, using quantitation by thin-layer chromatography (TLC). Sophisticated techniques, such as high-performance liquid chromatography (HPLC) and high-performance thin-layer chromatography (HPTLC), are gradually being introduced, and should soon enable a faster and more accurate analysis of samples.

Future Research

Future research has been approved by the national committee in the areas of:

- Continued work on inhibition of aflatoxin-producing fungi by chemical treatment;
- Aflatoxin detoxification;
- Evaluation of the UTP system for producing low-aflatoxin maize on a commercial scale;
- Determination of the feasibility of increasing the proportion of second-crop, dry-season maize, which is known to have a low aflatoxin content, and to determine where such changes might be most appropriate;
- Study of aflatoxin distribution in low-aflatoxin content batches in order to devise appropriate sampling plans for use throughout the maize marketing chain;
- Development and evaluation of analytical techniques, both fully quantitative and semiquantitative, for use in quality control;
- Reduction of the risk of aflatoxin contamination in unshelled maize, e.g., in crib storage and extended field drying; and
- Study of the risk of aflatoxin contamination associated with maize shipping, and the development of suitable control measures.

Cooperative Research

Much of the aflatoxin research in Thailand can now be considered to be coordinated and cooperative, due to the influence of the national committee. Assistance from other countries to provide funding, training and staff is still needed; such support has played a significant role in aflatoxin research in the past. Various foreign agencies have given support to the Department of Agriculture through bilateral or multilateral assistance.

The United Kingdom has provided training, equipment, staff and volunteers to join in collaborative projects with Thai researchers, at a value of approximately 15 million baht (US$ 600,000). The United States Agency for International Development in phase 1 of its contract, has approved a soft loan of approximately US$ 200,000 and a grant for research staff and overseas training and study tours for Thai scientists for 1985 and 1986. The United Nations Development Programme (UNDP) has approved funds of US$ 38,500 for 1985 and 1986. In addition, the Tropical Agricultural Research Centre (Japan) has approved a cooperative project with the Division of Plant Pathology and Microbiology of the Department of Agriculture on quality and preservation of maize by preventing aflatoxin contamination. The Tropical Agricultural Research Centre supplies senior researchers, training, analytical equipment and software.

Acknowledgements

Credit for the rapid progress in the battle against aflatoxin must go to the National Committee on Mycotoxin Control in Agricultural Commodities and its constituent organizations. Particular mention should be made of the research carried out by the Department of Agriculture, which through its own research and in collaboration with the United Nations and British and Japanese teams has provided the data reported in this paper. Scientists who deserve special mention are P. Tonboon-Ek, P. Siriacha, C. Lorsuwan and A. Wong-Urai; U.L. Diener, S. Nesheim, E. Smalley and G. Ware (members of the UNDP team); K. Jewers, M. Nagler, J. Meadley and S. Kenniford (British team members); and M. Saito, T. Goto and S. Kawasuki (the Japanese team).

References

1. Kenniford, S., J. Meadley, J. Morris and M. Nagler. 1985. Report on Aflatoxin in Maize in Thailand, vol. 2: Main Report. Overseas Development Administration, Government of the United Kingdom, London, England.

2. Nagler, M., *et al.* 1983. Paper presented at the Regional Grain Post-Harvest Workshop, Puncak Pass, Indonesia, May 3-6, 1983.

3. Siriacha, P., A. Wong-Urai, P. Tonboon-Ek and D. Buangsuwon. 1983. Paper presented at the Regional Grain Post-Harvest Workshop, Puncak Pass, Indonesia, May 3-6, 1983.

4. Wong-Urai, A., P. Siriacha, P. Tonboon-Ek and D. Buangsuwon. 1983. Paper presented at the Regional Grain Post-Harvest Workshop, Puncak Pass, Indonesia, May 3-6, 1983.

5

Aflatoxin Workshop Recommendations

The main objective of the aflatoxin workshop was the examination of areas of potential cooperative research among scientists of the USA, developing countries and CIMMYT. The following areas covered at the workshop are recommended for consideration.

General

I. Alert regulatory and administrative personnel in all countries, especially those located in the tropics and subtropics, about the potential hazards of aflatoxin-contaminated maize.

II. Develop ways of reducing the maize aflatoxin contamination problem by cooperative research, information dissemination and education of farmers, processors and consumers.

Potential Areas of Research

1. Post harvest

- Survey maize storage and handling procedures, both on the farm and in commercial facilities, to identify storage practices that contribute to aflatoxin contamination;
- Identify unique regional factors that contribute to aflatoxin contamination;
- Study the potential of new, regionally available preservatives that would reduce *Aspergillus flavus* development and insect damage in maize stored above safe moisture levels; and
- Identify the utility of biological control of *A. flavus* through the addition of competitive microbes.

2. Preharvest environment

- Study the different components of environmental stress to determine the relative importance of maximum and minimum temperatures, rainfall amounts and patterns, and time and amount of dew accumulation as related to aflatoxin contamination;
- Develop early warning systems based on weather and insect activity for use in forecasting potential aflatoxin contamination outbreaks and assess the value of remote sensing in defining high-risk areas; and
- Examine the interaction of various kernel rot fungi, such as *A. flavus*, *A. parasiticus*, *A. niger* and *Fusarium moniliforme*, during ear development at different latitudes and environments and under different cultural practices to determine the effects on aflatoxin contamination.

3. Inoculum sources

- Identify the sources of *A. flavus* and *A. parasiticus* inocula that cause preharvest infection, including such soil factors as pH, mineral content and organic matter, and associations with plant residues;

- Measure *A. flavus* and *A. parasiticus* spore levels and distribution during maize development at environmentally diverse locations;

- Determine the effect of cultural practices (rotation, irrigation and minimum tillage) on *A. flavus* conidia, sclerotia levels and their survival;

- Determine sources of *A. flavus* and *A. parasiticus* inocula in grain storage, handling facilities and processing plants; and

- Examine seasonal relationships between soil insects and inoculum distribution.

4. Infection process

- Study the *A. flavus* infection process in developing kernels, with particular attention to the role of pericarp, aleurone, embryo glumes, hilar layer and silk scars;

- Elucidate the sequential infection process in the kernel, i.e., which sections of the kernel exhibit infection during specific developmental stages;

- Improve methods for *A. flavus* inoculation of developing kernels;

- Characterize the extent of aflatoxin contamination produced by *A. flavus* in undamaged kernels without insect vectors;

- Determine the potential for *A. flavus* penetration of kernels via the maize vascular system; and

- Identify the role of physical damage of kernels by insects, birds and mammals, as well as weather factors, in increased susceptibility to *A. flavus* infection.

5. Genetic control

- Determine differences among genotypes in susceptibility to *A. flavus* infection and mechanisms of kernel inhibition to aflatoxin contamination;

- Establish the influence of variations in kernel substrate among genotypes and endosperm mutants on *A. flavus* infection;

- Elucidate the association of inherited resistance to insects with variations in *A. flavus* infection;

- Initiate studies to determine the contribution of husk cover and tightness, chemical content, drying and silk differences in resistance to *A. flavus*;

- Identify the role of kernel hardness, ratio of vitreous and nonvitreous endosperm, and chemical compositiosn of the endosperm in *A. flavus* resistance; and

- Evaluate selected accessions from the maize germplasm bank at CIMMYT for *A. flavus* resistance by two methods: Method I, incubating 1000 kernels with *A. flavus* inoculum for five days and then planting, propagating surviving plants, repeating the cycle five times, and retaining a remnant sample from each cycle for evaluation; and Method II, growing accessions in areas where natural aflatoxin contamination is high, saving only kernels with no visible infection at harvest, planting selected kernels and repeating the process, and saving a remnant sample from each cycle for evaluation of resistance to *A. flavus* infection.

6. Sampling methods

- Standardize sampling procedures at the international level, among research and regulatory agencies;

- Develop methods for field plot techniques, sample size and number of replications, sample preparation (grinding and subsampling), extraction methods, separation and toxin quantification;

- Train pertinent technical personnel in mycotoxin detection, with emphasis on rapid methods and updated information on new detection techniques and safety procedures; and

- Study the possibility of establishing an international aflatoxin determination laboratory for developing countries at CIMMYT.

7. Human and animal toxicoses

- Develop a network for reporting human illness and death attributed to aflatoxin toxicoses; educate and offer assistance to establish a questionnaire for reporting illness and death where aflatoxin may be suspected;

- Identify the nutritional aspects of aflatoxicosis in areas where high mortality occurs among infants and older people;

- Establish safety guidelines in respect to dust inhalation, handling and processing of aflatoxin-contaminated maize;

- Survey the extent of exposure of domestic animals to aflatoxin by milk/urine testing for toxin and residues; correlate exposure data with animal carcass quality (abattoir) or production information;

- Develop an information service among veterinarians on the diagnosis of mycotoxicoses associated with the consumption of specific toxin-contaminated feed by animals; and

- Establish guidelines for the consumption of aflatoxin-contaminated feed by domestic animals to avoid economic losses and low performance.

8. Detoxification

- Establish unequivocally whether or not detoxified aflatoxin-contaminated maize can be used for *tortilla* manufacture; examine detoxified maize for changes in amino acids, oil and starch;

- Conduct studies on the use of urea as a detoxicant of aflatoxin in the manufacture of *tortillas*;

- Compare the efficiency of varied detoxifying agents and measure the response in animals when fed treated feeds; and

- Identify the practicality of mechanically separating *A. flavus*-infected kernels as a means of lowering aflatoxin levels in maize products.

9. **Management strategies**
   - Encourage and recommend periodic regional meetings at strategic locations to focus on aflatoxin contamination problems; suggested locations might be in Asia, Africa, Central and South America;

   - Establish an umbrella committee consisting of scientists from Asia, Africa, Central and South America, plus technical people from CIMMYT and the USA, to encourage cooperative research, receive and evaluate proposals, and distribute research findings;

   - Encourage the education and training of scientists, technicians and food processors who deal with aflatoxin contamination problems (CIMMYT should be considered as a potential training site); and

   - Introduce surveillance procedures for aflatoxin in maize products in high-risk areas and develop a mechanism for rapid, international dissemination of results to pertinent scientists and regulatory agencies on a regular basis.

6

Closing Remarks

R.L. Paliwal, Associate Director, Maize Program, CIMMYT, Mexico

It is a pleasure for me to address this closing session of the US Universities-CIMMYT Aflatoxin Workshop. The workshop is the result of a cooperative project that was started four years ago between the University of Missouri, CIMMYT and the national programs participating in the international testing of maize germplasm for its response to aflatoxin in various environments of the tropical world.

The project has clearly indicated that aflatoxin in maize is a widespread and serious problem. It was decided to bring together scientists from US universities and developing-country institutions working on aflatoxin with CIMMYT maize program scientists to discuss the problem. This workshop has been very successful in bringing about this interaction.

I wish to express CIMMYT's appreciation to USAID and UNDP for making the workshop possible. I also wish to note CIMMYT's appreciation to the organizers of the workshop, Marcos Zuber, Eivind Lillehoj and Bobby Renfro, and to thank all of the scientists who have attended the workshop and participated in its deliberations. The discussions have been held in an exemplary spirit of cooperation and constructive criticism. The harmony in which the discussions have taken place is demonstrated by the unanimity of the recommendations drawn up by the participants.

Most of our sessions have gone beyond normal working hours. This has shown the real interest of the participants in the problem of aflatoxin in maize. These long hours have been very well rewarded by the important conclusions and recommendations that have come out of the workshop. The real success of the workshop, however, will be measured by the follow-up and the cooperative projects between institutions that emerge as a result of these deliberations and recommendations.

I thank you once again for your participation and your contribution to the success of this workshop.

J.C. Sentz, Coordinator, US Universities-CIMMYT Collaborative Maize Research Program, University of Minnesota, St. Paul, Minnesota

This Aflatoxin Workshop has been a major step in expanding collaborative research among scientists from CIMMYT, cooperating countries and the USA, with the aim of insuring the production and availability of maize that is safe for both human and animal consumption. The need for collaborative effort to reduce aflatoxin contamination of maize supplies, primarily in tropical and subtropical areas, was identified as a priority by maize scientists from CIMMYT, US universities and developing countries meeting at CIMMYT in August 1984 for the U.S. Universities—CIMMYT Maize Conference on Collaboration Toward Mutual LDC Maize Production Objectives. Research on aflatoxin was one of six cooperative activities identified at that conference which subsequently received partial support from the US Agency for International Development. The present conference was initiated as a first step in attacking the aflatoxin problem.

The information presented at the workshop, the questions raised and the discussions, particularly about exchanging information routinely and working together on common problems and general aspects of aflatoxin contamination, have confirmed the priority given to this problem and the need to advance research for its control. The meeting will have very little impact, however, unless scientists seize the opportunity to assign priorities to the issues identified and develop strategies for their solution.

In developing a strategy and program that will be effective and efficient, there is a clear need to differentiate between those activities which are country specific and those which are regional or perhaps global.

It is not enough to meet again in three or in five years to assess the progress which has been made. There should be a plan, a program of activities, identifying work areas that can be pursued both individually and cooperatively, not only to guide and pace research, but also to facilitate obtaining fiscal support for the work regionally and locally.

I sincerely hope that the information assembled and the ideas and suggestions generated will initiate a collaborative effort to resolve the problems that cannot be pursued effectively in individual units and locations. The planning of such work should be delegated to a working committee and pursued immediately.

I want to express appreciation, on behalf of the Collaborative Maize Research Working Group, to the participants in this workshop, to CIMMYT for their direct support and as hosts and to the US Agency for International Development for planning and travel support. Our working group will be most interested in the impact of research generated through this workshop and will be available to assist in the achievement of your objectives.

7

Appendix

Participants, US Universities-CIMMYT Maize Aflatoxin Workshop, CIMMYT, Mexico, April 6-11, 1986

Argentina
Dra. Emilia E. Planes de Banchero
Investigadora, Departamento de Microbiología
Instituto Nacional de Tecnología Agropecuaria (INTA) Castellas 1712
Casilla de Correos 25
Buenos Aires, Argentina

Prof. Carlos Marcelo Loffler
Profesor, Facultad de Ciencias Agrícolas
Universidad de Mar del Plata
Casilla de Correos 276-7620
Balcarce, Argentina

Colombia
Dr. Raul Cuero
Departamento de Biología
Universidad del Valle
Cali, Colombia

Costa Rica
Dr. Ronald Echandi Zürcher
Catedrático/Investigador
Centro para Investigaciones en Granos y Semillas
Universidad de Costa Rica
Instituto Interamericano de Cooperación para la Agricultura (IICA)
Apartado Postal 55
Coronado, Costa Rica

Dr. Miguel Mora Cordero
Director
Centro para Investigaciones en Granos y Semillas
Universidad de Costa Rica
San José, Costa Rica

Ecuador
Dra. Susana Espin de Rivera
Investigador, Residuos Tóxicos en Alimentos
Departamento de Nutrición
Instituto Nacional de Investigaciones Agropecuarias (INIAP)
Apartado Postal 340
Quito, Ecuador

Ing. Mario Galarza Silva
Coordinador Nacional de Maíz
Instituto Nacional de Investigaciones Agropecuarias (INIAP
Apartado Postal 2600
Quito, Ecuador

El Salvador
Lic. Gloria Ruth Calderón
Jefe, Laboratorio de Residuos Tóxicos
Centro de Tecnología Agrícola (CENTA)
Apartado Postal 885
San Salvador, El Salvador

Guatemala
Dra. Elizabeth de Porres
Instituto Centroamericano de Investigación y Tecnología Industrial (ICAITI)
Avenida La Reforma 4-47, Zona 10
Apartado Postal 1552
Guatemala, Guatemala

India
Dr. Modan M. Payak
Head, Division of Mycology and Plant Pathology
Indian Agricultural Research Institute
New Delhi 110012, India

Dr. Kaushal K. Sinha
Chief Investigator, IFS
Post-Graduate Department of Botany
Bhagalpur University
Bhagalpur 812007, India

México

Ing. Gloria Laura Anguiano Ruvalcaba
Asistente de Investigación
CINVESTAV
Instituto Politécnico Nacional, Unidad Irapuato
Popocatépetl 52
Apartado Postal 629
36620 Irapuato, Guanajuato, México

MSc. Jose Luis Arellano V.
Investigador, Programa de Maíz
INIFAP-SARH-CAEVAMEX-CIAMEC
Apartado Postal 10
56230 Chapingo, Estado de México, México

Dr. Hector Bourges
Jefe, Division de Nutrición Experimental y Ciencia de los Alimentos
Instituto Nacional de la Nutrición
Vasco de Quiroga 15
Tlalpan, 14000 México, D.F., México

MSc. A. Magda Carvajal Moreno
Instituto de Biología
Universidad Nacional Autónoma de México
Apartado Postal 70-233
04510 México, D.F., México

MSc. Oscar Cota
INIFAP-SARH-CAEVY-CIANO
Apartado Postal 515
85000 Cd. Obregón, Sonora, México

MSc. Genoveva García Aguirre
Departamento de Botánica
Instituto de Biología
Universidad Nacional Autónoma de México
Apartado Postal 70-233
04510 México, D.F., México

MSc. Noel Gomez Montiel
Investigador de Maíz
INIFAP-SARH-CAEIGUA-CIAPAC
Apartado Postal 29
Iguala, Guerrero, México

MSc. Alfredo González Pérez
Investigador, Laboratorio de Control de Calidad
CONASUPO
Insurgentes Sur 3696-B
Tlalpan, 14000 México, D.F., México

MSc. Doralinda Guzman de Peña
CINVESTAV
Instituto Politécnico Nacional, Unidad Irapuato
Popocatépetl 52
Apartado Postal 629
36620 Irapuato, Guanajuato, México

Ing. Francisco Javier Herrera Rodríguez
Investigador, Programa Granos Almacenados
INIFAP-SARH-CIAPY
Campo Agrćola Experimental Uxmal
Apartado Postal 50, Sucursal D
97000 Mérida, Yucatán, México

Dr. Luis César López Frías
Coordinador, Fitopatología Z. Centro
INIFAP-SARH-CAEPAB-CIANOC
Apartado Postal 20
20660 Pabellón de Arteaga, Aguascalientes, Mexico

MSc. Josefina Morales de León
Jefe, Departamento de Ciencia y Tecnología de Alimentos
Instituto Nacional de la Nutrición
Vasco de Quiroga 15
Tlalpan, 14000 México, D.F., México

Dr. Ernesto Moreno Martínez
Instituto de Biología
Universidad Nacional Autónoma de México
Apartado Postal 70-233
04510 México, D.F., México

MSc. Dora Alicia Ortega Zaleta
Investigador, CIAGOC-INIA-SARH
Campo Agrícola Experimental Cotaxtla
Carretera Veracruz-Córdoba Km 34
Apartado Postal 429
91700 Veracruz, Veracruz, México

Dr. José Ron Parra
Mejorador de Maíz
INIFAP-SARH-CAEZAP-CIAB
Apartado Postal 77, Sucursal A
Zapopan, Jalisco, México

MSc. Manuel Rosas Romero
Investigador Adjunto
Departamento de Fitopatología
Colegio de Post-Graduados
56230 Chapingo, Estado de México, México

Dr. Rene Rosiles M.
Laboratorio de Toxicología
Facultad de Medicina Veterinaria y Zootecnia
Universidad Nacional Autónoma de México
04510 México, D.F., México

MSc. Mauro Sierra Macías
Investigador de Maíz
INIFAP-SARH-CIAGOC
Campo Agrícola Experimental Cotaxtla
Apartado Postal 453
91700 Veracruz, Veracruz, México

MSc. Angel Torreblanca Roldán
Director, Departamento de Ciencias de la Nutrición y de los Alimentos
Universidad Iberoamericana
Avenida Cerro de las Torres 395
04200 México, D.F., México

Ing. Gustavo A. Velázquez C.
Investigador del Programa de Maíz
Campo Agrícola Experimental del Valle de México
Centro de Investigaciones de la Mesa Central
Campo El Horno
Apartado Postal 10
56230 Chapingo, Estado de México, México

Héctor Vilches Landin
Universidad Iberoamericana
Avenida Cerro de las Torres 395
04200 México, D.F., México

MSc. Miguel Villalobos Mancilla
Mejorador del Maíz
INIFAP-SARH-CAESICH-CIAN
Apartado Postal 554
31500 Cd. Cuauhtémoc, Chihuahua, México

Philippines

Dr. Rosalinda P. Garcia
Researcher and Assistant Professor
National Crop Protection Center
University of the Philippines at Los Baños
Laguna 3720, Philippines

Thailand

Dara Buangsuwan
Head, Seed and Post-Harvest Pathology Branch
Division of Plant Pathology and Microbiology
Department of Agriculture
Bangkhen 9, Bangkok 10900, Thailand

USA

Dr. J. Scott Angle
Assistant Professor
Department of Agronomy
University of Maryland
College Park, Maryland 20742, USA

Dr. J. Lawrence Apple
Director, International Agricultural Programs
North Carolina State University
P.O. Box 7112
Raleigh, North Carolina 27695-7112, USA

Dr. Dean Barry
Agricultural Reserach Service
US Department of Agriculture
1-67 Agriculture Building
University of Missouri
Columbia, Missouri 65211, USA

Dr. George A. Bean
Professor
Department of Botany
University of Maryland
College Park, Maryland 26742, USA

Dr. Larry L. Darrah
Research Geneticist
Agricultural Research Service
US Department of Agriculture
110A Curtis Hall
University of Missouri
Columbia, Missouri 65211, USA

James W. Dickens
Agricultural Engineer
Agricultural Research Service
US Department of Agriculture
North Carolina State University
P.O. Box 7610
Raleigh, North Carolina 27695-7610, USA

Dr. Urban L. Diener
Professor
Department of Plant Pathology and Microbiology
Auburn University
Auburn, Alabama 36849, USA

Dr. Bruce A. Fortnum
Associate Professor
Department of Plant Pathology and Physiology
Pee Dee Experiment Station
Clemson University
P.O. Box 271
Florence, South Carolina 29503, USA

Dr. Pat B. Hamilton
Professor
Department of Poultry Science
North Carolina State University
Raleigh, North Carolina 27609, USA

Dr. Stanley D. Jensen
Research Coordinator, Western Region
Pioneer Hi-Bred International, Inc.
Route 2, Box 1A
York, Nebraska 68467, USA

Dr. Chris J. Johannsen
Director, Agricultural Data Network and
Laboratory for Applications of Remote Sensing
105 Smith Hall
Purdue University
West Lafayette, Indiana 47907, USA

Dr. Roger K. Jones
Extension Plant Pathologist
Agricultural Extension Service
Texas A&M University
College Station, Texas 77843, USA

Dr. Eivind B. Lillehoj
Microbiologist
Southern Regional Research Center
Agricultural Research Service
US Department of Agriculture
1100 Robert E. Lee Boulevard
P.O. Box 29687
New Orleans, Louisiana 70197, USA

Dr. William W. McMillian
Research Entomologist
Insect Biology Laboratory
Agricultural Research Service
US Department of Agriculture
Georgia Coastal Plain Experiment Station
Tifton, Georgia 31793, USA

Dr. T. Everett Nichols, Jr.
Department of Economics and Business
North Carolina State University
P.O. Box 8109
Raleigh, North Carolina 27650, USA

Dr. Gary A. Payne
Associate Professor
Department of Plant Pathology
North Carolina State University
P.O. Box 7616
Raleigh, North Carolina 27650, USA

Dr. Allan C. Pier
Professor, Department of Veterinary Science
University of Wyoming
Laramie, Wyoming 82017, USA

Dr. David Sauer
US Grain Marketing Research Laboratory
Agricultural Research Center
US Department of Agriculture
1515 College Avenue
Manhattan, Kansas 66506, USA

Dr. Gene E. Scott
Supervisory Research Agronomist
Agricultural Research Service
US Department of Agriculture
P.O. Box 5248
Mississippi State University
Mississippi State, Mississippi 39762, USA

Dr. James Sentz
International Agricultural Programs
320 Vocational and Technical Building
University of Minnesota
1954 Buford Avenue
St. Paul, Minnesota 55108, USA

Dr. Phillip F. Sisson
Director, Agricultural Economics
Quaker Oats Company
345 Merchandise Mart
Chicago, Illinois 60654, USA

Dr. Jack R. Wallin
Research Plant Pathologist
University of Missouri, retired
Route 5
Fulton, Missouri 65251, USA

Dr. Neil W. Widstrom
Research Geneticist
Agricultural Research Service
US Department of Agriculture
Georgia Coastal Plain Experiment Station
P.O. Box 748
Tifton, Georgia 31793, USA

Dr. David M. Wilson
Professor, Department of Plant Pathology
Georgia Coastal Plain Experiment Station
University of Georgia
Tifton, Georgia 31793, USA

Dr. Marcus S. Zuber
Professor Emeritus
Department of Agronomy
109 Curtis Hall
University of Missouri
Columbia, Missouri 65211, USA

CIMMYT

Dr. D.L. Winkelmann
Director General

Dr. Ronald P. Cantrell
Director, Maize Program

Dr. R.L. Paliwal
Associate Director, Maize Program

Dr. Bobby L. Renfro
Pathologist, Maize Program

Dr. John A. Mihm
Entomologist, Maize Program

Dr. Magni S. Bjarnason
Breeder, Quality Protein Maize

Dr. James A. Deutsch
Breeder, Advanced Unit
Maize Program

Dr. Margaret Smith
Breeder, Back-up Unit
Maize Program

Dr. Gregory Edmeades
Physiologist, Maize Program

Dr. David C. Jewell
Wide Crosses, Maize Program

Dr. James E. Lothrop
Breeder, Highland Maize

Dr. Hiep Ngoc Pham
International Testing
Maize Program

Dr. Suketoshi Taba
Germplasm Bank
Maize Program

Dr. Surinder K. Vasal
Breeder, Hybrid Program

Dr. H. Garrison Wilkes
Visiting Research Fellow
Maize Program

Dr. Dana L. Eaton
Associate Scientist
Maize Program

Dr. Narceo B. Bajet
Postdoctoral Fellow
Maize Program

Dr. Carlos de León
Asian Regional Maize Program
Department of Agriculture
Bangkhen, Bangkok 10900, Thailand

Dr. Alejandro Ortega C.
Regional Maize Program,
Mexico, Central America
and Caribbean

Dr. Evangelina Villegas M.
Head, General Laboratories

Dr. J. Michael Prescott
Head, Seed Health
Wheat Program

Elizabeth S. Cuéllar
Editor